Futures of the Study of Culture

Concepts for the Study of Culture

Edited by
Doris Bachmann-Medick, Horst Carl,
Wolfgang Hallet and Ansgar Nünning

Volume 8

Futures of the Study of Culture

Interdisciplinary Perspectives, Global Challenges

Edited by
Doris Bachmann-Medick, Jens Kugele,
and Ansgar Nünning

DE GRUYTER

ISBN 978-3-11-065509-4
e-ISBN (PDF) 978-3-11-066939-8
e-ISBN (EPUB) 978-3-11-066954-1
ISSN 2190-3433
DOI https://doi.org/10.1515/9783110669398

Library of Congress Control Number: 2020934399

Bibliographic information published by the Deutsche Nationalbibliothek
The Deutsche Nationalbibliothek lists this publication in the Deutsche Nationalbibliografie; detailed bibliographic data are available on the Internet at http://dnb.dnb.de.

The book is published with open access at www.degruyter.com.
Typesetting: Integra Software Services Pvt. Ltd.
Printing and binding: CPI books GmbH, Leck

www.degruyter.com

Preface and Acknowledgements

The exploration of possible futures of the study of culture is more than a prognostic effort, diagnosis of trends, or progressive elaboration of theories and methods. It also requires the critical consideration of possible future topics, transformations, and potentials within an interdisciplinary and international research field that faces contested futures in a rapidly changing global world. This volume discusses recent developments, emerging directions, and concerns for the study of culture from a wide range of national and disciplinary contexts, while addressing pressing challenges and crucial issues found in contemporary public discourse.

The articles in this volume have been written and edited well before there were any signs of the current global Covid-19 pandemic that has rapidly brought death, fear, and unforeseen challenges to individual lives and cultural systems. We, of course, do not know what the future will bring or hold in store for our world, but we sincerely hope that we will find ways to cope with all the challenges resulting from this global pandemic. What the corona crisis shows us already, however, is that we depend not only on political and economic systems, but also on ideas, common values, and cultural practices to shape a common future. We need the perspectives of the humanities and social sciences to understand and to create our society, culture, and global world.

We have rarely experienced this fragility of our globalized world and such uncertainty of any future outlook. In times when human lives, economies, and political systems are at stake, we grope our way forward taking very small steps at a time as the very foundations of future expectations seem radically shaken. Yet, although written well before this global crisis, the articles in this volume have approached the topic of 'futures' rather cautiously and with nuance. Instead of generating a global prognostic vision, this collection pursues incipient approaches that try to expand the limits of our established but often ill-suited conceptual settings and disciplinary and institutional arrangements. It aims to open up new horizons for the study of culture by bringing changed conceptual tools and research practices in sight that could perhaps make us better equipped for dealing with urgent concerns and future issues yet unknown.

With the generous support of the University Library Giessen, we have made the book available through Open Access to maximize the accessibility and potential of its contributions to spark debates worldwide. Our goal was to produce a collection that is not only multidisciplinary but also multi-voiced, as exemplified by our two-perspective introductions and an interview with Peter L. Galison.

Most of the contributions originated at the international symposium held in 2016 to celebrate the 15th anniversary of the Giessen Graduate Centre for

https://doi.org/10.1515/9783110669398-202

Humanities (GGK) and the 10th anniversary of the Excellence Initiative-funded International Graduate Centre for the Study of Culture (GCSC).

We would like to extend special thanks to our colleague and PhD candidate, Simon Ottersbach, and to our student assistant, Franziska Eick (both at the GCSC). Their support in formatting the manuscript has been invaluable. Anne Wheeler, Marie Schlingmann, and Elizabeth Kovach were of tremendous help as English language proofreaders of the manuscript. Finally, our thanks go to De Gruyter, in particular Manuela Gerlof, Stella Diedrich, Lydia White, Myrto Aspioti, and Dipti Dange for seeing the project swiftly through the publication process; to the series editors for their support; and last but not least to the GCSC, not only for generously supporting the Open Access publication of this volume but also for providing an intellectually stimulating environment that has been most fruitful for this endeavor.

Giessen, April 2020

Doris Bachmann-Medick, Jens Kugele, Ansgar Nünning

Contents

Preface and Acknowledgements —— V

Introductions: Futures of the Study of Culture

Doris Bachmann-Medick
Futures of the Study of Culture: Some Opening Remarks —— 3

Jens Kugele
Collaborative Research in the Study of Culture —— 17

I Horizons for Future Reflections

Ansgar Nünning
Taking Responsibility for the Future: Ten Proposals for Shaping the Future of the Study of Culture into a Problem-Solving Paradigm —— 29

Andreas Langenohl
The "Future Sense" and the Future of the Study of Culture —— 66

II No Future? Politics and Concerns

Nicole Anderson
Pre-Post-Apocalyptic Culture: The Future(s) of the Humanities —— 83

Isabel Capeloa Gil
The Global Eye or Foucault Rewired: Security, Control, and Scholarship in the Twenty-first Century —— 94

Richard Grusin
No Future: The Study of Culture in the Twenty-first Century —— 110

Hubertus Büschel
Beyond the Colonial Shadow? Delinking, Border Thinking, and Theoretical Futures of Cultural History —— 123

III Theorizing Pasts, Presents, Futures

Andreas Reckwitz
The Society of Singularities — 141

Frederik Tygstrup
After Literature: The Geographies, Technologies, and Epistemologies of Reading and Writing in the Early Twenty-first Century — 155

Andressa Schröder
The Integrative Potentials of Arts-based Research for the Study of Culture: A Reflection on *The Lagoon Cycle* by Helen Mayer Harrison and Newton Harrison — 169

Uwe Wirth
After Hybridity: Grafting as a Model of Cultural Translation — 182

Dirk van Laak
Liquid Spaces in Modern Historiography — 203

IV Future Connectivities: Economy, Natural Sciences, Ecology

Tom Clucas
Culture in the Marketplace — 223

Laura Meneghello
Cultural History, Science Studies, and Global Economy: New and Future Approaches — 236

Silke Schicktanz
Normativity and Culture in the Context of Modern Medicine: A Prospective Vision of an Elective Affinity — 250

Ursula K. Heise
Multispecies Futures and the Study of Culture — 274

Peter L. Galison and Jens Kugele
Future Trading Zones for the Study of Culture: An Interview with Peter L. Galison — 288

Notes on Contributors — 299

Index — 305

Introductions: Futures of the Study of Culture

> Have we now reached a plateau in which the future is likely to be one of consolidation, refinement, and continuity? Or are we at the threshold of new developments, whether reactive rollbacks to earlier paradigms or dimly foreseen revolutions and emergent innovations?
> (Mitchell 2004, 330)

> For us, the future no longer presents itself as an open horizon of possibilities; instead, it is a dimension increasingly closed to all prognoses – and which, at the same time, seems to draw near as a menace. (Gumbrecht 2014, xiii)

Doris Bachmann-Medick

Futures of the Study of Culture: Some Opening Remarks

In his book *The Future as Cultural Fact* the anthropologist Arjun Appadurai claims that the "orientation to the future" should be revalued as a main dimension of culture – though a dimension "that is almost never explicitly discussed" (Appadurai 2013, 179). In cultural anthropology, he observes, the future has so far been repressed in favor of tradition, heritage, and other past-oriented concepts. Cultural anthropology thus sharply contrasts with economics, a science explicitly of the future, of forecasting, prognoses, and expectations.[1] Appadurai calls on cultural anthropology to redefine culture as the "capacity to aspire" (195), i.e., to view culture as that which strengthens impoverished or marginalized groups and social classes and allows them to develop. In Appadurai's view, this would also strengthen cultural anthropology as a discipline, and allow it to unfold in the future. But what about the interdisciplinary study of culture? Has it perhaps been more open to "futurity as a cultural capacity" (180) that is based on anticipations, aspirations, and imaginations (286) from the beginning?[2] Is it more future-oriented than the discipline of cultural anthropology?

Before answering this question, we must differentiate between futurity as a cultural activity on the one hand, and the multidirectional future potential of cultural research on the other (see Andreas Langenohl in this volume). How closely are these two understandings of "futurity" related? Do they stand for two sides of the same coin, as they shape the entire "social formation as a configuration of unequal positions and relations" (Grossberg 2006, 3)? As Lawrence Grossberg claims in reference to Stuart Hall, engaging with this "social formation" in its entirety and contextualizing instead of isolating categories and concepts is essential for a socially relevant cultural studies. This leads to "conjunctural analyses" (5) of conflictual social formations that combine first- and second-order observations. Appadurai, however, engages mainly with first-order observation and the future-oriented capacities of culture itself. But the study of culture also needs to connect cultural aspirations much more strongly to new analytical research categories and a conceptualization

1 For a discussion of economics as a science of imagined futures, based on cultural tools such as "fictional expectations" and narratives that cope with the uncertainty of the future, see Beckert 2016, 3.

2 See Andreas Langenohl in this volume for a more detailed interpretation of Appadurai's concept of a cultural "capacity to aspire."

https://doi.org/10.1515/9783110669398-001

of main entry points that structure future cultural research: risk, imagination, affects and anxieties, media representation, ecological crises, public health crises, etc.

1 Changing Positions, Changing Concepts, Changing Frames

What is the state of the art? Since the nineteenth century, we can no longer assume that the study of culture and other fields of the humanities and social sciences use the 'future' as a fixed frame of reference, let alone as a category of progress (see Freitag and Groß 2017, 8). Conceptions of the future today instead seem to oscillate between an evocation of crisis, a continuation of contemporary theory dynamics, and the generation of fundamentally new paradigms in the face of a "future as catastrophe" (Horn 2018, 5). Often these conceptions diagnose a massive disruption through unforeseeable destabilizing "tipping points" of social and theoretical processes (Horn 2017, 11, 2018, 5). Alternatively, they identify long-term transitions in the humanities, such as "a movement away from 'signification' and 'meaning' toward 'communication and affect'" (Venn 2007, 51); a shift from constructivist to non-constructivist approaches culminating in evidence, presence, and materiality (Gumbrecht 2010, 2014); or a technological transformation of literary representations into new sorts of texts and new forms of reading (see Frederik Tygstrup in this volume). Another strand of future research has extended the familiar pathways of humanist thinking in a post- or non-humanist direction – following explicitly programmatic ideas and critical-ethical aims for the humanities in the twenty-first century (see Braidotti 2013; Grusin 2015). But in the end, do all of these diagnoses of future transitions not remain within the framework of 'change,' do they not evoke a chain of developments and a linear projection into the future? It seems worth mentioning at this point that institutional prerequisites for the development of the humanities and social sciences such as strategic financing schemes and research collaborations have played a key role in shaping such theories of the future according to the logic of their own project proposals. Questioning the frameworks that currently underpin theories of the future, however, could open up new ways of understanding and theorizing the future. We are not talking here about new ways of speculating on future possibilities, problems, anxieties, key concerns and scenarios, cutting-edge research, and emerging topics – such as, for instance, living in or constructing future cities, developing or applying future technologies, coping with surveillance cultures, etc. (see Folkerts, Lindner, and Schavemaker 2015). Nor are we talking about reframing how we acquire knowledge through distinct methods of scaling history, for example, by turning our attention to the Anthropocene or epochal microsections and upheavals as the late Ulrich Beck does (2016, 51–60). After pointing to the Axial Age, the French Revolution, and colonial transformation, he emphasizes the current all-encompassing metamorphosis of the world. Our approach suggests something different: It encourages paying attention to the methodological suppositions underlying the various conceptions of the future, which involves digging out and differentiating shared points of reference that could highlight significant issues for social action as well as for futures of cultural research in explicitly plural terms.[3]

Before outlining this new approach, however, we should first consider the study of culture in its dynamic unfolding, in its own theoretical and methodological development. This unfolding or *Eigendynamik* is inflected by the cultural conflicts and asymmetries of global society, which is why a consideration of futures of the study of culture never only concerns prospective theories and methods. It demands engagement with the emerging futures of cultures and societies in their global conditions. Following this premise, Richard Grusin in this volume ties the futures of the study of culture to "the study of key concerns of the twenty-first century." Referencing Ulrich Beck's notion of a global risk society, which, in his view, we are increasingly becoming, Grusin contends that the study of culture can no longer be left to the traditional humanities alone. It should explicitly be blurred with scientific and public debates on the geological scale of the Anthropocene and the environmental threats facing it (see Chakrabarty 2018), and with studies of media technologies, digitalization, and surveillance – to name but a few challenging fields of research. In the spirit of enriching cultural research with such diverse paradigms, Isabel Gil in this volume focuses on surveillance, showing that the practice or even the system of surveillance not only shapes present and future cultural conditions but also changes the entire framing of the study of culture itself. This approach to surveillance indicates that the future study of culture will be obliged to address pressing problems within society.

Can this reference to the social sphere be seen as a moral-political common denominator for the study of culture? Is the familiar practice of working with 'concepts' as analytical tools giving way to a deeper engagement with 'concerns' (on matters of concern, see Latour 2014, 231–232)? This question does not necessarily call for a normative basis for the study of culture, but increasingly for

---

3 On the significant shift at the end of the twentieth century toward reconceptualizing 'the future' as a multiplicity of futures, see Gidley 2017 (ch. "The Future Multiplied") and Seefried 2014, 2015.

a commitment to responsibility, to rethinking the common denominators and points of reference of our work with concepts in the study of culture – rethinking 'humanity,' 'the world,' 'climate,' 'public health,' 'global justice,' 'human rights,' or 'humans' as a species. Humans are no longer considered to be autonomous from the rest of being but are rather regarded as relationally woven into a network which includes non-humans, technologies, resources, objects, etc. (Horn 2017, 9). As "re-thinking key categories like subjectivity and affect, the environment and technology" (Venn 2007, 49) is the challenge of the day, it is important to also consider the categories with reference to which we analyze pressing global problems.

But where might potential research in the future of the study of culture take place? Though it would be naive to neglect the important institutional dimension of academic work, we should not confine research to the corporate, "entrepreneurial" university. However, the academic environment requires researchers to strategically position themselves in multiple competitive contexts. To position oneself in this field means to distinguish oneself by exploiting "ever more specialised niches" (Angermuller 2013, 265) within the academic market (on academic and financial markets in their potential of shaping research futures, see Tom Clucas in this volume). Alongside this established social-academic trajectory toward marketable professional futures, one could identify a trend in the signature areas of Western research. I am referring to the increased relevance of a culture of singularities such as that outlined by Andreas Reckwitz both in his contribution to this volume and in his provocative book *The Society of Singularities* (2020). Does the tendency to find one's place in society by choosing a position of singularity and uniqueness apply to the field of theory, too? Are we perhaps running into a multitude of singular approaches, "a canon of singularities, a collection of intellectual incursions that were, by definition, without precedent" (Potts and Stout 2014, 2) – not a traditional canon based on "singular names" (2) of outstanding theorists, but rather a new canon of singular approaches? A symptom of this trend could be the contemporary turning away from schools and key theorists in favor of transformative theoretical breaks and new orientations such as the "cultural turns" of the past two decades (see Bachmann-Medick 2016). These "turns" suggest that there is a tendency amongst researchers to carve out and occupy specific research fields exclusively: "Working academics struggle to publish before the flag under which they began their research has been captured and replaced with another" (Potts and Stout 2014, 3). The quick turnover rate that comes with the flagging of claimed research fields seems to be accountable for an almost never-ending compulsion to produce newness. But what about already existing conversations and debates? Why should they be overrun by the obsession with newness that governs current research dynamics? Reflecting on the future of the study of culture must not necessarily repeat this entanglement between linear theory developments and the obsession with (their) newness. Perhaps it would be more effective to employ practical-theoretical tools that follow innovative and future-oriented paths by focusing on new ways of synthesis and linking, critical revision and delinking.

2 Changing Turns or a Grand Paradigm Shift?

Will the emergence of ever-new theoretical turns make the future of cultural research more diverse, more pluralistic? The range of recent turns has drawn attention to a number of emerging topics or concerns which show and demand a deeper involvement with cultural realities (such as global migration, pandemics, climate change, the Anthropocene, etc.). Do we need to rethink our key research categories in light of increased involvement of research with cultural realities and the resulting 'turns' or transdisciplinary 'studies' – such as the ontological turn, or posthumanist, animal, disability, sustainability, etc. studies – before we can even speak of the future of the study of culture? Or will it become inevitable to break entirely with familiar theories and concepts, the *longue durée* constellations of interwoven turns and their increasing differentiation into a prolonged series of sub-sections and studies? In the end, any linear trajectories of theory might prove to be inadequate to analyze and address the contemporary dynamics of newly emerging global problems and systemic disruptions. Will it thus become unavoidable to suggest a hitherto unheard-of paradigm shift in a Copernican sense? In any case, the overarching question is: Are we forced to leave familiar theoretical frameworks behind and adjust our terms and concepts to a world that is "fundamentally different" (Beck 2016, 9) from what we have experienced so far? Ulrich Beck takes a clear position towards this question, claiming that we will be forced to carry out "epochal change" (5) in how we think about the future, to conceptualize a void that until now was never thought to be thinkable at all (see 28–30).

What, then, is the starting point for reflecting on the future of the study of culture? A good starting point would be a new conceptualization of the past. We need to historicize the key concepts that guide our engagement with the future. Historian Dirk van Laak maintains that it is a precondition for dealing with the future: We need historians to act as "prophets of the past" (van Laak in this volume, 215) and reject the assumed continuity between past, present, and future in favor of an openness for "different rhythms and paces of change" (van Laak, 215). But would the reflection on the future not go even further if we started with a new conceptualization of the present? In any case, Hans Ulrich Gumbrecht is convinced that the "broad present," in order to be grasped, requires a new epistemological framework. This unaccustomed framework has to be developed out of an enlarged notion and awareness of a present that is no longer informed by the persistent concept of "historical consciousness" and temporal sequence, but which instead suggests a new "chronotope," one that is shaped by simultaneity and oscillation (Gumbrecht 2014, 75–76). Gumbrecht's postulate of a "broad present" thus implies that before we can even begin to reflect on future developments, we must question the adequacy of our temporal mindset by asking whether we can still rely on our familiar "epistemological habitat" (xiii).[4] Such skeptical interrogation is all the more necessary in view of the global simultaneities of uneven cultural and political conditions that are a challenge to any linear projections of the future. In this context the epistemological lens could also be an eye-opener for the multiple pathways of future research that should no longer be confined to Western scholarship (see Schulz 2019, 4–5), but rather exposed to cross-cultural efforts to address the complexities, diversity, and unevenness of the contemporary world. An interdisciplinary switch to cultural anthropology/ethnology might be a productive starting point for grasping such complexities, as this discipline of complex entanglements has been critically taking up the issue that "new forms of globalization and modernization are bringing all parts of the earth into greater, uneven, polycentric interaction" (Fischer 2003, 3).

3 Changing Points of Reference, Grasping Various Futures

The complex cultural entanglements of the present and the increasing experience of the present as uneven and multiple are good starting points for reflecting upon the future in the plural. This does not mean we should project specific frames of reference onto an unknown future. It means seeking, encouraging, questioning, and critically developing new frameworks in contemporary cultural theory. It means engaging in practice-related knowledge production, not least through the work of above-mentioned multifaceted transdisciplinary 'studies' "that are currently cross-breeding nomadically" (Braidotti 2018, 10), with their broad range of disciplinarily hybrid critical terms. Contributors to this volume exemplify such

4 "... the narrow present of 'history' was the epistemological habitat of the Cartesian subject, another figure of reference (and self-reference) must emerge in the broad present" (Gumbrecht 2014, xiii).

transgressive approaches: Silke Schicktanz draws new ethical inspirations from biomedicine; Hubertus Büschel critically exposes entanglements between ethnography/anthropology and new cultural history; Andressa Schröder outlines arts-based cultural research on ecological issues; and Laura Meneghello offers a new cultural perspective on global economy. In cultural life itself, cross-border perspectives have been made productive for elaborating critical frames or shared points of reference. To name an example: The polyphonic negotiation of 'universal' human rights in the context of local social conflicts or clashes shows that such conflictual scenarios can often be mastered only by seeking shared frames of reference. To look for common frames of reference with regard to future cultural and sociological research is certainly challenging as well. However, it demands a new epistemological starting point: a fundamental "transformation of the reference horizon" (Beck 2016, 17).

As Ulrich Beck, among others, maintains, the future can only be approached if we break down our certainties and, above all, leave behind our traditional perceptions of social 'change' and 'transformation.' Instead, the future opens up in a world where change, with its reference to existing orders and institutions, is replaced by the emerging concept of 'metamorphosis' (Beck 2016, 29). Beck identifies a radical shift and break between the age of change, up until the present, and a coming age of metamorphosis. Global turmoils and global problems have become so complex that they can no longer be grasped and analyzed with familiar concepts. Even the concept of culture itself has to undergo massive transformation. More than ever before, culture is about to be re-envisioned as "more-than-human" (see Ursula Heise in this volume), critically engaging with the rapid developments of artificial intelligence located at complex intersections between fields of the material and 'non-human,' technology, medicine, ecology, computer science, biopolitics, design, and the environmental humanities. Climate change is only one significant reason for this new cultural assemblage. The familiar nature-culture divide is no longer valid; the traditional human subject has been mutated into a "controllable consumer" (Beck 2016, 9); human life has turned into "manufacturability" (25). In these terms Beck outlines a new paradigm which he calls – quite loftily – the ensuing "metamorphosis of the world." This metaphoric phrase points to a complete change of worldviews: a "new way of generating critical norms" (39), new concepts, frameworks, and conditions, "creating a cosmopolitan frame of reference" (40). It represents the acute sense that we can no longer stick to the familiar horizon and extrapolate possible future developments from this present situation.

And yet we can only approach the future by working in the present. Indeed, metamorphosis is for Beck a "characteristic feature of the present age" (20). Finding ways to implement such grand Copernican paradigms, to put them into practice, helps us not to get overwhelmed by them. But how should we implement these new conceptual frames in our investigations? A first "point of entry" might be to focus on a "future sense" already at work within the present, as Andreas Langenohl suggests in his contribution. Langenohl refers to Leslie Adelson's conceptual elaboration of "futurity"[5] but focuses on its applicability in the practical sphere of "prefigurative politics." In the face of new radical transformations in medicine and biotechnology, global risks and catastrophes, and – last but not least – the digital revolution, the focus must shift onto new forms of collaboration and competition, new global climate and health alliances, strategies for traffic and transportation, efforts of global social justice, and increased attention to urban rights in world cities. Focusing on such mobilizing pivot points for analysis enables us to develop the paradigm of metamorphosis in concrete fields of action. Thus, the cultural and cultural-analytical reflection on futures of the study of culture could affect more important areas of investigation and research than just the isolated sequence of theories, turns, and paradigms. It would focus the study of culture on the emergence and elaboration of rather practice-oriented approaches, more so than has been done in the past.

Concerns with the dynamic of 'turns' and transdisciplinary 'studies' in their "hybrid cross-fertilization" (Braidotti 2019, 43) have already created a "nomadic expansion of multiple practices and discourses" (44). These practices across and beyond disciplines (that also find expression in this volume) have already paved the way in such a pragmatic or practical-theoretical direction. But one of the main challenges from now on is to come up with new operative tools and practices for "making" futures. Translation may well constitute such an operative tool – be it the translation of cultural-analytical concepts into societal-political concepts, the translation of academic issues into the public humanities, the study of culture as a capacity to translate between disciplines and cultures,[6] or – facing the Anthropocene – the "displacement-translation of 'force' into 'power'" (Chakrabarty 2018, 13), especially when it comes to the translation of physical-geological categories into social categories of action and responsibility. This translational or, more generally speaking, operative concern goes hand in hand with a new sensibility for processes of transition and lenses of liminality, contact zones, and mechanisms for coping with passages and context shifts. It could encourage the creation of liminal third spaces as possible junctions for giving terms such as

5 Leslie Adelson elaborated the "future sense" as a disjunctive, counterfactual, "long-distance sense organ of temporal perception" (2017, 200, also 40–41).

6 On the study of culture/the humanities as translation studies, see Bachmann-Medick 2012, 35–40; on complexities of 'cultural translation' seen through the lens of "grafting," see Uwe Wirth in this volume.

'humanity,' 'the world,' or 'cosmopolitanism' substance. In other words, all new theoretical tools which will be developed must still be made relevant to practice. Sociology and the study of culture could once again learn how to do so from cultural anthropology: "anthropology operates in a set of third spaces (...) where new multicultural ethics are evolving (...)" and its "challenge is to develop translation and mediation tools for helping make visible the differences of interests, access, power, needs, desire, and philosophical perspective," as Michael Fischer writes in *Emergent Forms of Life and the Anthropological Voice* (2003, 3). And yet, as Hubertus Büschel argues in this volume, one has to be aware of the colonial roots of cultural anthropology and their lasting impact, especially on the recently reconceptualized new cultural history. Referring critically to examples of the entanglement between new cultural history and modes of colonial knowledge production, he also reflects on important operative tools: on practices of provincializing and decentering, delinking with Western mindsets, and border thinking.

Such operative approaches do not have prophetic qualities, nor do they advocate big paradigm shifts or make predictions about future developments. Rather they allow us to start at a different point, perhaps at some impasse or rupture of seemingly continuous trajectories of theory, with stronger regard to breaking and groundbreaking practices of agency and theoretical (trans)formations as well – with a special focus on their non-linear, network-like, translational modes. If we follow Beck, for example, and consider metamorphosis to be a new paradigm, we still need to employ the "jeweler's-eye work of ethnography and social anthropology, the back and forth of detail work and sitting back to view the settings" (Fischer 2009, 270). What is meant here is a fine-tuning of context-related and situation-adequate research attitudes. In the end, such nuanced attitudes will lead to a reconceptualization of the study of culture itself: The study of culture thus turns into a mode of translation studies. As I have tried to explain in other contexts, the study of culture could in a fundamental sense be considered translation studies, since it also strives to pluralize relations and phenomena precisely through the disruption of concepts of wholeness and unity that each translation process inevitably accretes (see Bachmann-Medick 2009, 12). Returning to an idea discussed above, translation as an analytical category could be made fruitful and future-promising if we further connected it with a practical-theoretical, translational mode of acting and agency. Peter Galison's concept of "trading zones" could be especially useful in such an effort (see Galison in this volume). This concept suggests that we can ensure the collaboration of seemingly incompatible language and knowledge communities in interdisciplinary academic contexts and heterogeneous social encounters by establishing a "restricted" exchange language that allows a coordination of action. In a broader historical sense, however, translation as a mode of action could have strategic potential for "making" futures: Past experiences can be reinterpreted and translated by taking up their symbols and shapes and by inserting them into new contexts. In this process new meanings are made more acceptable in traditional forms. Thus, by such innovative translations new horizons can be opened up (see Bachmann-Medick 2017).

To conclude: Discussing the future of the study of culture means much more than elaborating on emerging concepts or even paradigms. It means engaging with innovative methodological infrastructures – such as scaling, zooming (Hannerz 2016, 5), translation, grafting (see Uwe Wirth in this volume), linking or delinking, and other practical efforts to find "strategic switches and pressure points" (Fischer 2009, 270) that have the potential to transform entire research scenarios. But paying attention and fostering new methodological approaches or developing practice tools in fields of action has its limitations, too. It in no way makes the trajectories towards a plural future more manageable. The future of the study of culture is by no means to be understood as a matter of management (see Bachmann-Medick 2017a). The illusion that the future of the study of culture can be managed is maintained, on the level of individual scholars, by activities such as continuously writing reports, peer-reviewing, forming working groups, taking part in evaluation processes, participating in appointment committees, and – on the level of academic organizations and university institutions – by building research associations, making decisions about university rankings, or setting priorities in funding. Thinking about the futures of the study of culture may well lead to a dead end as long as it overestimates such strategic calculations that point to a mere technocratic image of the future of scholarship.

The subject of these introductory remarks has been neither speculation nor prophecy on possible developments in cultural theory and research. The intention was rather to outline a way to future research by mapping out new practical methods of inquiry and point to "critical thinking tools" of the humanities (as Nicole Anderson claims in this volume) and shared, transdisciplinary points of reference that make cultural analyses translatable onto the field of action. But can an approach like this prevent futures from "draw[ing] near as a menace" – as one of the themes of this essay evokes? It can conceivably help us consider the openness of the future and the opportunities for an emergence of unplanned and new perspectives, by admitting the limited manageability of future developments in the study of culture – and by suggesting instead the use of critical analytical, communicative, and ethical skills of the study of culture. Even with all these practical possibilities of knowledge in mind, the theoretical epicenter for orbiting the future resides in the present. In this sense, the words of Teresa de Laurentis could provide further food for thought: "The time of theory, as articulated thought, is always the present" (Laurentis 2004, 365).

Postscript April 2020

We are confronted at this moment with a reality that is dominated by the global coronavirus disease (Covid-19) pandemic. How will the futures of the study of culture emerge from this crisis? At a time when bare survival is at stake, all projections into the future become more uncertain than ever. Even if it is still unknown today how enduring the repercussions of this catastrophe really will be – whether one can truly speak of an epochal 'turn of an era' – this situation is likely to have considerable consequences for the study of culture. It is to be feared that this worldwide unsettling of our survival conditions will continue to entail massive global challenges in the future. Alongside issues such as climate change, migration, war, terrorism, and human rights, the present crisis is bringing inherent dangers in the production and reproduction of human societies to the foreground. And with this new momentum, issues of 'biopolitics' come to the center of public attention: the increased urgency of public health and global health policy, coupled with the prominent role of scientific medical experts as the right arm of political decision-makers. This situation could massively accelerate a development that has been emerging for some time now: The public importance of the humanities and cultural sciences could rescind even further.

If it is currently the virologists, biologists, physicians, pharmaceutical chemists, biomedical technologists, and specialists in digital surveillance who set the tone and determine the agendas of research, and to whom the political decision-makers defer, they will also most likely collect the lion's share of research funds in the future. But can virologists, biomedical pharmacists, technologically competent physicians, and big data specialists solve social and cultural problems? Who will be dealing with the obvious social downsides of this global crisis?

In all this, a new hour for the study of culture, for the humanities, and the social sciences could arrive. Future efforts to solve the problems caused by the pandemic will have to concern themselves with counteracting a hitherto one-sided orientation towards 'economic globalization.' The study of culture will have to consider the cultural, symbolic, experiential, affective, and discursive implications of the crisis. It could also furnish conceptual tools to handle the greater need for cross-border networking, solidarity, and collaborations on a global scale. But in all these developments, the cultural power asymmetries and economic inequalities will have to be assessed anew, leading to a critical analysis of the seemingly unavoidable reshaping of the global order. But there will be additional impulses to reposition the study of culture in the present turmoil of the world. They will arise from new concerns – the necessity to uncover, problematize, and counteract the massive restrictions of democratic rights and liberties; the obligation to deal with the symptoms of increasing racism and populism; but also the need to question the forceful interventions of crisis measures into our ways of life and sociality.

One thing seems certain: The futures of the study of culture will surely be "infected" by this pandemic crisis. They will be confronted with new fundamental problems and their consequences for our everyday lives. To name but a few important ones: the changing relationship between the generations, the new rules of physical and spatial distance, the intensified mediatization and digitalization of our communication through 'social distancing' and its virtual tools, the transformation of our mobility in public spaces and of our concepts, practices, and relations of work. How are these new social conditions to be analyzed with a differentiating vocabulary?

There may be other and more encompassing components that demand new framings for the study of culture: How can we define 'systemic relevance' in our societies under the conditions produced by the crisis? Should the study of culture be further opened up to economics (see Tom Clucas in this volume)? How can we develop ethical and bioethical norms that are adequate to our needs and at the same time responsive to different cultural frameworks (see Silke Schicktanz in this volume)? Last, but not least: How can the distinctions between the spheres of the human and the non-human – in view of the present challenge from the viral world – be reconsidered and the necessary recognition of multispecies cultures be newly assessed (see Ursula Heise and Richard Grusin in this volume)?

In addition to the challenges posed by such newly pressing issues, coping with changed practices and forms of communication will be of fundamental importance. It is here that translation as a 'methodological concept' promises to gain further importance. When experts and politicians collaborate in an entangled way, and scientific studies and findings more increasingly become the basis for political decision-making, then the refined and critical translational capacities of the study of culture are needed all the more. It is these translational capacities that might help to steer and control booming practices of mediation management and to develop communication strategies that include democratic public participation. In this way further 'trading zones' for the collaboration between different knowledge and decision-making systems could be implemented (see Peter Galison and Jens Kugele in this volume). Other cultural and social practices that make a study of culture approach indispensable will gain in importance: developing modes of resilience, coping with existential liminal situations, and the modes of cultivating global social solidarity and responsibility (see Ansgar Nünning in this volume). Giving increased attention to such novel forms of action could perhaps lend a practical dimension to the rather abstract concept of a "metamorphosis of the world," which Ulrich Beck coined to describe a radical disruption of all familiar certainties, conditional frameworks, and analytical competences in grasping possible futures. Focusing on a practical approach like this could help the study of culture break down such overarching concepts into the operative levels of our capacities to act. Moreover, it could also help to inflect our analytical research in directions that have not yet been illuminated, since they have been almost unthinkable so far.

References

Adelson, Leslie A. "Futurity Now: An Introduction." *The Germanic Review* 88.3 (2013): 213–218.
Adelson, Leslie A. *Cosmic Miniatures and the Future Sense: Alexander Kluge's 21st-Century Literary Experiments in German Culture and Narrative Form.* Berlin/Boston: De Gruyter, 2017.
Angermuller, Johannes. "How to Become an Academic Philosopher: Academic Discourse as a Multileveled Positioning Practice." *Sociología Histórica* 2 (2013): 263–289.
Appadurai, Arjun. *The Future as Cultural Fact: Essays on the Global Condition.* London/New York: Verso, 2013.
Bachmann-Medick, Doris. "Introduction: The Translational Turn." *Translation Studies* 2.1 (2009) (special issue *The Translational Turn*): 2–16.
Bachmann-Medick, Doris. "Translation: A Concept and Model for the Study of Culture." *Travelling Concepts for the Study of Culture.* Eds. Birgit Neumann and Ansgar Nünning. Berlin/Boston: De Gruyter, 2012. 23–43.
Bachmann-Medick, Doris. *Cultural Turns: New Orientations in the Study of Culture.* Transl. Adam Blauhut. Berlin/Boston: De Gruyter, 2016.
Bachmann-Medick, Doris. "Cultural Turns: A Matter of Management?" *ReThinking Management: Perspectives and Impacts of Cultural Turns and Beyond.* Eds. Wendelin Küpers, Stephan Sonnenburg, and Martin Zierold. Wiesbaden: Springer VS, 2017a. 31–55.
Bachmann-Medick, Doris. "Übersetzung zwischen den Zeiten – ein *travelling concept*?" *Saeculum. Jahrbuch für Universalgeschichte* 67.1 (2017): 21–43.
Beck, Ulrich. *The Metamorphosis of the World.* Cambridge/Malden, MA: Polity, 2016.
Beckert, Jens. *Imagined Futures: Fictional Expectations and Capitalist Dynamics.* Cambridge, MA: Harvard University Press, 2016.
Braidotti, Rosi. *The Posthuman.* Cambridge/Malden, MA: Polity, 2013.
Braidotti, Rosi. "A Theoretical Framework for the Critical Posthumanities." *Theory, Culture & Society* 36.6 (2019) (special issue *Transversal Posthumanities*): 31–61.
Chakrabarty, Dipesh. "Anthropocene Time." *History & Theory* 57.1 (2018): 5–32.
Fischer, Michael M. J. *Emergent Forms of Life and the Anthropological Voice.* Durham, NC/London: Duke University Press, 2003.
Fischer, Michael M. J. *Anthropological Futures.* Durham, NC/London: Duke University Press, 2009.
Folkerts, Hendrik, Christoph Lindner, and Margriet Schavemaker, eds. *Facing Forward: Art & Theory from a Future Perspective.* Amsterdam: Amsterdam University Press, 2015, open access: https://www.oapen.org/search?identifier=576931 [accessed: 16 November 2019].
Freitag, Klaus, and Dominik Groß, eds. *Zurück in die Zukunft. Die Bedeutung von Diskursen über „Zukunft" in der Wissenschaftsgeschichte.* Kassel: Kassel University Press, 2017.

Gidley, Jennifer M. *The Future: A Very Short Introduction*. Oxford: Oxford University Press, 2017.
Grossberg, Lawrence. "Does Cultural Studies Have Futures? Should It? (Or What's the Matter with New York?)." *Cultural Studies* 20.1 (2006): 1–32.
Grossberg, Lawrence. *Cultural Studies in the Future Tense*. Durham, NC/London: Duke University Press, 2010.
Grusin, Richard, ed. *The Nonhuman Turn*. Minneapolis/London: University of Minnesota Press, 2015.
Gumbrecht, Hans Ulrich. *Our Broad Present: Time and Contemporary Culture*. New York: Columbia University Press, 2014.
Hannerz, Ulf. *Writing Future Worlds: An Anthropologist Explores Global Scenarios*. London: Palgrave Macmillan, 2016.
Horn, Eva. "Jenseits der Kindeskinder. Nachhaltigkeit im Anthropozän." *Merkur* 71.814 (2017): 5–17.
Horn, Eva. *The Future as a Catastrophe: Imagining Disaster in the Modern Age*. New York: Columbia University Press, 2018.
Latour, Bruno. "Why Has Critique Run out of Steam? From Matters of Fact to Matters of Concern." *Critical Inquiry* 30.2 (Winter 2004) (special issue *The Future of Criticism*): 225–248.
Laurentis, Teresa de. "Statement Due." *Critical Inquiry* 30.2 (Winter 2004) (special issue *The Future of Criticism*): 365–368.
Mitchell, W.J.T. "Medium Theory: Preface to the 2003 *Critical Inquiry* Symposium." *Critical Inquiry* 30.2 (Winter 2004) (special issue *The Future of Criticism*): 324–335.
Potts, Jason, and Daniel Stout, eds. *Theory Aside*. Durham, NC/London: Duke University Press, 2014.
Reckwitz, Andreas. *The Society of Singularities*. Cambridge/Malden, MA: Polity, 2020 (German original: *Die Gesellschaft der Singularitäten. Zum Strukturwandel der Moderne*. Berlin: Suhrkamp, 2017).
Schulz, Markus S., ed. *Global Sociology and the Struggles for a Better World: Towards the Futures We Want*. London/Thousand Oaks, CA/New Delhi/Singapore: Sage, 2019.
Seefried, Elke. "Steering the Future: The Emergence of 'Western' Futures Research and Its Production of Expertise, 1950s to Early 1970s." *European Journal of Futures Research* 2.1 (2014): 1–29. https://doi.org/10.1007/s40309-013-0029-y [accessed: 29 August 2019].
Seefried, Elke. *Zukünfte. Aufstieg und Krise der Zukunftsforschung 1945–1980*. Berlin/Boston: De Gruyter, 2015.
Thorkelson, Eli. "The Infinite Rounds of the Stubborn: Reparative Futures at a French Political Protest." *Cultural Anthropology* 31.4 (2016): 493–519.
Venn, Couze. "Cultural Theory and Its Futures: Introduction." *Theory, Culture & Society* 24.3 (2007): 49–54.

Jens Kugele

Collaborative Research in the Study of Culture

1 Thinking about Futures

On September 5, 2019, the "Futurium" opened its doors to the public in Germany's capital, Berlin, and extended an invitation to reflect on the possible futures we imagine for our world. This new building illustrates several key characteristics of our thinking about the future, futures, and futurity in this volume as well. First, in its spatial interplay of exhibition, forum, and lab, the Futurium demonstrates that thinking about futures requires a variety of dynamic spaces. Second, as Stefan Brandt, director of the Futurium emphasizes, it invites us to think about the future in the plural (Checchin 2019). Third, located in the government quarter of the capital and sponsored by the German Federal Ministry for Education and Research as well as by several foundations and companies, this 60-million-Euro project reminds us of the role that infrastructure, politics, and economics play in our thinking about futures. Fourth, its architecture, a result of a 20-year planning and building process, presents a fundamental dilemma that all collective and institutional thinking about possible futures faces: Behind its concrete walls and its glass façade, this edifice, built with today's materials and envisioned by yesterday's architects, hosts visions of tomorrow. While limited by its conventional materiality, it displays in its interior exhibitions on an envisaged future architecture that uses crab shells, bamboo, fungi cultures, brick clay, and recycled materials (see Richter 2019). Fifth, inside the Futurium, visitors find a space

Note: As mentioned above in the Preface and Acknowledgements to this volume, our texts were conceptualized, written, and edited well before there were any signs of the current global covid-19 pandemic that has rapidly brought death, fear, and unforeseen challenges to individual lives and cultural systems. In light of the current global pandemic, experts are expecting that the covid-19 crisis will change the future of our health systems, our political systems, and more generally, our culture. Although we are only at the very beginning of this pandemic, it can be predicted, that, in many ways, these developments will also have unforeseeable consequences for the higher education system in general and the study of culture more specifically. Just as the crisis already has changed our perspectives on health, social interaction and distance, our notions of home, our organization of the private and public sphere, it will change the ways we organize our classrooms, our research, travels, meetings, and conferences, our interactions with colleagues, fellow researchers, and students. As leading economists at the I.M.F. expect the global economy to face the worst slump since the Great Depression, many higher education institutions and humanities departments might have to deal with major budget cuts in the near future.

https://doi.org/10.1515/9783110669398-002

of interaction and active participation that moves beyond mere representation and descriptive texts. Sixth, at its conceptual core, the Futurium features creative collaboration that reaches across institutional contexts and fields of expertise and engages in an exchange with citizens. The future "lab" inside the Futurium is thus not only an attraction for family excursions on rainy Sunday afternoons, but it enables the conceptual interaction between academic research, exhibition space, participating visitors, and the general public.

These elements are central in our thinking about futures of the 'study of culture' as well, which requires dynamic spaces that allow for creative reflection about the future in the plural, always with an awareness of and consideration for its political dimensions. Most centrally, exchange, in the form of collaborative research, lies at the heart of the scholarly study of culture, which imagines the possible futures of its field as well as possible futures of culture more generally.

2 Collaborative Research

At its core, an integral element of the interdisciplinary study of culture is such collaborative research across various borders. This is the case, at least, if we conceive of the study of culture not as resorting to one particular tradition such as the British *Cultural Studies*, the North American *Cultural Studies*, or German *Kulturwissenschaft* in the singular form (see also Ansgar Nünning's contribution to this volume), but instead as an attempt to foster a non-ideological intellectual exchange among all scholarship on culture that employs theoretical and conceptual tools and takes into account its historical dimensions. In what follows, I will highlight five aspects of such collaborative research: first, developing knowledge through the work of thought collectives in the Fleckian sense; second, exchange across various boundaries, including training future generations of researchers for the study of culture; third, forms and formats that allow this collaboration including administrative imagination and structures; fourth, the academic status of collaborative work; and fifth, inextricably linked to the latter, the status of the study of culture as an academic field in the context of disciplinary formations and degree-awarding institutions.

3 Collective Knowledge Construction

Rumor often has it that academic work in the humanities and in the social sciences is the solitary work of individual scholars. The prevailing myth of the individual, independent, and solitary genius scholar goes hand in hand with the fictions of individual talents and skill sets, independent decision-making and selection of research topics, solitary research and problem-solving, as well as single-authored publications. What this myth of the individual genius scholar does *not* account for is best captured in the notion of "thought collective" ("Denkkollektiv"), a term coined by Ludwik Fleck in the 1930s. As the title of his work *Entstehung und Entwicklung einer wissenschaftlichen Tatsache* (1980 [1935]) indicates, Fleck points us to his fundamental notion of a collective "development" and "formation" of scholarly facts. These scholarly facts, in Fleck's view, are in their essence shaped and constructed by a collective of people, inherently linked to language, and instantiated primarily in different forms of scholarly texts. Thus, they bear witness to interrelations among individuals as well as across time and place. Any individualistic accounts of knowledge and of independent genius scholars must therefore be interpreted as mere fiction. While the thought collective in Fleck's sense might often be silent in individual publications, collaborative research offers ways to make it explicit (see Wray 2002, 152). This is certainly not intended to debase individual work entirely, but to explore ways of combining solitary work with collaborative work, and to make the thought collective more explicit in the social-linguistic utterances that, in combination with academic practice, create the development of knowledge.

4 Crossing Boundaries

As Peter L. Galison suggests in this volume, "collaborating across boundaries requires a certain kind of attentive listening." Such active engagement with the work, motivations, values, and goals of others may question established structures, hierarchies, and epistemic regimes; yet it also forms the foundation for collaboration across disciplinary, regional, national, institutional, and linguistic boundaries. Such boundary-crossing includes collaboration across status groups in academia. Integrating students and early-career researchers at a doctoral and postdoctoral level using this notion of collaboration creates opportunities to train future generations of researchers in the study of culture to enter the profession equipped with competences beyond their specific fields of expertise and beyond their individual thesis work. Lawrence Grossberg, in his preface to his *Cultural Studies in the Future Tense*, highlights the value of such collective work when he thanks his students "who have helped shape cultural studies at the University of North Carolina, in my seminars [...] and in the various working groups of the University Program in Cultural Studies," as well as his graduate students, "past and present [...] for their collaborative and collective labors" (Grossberg

2010, xi). Grossberg goes on to thank his translator, his audiences, those who extended speaker invitations to him as well as his junior faculty colleagues, i.e., multiple participants in the (academic) thought collective and the construction of knowledge behind Grossberg's own single-authored publications. While it is encouraging to see esteemed scholars like Grossberg acknowledge the value of collaborative research across boundaries and status groups in their prefaces, such research needs to be acknowledged and fostered every step of the way. To use Peter Galison's words from this volume again: "There are substantive things one can do to promote the visibility and recognition of rising PhDs, postdocs, and assistant professors: They can be promoted to give academic and public talks, they can take on recognized roles in working groups, they can report at collaboration meetings, they can be leads on white papers. We ought to be thinking now about ways to do such things in the growing number of interdisciplinary collaborations in the human sciences."

5 Forms and Formats

Successful collaborative research requires appropriate forms and formats of collaboration. It requires administrative imagination, visionary institutional formations, and innovative structures. Research centers such as the Center for 21st Century Studies (C21) at the University of Wisconsin-Milwaukee is a case in point (see Richard Grusin's contribution to this volume). In light of developments in higher education over the past decades, including financial as well as technological transformations and its increasing professionalization and institutionalization, R. Eugene Rice and others have observed that their fellow faculty members increasingly turn their thoughts inward (see Rice 1996). Collaborative research, by contrast, requires a reflection on academic genres, on both well-tested and alternative formats for research events, on enabling spaces inside and outside of buildings, and on the accessibility of research results, open access publications, and open science more generally. Scholars in the study of culture will need to become adept at using multiple modalities to present their work beyond the conventional genres and media as they expand their work into the realms of film, exhibitions, newspaper articles, community work, etc. Mary Frank Fox and Catherine A. Faver (1984) point to the advantages of such collaborative work and highlight its potential to foster efficiency, sustained motivation, and interpersonal commitment. At the same time, they also draw our attention to its costs and risks such as logistical efforts, travel costs, energy-consuming social conflicts, evaluation of publications, and ethical standards.

6 The Status of Collaborative Work

The risks and potential of collaborative research are inextricably linked to its academic status. To achieve sustained success for collaborative research in the study of culture, we ought to reevaluate our hiring practices and reconsider our idea of academic careers. Our perception of academic institutions would benefit from continued exchange amongst scholars about our conceptions of the study of culture as a research field; its relation to analyses of cultural systems, representations, historical dimensions, prognosis, and citizenship; its positionalities; and its relationship with artists and activists. One of the central questions for scholars in the study of culture will be how to situate their scholarship and thus the enterprise of the study of culture more broadly vis-à-vis the issues debated in a changing world. Topics such as climate change (see Ursula Heise's contribution to this volume), big data and surveillance (see Richard Grusin's contribution to this volume), artificial intelligence, public health and, most recently, global pandemics are major concerns in public as well as academic discourse. In light of the developments in the field of artificial intelligence and as far as the participants in our collaborative research are concerned, a new idea of "the machine" might even be needed (McCarty 2012, 7). A value-neutral version of the study of culture is unachievable, not only for epistemological reasons, but also in light of the increasing commercialization of higher education that forces the humanities to emphasize values other than those of the market, as Martha Nussbaum (2012) argues (see also Tom Clucas in this volume). Scholars in the study of culture will thus have to debate, for example, how to address political issues without resorting to the programmatic positions of British cultural studies, or how the "Heart of Cultural Studies" (Grossberg 2010) relates to the heart of the study of culture. If these discussions include a vision of collaborative research with participants from outside of academia, the study of culture might be able to realign the priorities of the professorate with democratic imperatives, thereby creating more public space in higher education (see Mathews 1998; Checkoway 2001).

In several influential articles, Clifford Geertz points to the important political role scholarly work on culture plays, particularly because of its emphasis on the constructedness of knowledge. At its core, Geertz's essay "Blurred Genres" makes a statement on the epistemological independence of the humanities. By reviewing their proper area of inquiry as well as their substantial theoretical tools, Geertz emphasizes the prominent status of the humanities in the academic construction of knowledge. Geertz's renunciation of "facticity" does not negate the possibility of substantial arguments. Rather, he invites us to ask different questions and to address emerging topics and concerns in academic and social discourse while reflecting on the methodological questions with which we are presented. Against attempts to mimic physics in order to reach higher predictability and therefore seem more legitimately scientific, Geertz's approach favors, for example, the interpretation of dynamic variation over the quest for generalizing laws or definitions (Geertz 2000 [1980]). The latter runs the risk of violating the fundamental flexibility, nuance, and variability in the interrelations between the individual and the environment. In Geertz's view, the social sciences, having just freed themselves from "dreams of social physics" (Geertz 2000 [1980], 23), can self-confidently claim a voice in the process of academic knowledge-construction, not least because they are well equipped and much needed in times of a general "muddling of vocational identities" (Geertz 2000 [1980], 23). Geertz stresses the historical, sociological, comparative, interpretive, and "catch-as-catch-can enterprise" of rendering matters understandable as well as the importance of context. Recognizing the *grande peur* of relativism, Geertz emphasizes diversity not so much in an act of exaltation, but rather to argue that we need to take diversity seriously as an object of analysis. In regarding pluralism as an entity in and of itself, the particularities would risk being subsumed in the generalizations, which translates to a threatening of social cohesion (values, beliefs) and an endangering of the ability to understand each other. The interplay between flexibility and stability or, as Mikhail Bakhtin describes it, the tension between centripetal and centrifugal forces, needs to be balanced. Thus, pluralism should be taken seriously, and intellectual and social work will need to be vigilant about the balance between these tensions as we follow Arjun Appadurai's call to "collaboratively envisage and build a robust anthropology of the future" (Appadurai 2013, 4).

7 The Status of the Study of Culture

As the contributions to this volume demonstrate, imagining possible futures makes critical reference to the present and, at the same time, makes us more attuned to characteristics of the present (see Katharina Martin and Christian Sieg 2016). This also applies to thinking about the future of the study of culture as an academic field in the context of disciplinary formations and degree-awarding institutions. It is clear that people and ideas are always on the move, and we might agree that there are no strict borders between previously separate disciplines and subdisciplines: that, for example, string theory shares techniques with what used to be called condensed matter physics (Peter Galison in this volume). At the same time, it has been argued that the interdisciplinary research perspectives constituting the research field 'study of culture' should be transformed into an academic discipline of its own (see Böhme 2016). What is at stake in these discussions about disciplinarity (see Assmann 2016, ch. 2 and 5), interdisciplinarity (see

Bachmann-Medick 2016; Nünning 2016), and transdisciplinary collaboration, is the very fabric of the study of culture, including questions of assessment, hiring practices, tenure review processes, translatability of research questions, degrees, standards, review and assessment cultures, publication cultures (see Endersby 2016), and notions of best practice across national contexts: in short, the central institutional dimensions of the construction of academic knowledge and power.

Including work in the study of culture ranging from institutionalized forms of disciplinary formations to the work of (and with) independent scholars, artists, activists, and citizens, collaborative research in the study of culture offers us opportunities to rethink academic careers, reconceptualize our notions of excellence, reconceptualize our notions of research, and rethink our visions for scholarship. We should aim to design administrative and departmental structures that recognize diverse forms of scholarship (Bringle, Games, and Malloy 1999) and diverse roles in departmental contexts in higher education; that integrate the different phases in academic careers; and that recognize scholars who feel a responsibility towards communities, civic life, and democratic discourse more generally. This might also lead to a rethinking of our curricular designs in the context of the study of culture: We should aim to create an interdisciplinary horizon for the research field 'study of culture' by addressing the very issues of translating scholarship across disciplinary, national, and linguistic boundaries, and by engaging in an exchange on them together. As Fox and Faver postulate, "[i]n the future, collaborations should be used systematically, rather than haphazardly, not only to fulfill the needs of individual researchers, but also to advance science and scholarship as a whole" (Fox and Faver 1984, 356).

As Arjun Appadurai reminds us in *The Future as Cultural Fact*, it is "vital to build a picture of the historical present that can help us to find the right balance between utopia and despair" (2013, 3). Grouped in four clusters, the contributions in our volume attempt to build this picture as they first point to the horizons for our future reflections; second, discuss the political dimensions of possible futures of the study of culture; third, rethink inter/disciplinary perspectives, heuristics, and epistemologies; and, fourth, invite us to consider future connectivities, and emerging topics and concerns.

References

Adelson, Leslie A. "The Future of Futurity: Alexander Kluge and Yoko Tawada." *The Germanic Review* 86.3 (2011): 153–184.

Adelson, Leslie A. "Futurity Now: An Introduction." *The Germanic Review* 88.3 (2013) (special issue *Futurity Now*): 213–218.

Appadurai, Arjun. *The Future as Cultural Fact: Essays on the Global Condition*. London/New York: Verso, 2013.
Assmann, Aleida. "Die Grenzenlosigkeit der Kulturwissenschaft." *Kulturwissenschaftliche Zeitschrift* 1.1 (2016): 30–48.
Babich, Babette. "Kuhn's Paradigm as a Parable for the Cold War: Incommensurability and Its Discontents from Fuller's Tale of Harvard to Fleck's Unsung Lvov." *Social Epistemology* 17.2/3 (2003): 99–109.
Bachmann-Medick, Doris. "Kulturwissenschaft in der Ermüdung? Anmerkungen zu einer Neuorientierung." *Kulturwissenschaftliche Zeitschrift* 1.1 (2016): 49–55.
Böhme, Hartmut. "Perspektiven der Kulturwissenschaft in historischer und gegenwartsanalytischer Perspektive." *Kulturwissenschaftliche Zeitschrift* 1.1 (2016): 17–31.
Bringle, Robert G., Richard Games, and Edward A. Malloy, eds. *Colleges and Universities as Citizens*. Boston: Allyn & Bacon, 1999.
Bühler, Benjamin, and Stefan Willer. *Futurologien. Ordnungen des Zukunftswissens*. Paderborn: Fink, 2016.
Checchin, Luise. "Interview am Morgen: 'Futurium-Eröffnung.'" *Süddeutsche Zeitung* (5 September 2019). <www.sz.de/1.4587554> [accessed: 5 September 2019].
Checkoway, Barry. "Renewing the Civic Mission of the American Research University." *The Journal of Higher Education* 72.2 (2001): 125–147.
Eagleton, Terry. "The Death of Universities." *The Guardian* (17 December 2010). <https://www.theguardian.com/commentisfree/2010/dec/17/death-universities-malaise-tuition-fees> [accessed: 15 August 2019].
Endersby, James W. "Collaborative Research in the Social Sciences: Multiple Authorship and Publication Credit." *Social Science Quarterly* 77.2 (1996): 375–392.
Fleck, Ludwik. *Entstehung und Entwicklung einer wissenschaftlichen Tatsache*. Frankfurt a.M.: Suhrkamp, 1980 [1935].
Fox, Mary Frank, and Catherine A. Faver. "Independence and Cooperation in Research: The Motivations and Costs of Collaboration." *The Journal of Higher Education* 55.3 (1984): 347–359.
Geertz, Clifford. "Blurred Genres: The Reconfiguration of Social Thought." *Local Knowledge: Further Essays in Interpretive Anthropology*. New York: Basic Books, 1983 [1980]. 19–34.
Gibbs, Robert. "Meeting Our Publics: A Search for the Right Questions in Public Humanities." *University of Toronto Quarterly* 85.4 (2016): 1–5.
Grossberg, Lawrence. *Cultural Studies in the Future Tense*. Durham, NC/London: Duke University Press, 2010.
Martin, Katharina, and Christian Sieg, eds. *Zukunftsvisionen zwischen Apokalypse und Utopie*. Würzburg: Ergon, 2016.
Mathews, David. "Character for What? Higher Education and Public Life." *Educational Record* 78 (1997): 11–17.
Mathews, David. *Creating More Public Space in Higher Education*. Washington, DC: Council on Public Policy Education, 1998.
McCarty, Willard, and Marilyn Deegan, eds. *Collaborative Research in the Digital Humanities*. London/New York: Routledge, 2012.
Nünning, Ansgar. "Perspektiven der Kulturwissenschaften im internationalen Kontext." *Kulturwissenschaftliche Zeitschrift* 1.1 (2016): 70–75.
Nussbaum, Martha C. *Not for Profit: Why Democracy Needs the Humanities*. Princeton: Princeton University Press, 2012.

Rice, R. Eugene. *Making a Place for the New American Scholar*. Washington, DC: American Association for Higher Education, 1996.
Richter, Peter. "Neueröffnung 'Futurium.' Messe der Meister von morgen." *Süddeutsche Zeitung* (4 September 2019). <www.sz.de/1.4587251> [accessed: 5 September 2019].
Wray, K. Brad. "The Epistemic Significance of Collaborative Research." *Philosophy of Science* 69.1 (2002): 150–168.

I Horizons for Future Reflections

Ansgar Nünning

Taking Responsibility for the Future: Ten Proposals for Shaping the Future of the Study of Culture into a Problem-Solving Paradigm

1 On the Need for Rethinking, Reframing, and Reinventing the Study of Culture for the Twenty-first Century

There is a curious lack of alignment between the challenges and problems that we face in the twenty-first century and the established ways in which academic disciplines and institutions have been organized since the nineteenth and twentieth centuries. At an early meeting of the International Advisory Board of the International Graduate Centre for the Study of Culture, which is the institutional site of knowledge production that has both shaped the observations and proposals in this essay, the renowned literary and cultural theorist Ursula Heise from UCLA once remarked that there is an unfortunate disparity between many of the concerns and issues with which the study of culture tries to come to terms and the disciplinary matrix and institutional frameworks within which we operate.

Note: For this essay I have drawn on and adapted some ideas and formulations that were first broached or developed in earlier articles (see especially Nünning 2010, 2012, 2014) and in passages that I contributed to introductions of co-edited volumes (see, e.g., Baumbach, Michaelis, and Nünning 2012; Nünning and Nünning 2010, 2018). Sections 7 and 10 are largely based on a reframed summary of Nünning (2014), from which several ideas and passages have been adapted and only slightly rephrased. Now that the extramural funding from the Excellence Initiative and the funding line called 'Graduate Schools' have unfortunately come to an end, I have also for the first time incorporated some of the ideas that were developed for the original proposal for the establishment of the "International Graduate Centre for the Study of Culture" submitted in 2006 and the renewal proposal of the GCSC from 2012 into an article.

On behalf of all the colleagues who have worked together at the GCSC, the three editors of this volume would like to express our and their tremendous gratitude for the generous financial support and extramural funding that the GCSC has received through the German Excellence Initiative from 2006 until 2019. I am also very grateful to my two co-editors, both for doing the lion's share of the editorial work and the fruitful collaboration over the last ten years or so. I would also like to thank my secretary Rose Lawson for her careful proof-reading, and Elizabeth Kovach for her copy-editing and for making valuable suggestions for improvement.

https://doi.org/10.1515/9783110669398-003

Although Ursula Heise no doubt phrased her observations much more elegantly, the gist of it was that we are trying to solve twenty-first-century problems with theories, concepts, and methods developed in the twentieth century, while working within a disciplinary matrix and the constraints of institutions that largely emerged in the nineteenth century.

Quite a few scholars and commentators from diverse disciplinary and institutional backgrounds have recently made similar remarks. The well-known German social scientist Harald Welzer, for instance, observed that both the programmatic political blueprints and the economic and industrial policies that were developed in the twentieth century are ill-suited for finding adequate answers to the challenges of the twenty-first century, especially to the question of how the current level of civilization could be maintained with a radically reduced use of resources (see Welzer 2013, 220, 288). In a somewhat similar vein, Cathy Davidson also attributes current problems to the disparity between the challenges they pose and the outdated strategies we use in trying to cope with them: "If we're frustrated at the information overload, at not being able to manage it all, it may well be that we have begun to see the problems around us in a twenty-first century multifaceted way, but we're still acting with the individualistic, product-oriented, task-specific rules of the twentieth" (Davidson 2011, 7). In short, we seem to be facing a wide range of new challenges and twenty-first-century global (or postmodern) problems for which there are no ready-made, traditional, or modern solutions.

The disparity between the somewhat ill-suited concepts and methods as well as the equally outdated institutional and disciplinary arrangements and the most pressing challenges, concerns, and issues that the study of culture is faced with in the twenty-first century may be one of the reasons for a loss of faith in both the relevance of academic work and our ability to imagine or shape the future at large. This is exactly what Lawrence Grossberg, one of the most distinguished pioneers of cultural studies in the United States, suggested in his seminal book *Cultural Studies in the Future Tense*: "People seem to be losing their faith in their ability to shape the future. It is not that they do not care about the future, but that they no longer feel that their caring can shape the future. We take no responsibility for the future" (Grossberg 2010, 62, see 284). Although many activists and campaigns like 'Fridays for Future' serve to show that there are indeed people who still retain some faith in their ability to shape the future, I tend to agree with Grossberg that the latter have become the exceptions rather than the rule. While engaged activists from the younger generation have begun to hold politicians and established institutions accountable for the proliferation of ecological crises and natural disasters that threaten the future of our planet, attempts to respond to cultural issues that matter globally have largely been few and far between.

Taking my cue from these observations, I would like to argue that we need to rethink, reframe, and even reinvent the study of culture in such a way as to ensure that there will be a better fit between current concerns of the twenty-first century (henceforth often abbreviated as C21 concerns or issues), on the one hand, and the conceptual, theoretical, and institutional frameworks with or within which we operate in the field of the study of culture, on the other hand. Moreover, as the title and subtitle of this essay (i.e., an essay in the original sense of the word, viz. an attempt) serve to emphasize, this essay argues that we should take responsibility for the future of both the study of culture and the world we live in by actively shaping the study of culture into a problem-solving paradigm, thus substantially reframing and even reinventing the field, the concepts and frameworks, and the projects and practices with which we are engaged. The emphasis in what follows will be on making some proposals for how the study of culture can be developed in such a way that it will not only have a future but also become an important interdisciplinary paradigm for coming to terms with key C21 issues.

Although the emphasis in both this volume and my essay is on the study of culture rather than the Anglo-American kind(s) of cultural studies, a contribution to a volume entitled *Futures of the Study of Culture* should offer clear answers to the two questions raised in the title of one of Grossberg's many important essays: "Does Cultural Studies Have Futures? Should It?" (2006). I am inclined to answer both questions in the affirmative, but I hasten to add that the study of culture will only have futures if we take full responsibility for them, i.e., if we are able and willing to develop relevant research agendas, reframe the concepts, theories, and methods we work with, and to state as clearly as possible why the study of culture matters in the twenty-first century more so than ever.

Since the title of this essay may sound grandiose and promise more than what I will be able to deliver, three brief provisos seem necessary. First, although it might go without saying, what follows is much more modest in scope than the comprehensive account that Grossberg delineates in his wide-ranging monograph *Cultural Studies in the Future Tense*, to which some of the suggestions made below are much indebted. Second, what follows does not claim to be about 'the' challenges, concerns, and issues that both contemporary societies and the study of culture are currently facing, but will instead only be able to cover a fraction of the wide array of aspects, dimensions, and issues that the topic of this volume addresses, focussing on some of the challenges and concerns that have not received as much attention as they arguably merit. Third, despite its emphasis on the future, this essay is neither an exercise in the emerging field of futures studies (aka futurology), nor an attempt to follow in the footsteps of Jonathan Swift. Apart from the intertextual reference in the subtitle to Swift's Juvenalian satirical essay, no irony or satire is intended.

2 Figuring out what is Going on, or: Challenging the Hegemonic Master Narratives and Fictions that Capitalist Cultures Live by, and Reframing Concerns, Conjunctures, and Contexts for the Study of Culture in the Twenty-first Century

Although limitations of space preclude the possibility of comprehensively gauging what the most pressing challenges and concerns are at present or taking stock of the state of the art in the study of culture, taking responsibility for the future presupposes understanding "what's going on in the worlds in which we live" (Grossberg 2010, 1) and the institutions in which we work. In his best-selling book *21 Lessons for the 21st Century*, the historian Yuval Noah Harari (2018) provides a wide-ranging overview of some of the most important technological and political challenges that we are faced with in an age that has variously been dubbed the age of acceleration, the age of bewilderment, the age of climate change, the Anthropocene, and the digital age. One of the few things that most commentators still seem to agree on is that our contexts, technologies, forms of life and work, and the concerns that emerge in relation to them have changed so drastically and rapidly that most people no longer know what is going on: "it is hard to maintain a clear vision. Frequently, we don't even notice that a debate is going on, or what the key questions are" (Harari 2018, ix).

If we accept this general diagnosis, it follows that such a lack of clarity and consensus poses a real challenge for the study of culture. It results in not only a radically altered context in which there is not even agreement about what the agenda or the priorities should be but also new concerns, issues, and research questions. As Cathy Davidson observes, "When suddenly, abruptly, our context changes, we are forced then to pay attention to all the things we didn't see before" (Davidson 2011, 206). It is anything but clear, however, what we should really pay attention to as scholars of culture, because there is no consensus as to what the greatest challenges and concerns are in the twenty-first century (see Harari 2018, 1–2). It is hardly controversial, however, that not only the contexts but also the 'conjunctures' (see Grossberg 2006, 4) of today's societies have changed so much that it is anything but clear what the most important "Problem-Spaces of Cultural Studies" (Grossberg 2010, 43) actually are.

Since any attempt at defining these new conjunctures and problem-spaces in a single essay would be doomed to failure, my first two proposals are that we should challenge the hegemonic master narratives and fictions that capitalist cultures live by, and that we should then try to reframe the main concerns and conjunctures for the study of culture in the twenty-first century accordingly. We are currently witnessing a crisis of many of the hegemonic master narratives and stories that we (or Americans) live by (see McAdams 2013), and a concomitant emergence of alternative cultural, medial, and narrative ways of sense- and worldmaking. The study of culture (or narrative theory, for that matter) has hardly begun to address the loss of faith in the master narratives of growth and progress (see Hänggi 2015), the proliferation of crisis narratives, broken narratives, and new kinds of fictional storytelling like fragmented essay-novels. Being as much shaped by contemporary culture as shaping it in turn, this rise of new kinds of narratives in twenty-first-century storytelling presents a challenge to both the study of culture and the study of narratives, questioning some of their most cherished premises and concepts. In the Appendix of his book *Living to Tell About It*, James Phelan observes that "the living of our lives affects the way we tell our stories, where the telling of our stories affects the way we go on living, and where part of our living is given over to talking about our telling" (Phelan 2005, 205). This observation not only suggests that the way we live and the way we tell our stories mutually shape each other; it also implies that wide-ranging changes in the way we live will have a profound impact on how we tell stories and what kinds of narratives we choose.

Taking my cue from Phelan's emphasis on the mutual constitution of living and storytelling, I should like to venture the hypothesis that we are currently witnessing a crisis of many of the hegemonic master narratives and stories that capitalist, or western, cultures live by and a concomitant emergence of alternative ways of knowing, sense-, and worldmaking. For worse rather than for better, we seem to live in an age in which disrupted lives (see Becker 1997) and broken narratives have become the rule rather than the exception. I would even go so far as to maintain that the plethora of broken narratives across a broad range of domains, genres, and text-types may suggest that, in a digitally enhanced and fragmented age like ours, there may be a change of dominant between the hitherto prevailing forms of coherent and linear stories that have served as cultural templates and new kinds of broken narratives, fragmented novels, and other fragmentary and often multimodal hybrids that combine heterogeneous genres and text-types. In a stimulating online-essay, "Fragmentary: Writing in a Digital Age" (2012), Guy Patrick Cunningham suggests that "works that deal with fragmentation, that eschew not only a traditional narrative structure but the very idea of a work comprising a single, unified whole [...] take on a special kind of relevance."

Moreover, in an age of ongoing crises, it is probably no coincidence that broken narratives seem to have proliferated in various contexts beyond literature and the arts. Salient cases in point include the realms of the economy, banking and finance, as well as politics. As a result of the financial crises (see Lanchester

2010) and soaring national debts, more and more banks, companies, and even states have been confronted with the fact that their cherished corporate or collective narratives no longer match recent developments. When financial dangers, debts, distress, and disorganization prevail, the collective narrative that tells the story of a firm or a nation may well be broken beyond repair. Moreover, we currently seem to be witnessing a crisis or even a breakdown of some of the master narratives that late capitalism has lived by, and it may well just be a question of time until the brokenness and obsolescence of master narratives such as those of economic growth, innovation, progress, and ever-growing prosperity will become impossible to ignore or overlook.

Being as much shaped by contemporary culture and altered forms of living as shaping them in turn, this rise of new kinds of narratives in twenty-first-century storytelling presents a challenge to both the study of culture and narrative theory, questioning some of their most cherished premises, concepts, and cultural templates. The phenomena that have been subsumed under the umbrella terms of 'broken narratives' and 'fragmented novels' challenge and even undermine key assumptions, i.e., that there is such a thing as a logic of narrative and that stories are endowed with orderly structures, coherence, and causality. Although it is definitely much too early to venture any hypotheses about whether such emergent narrative forms constitute a change of dominant in contemporary storytelling, their emergence and recent proliferation testifies to the fact that narratives not only shape cultures and world-models but are also very much shaped by them. Broken narratives and fragmented novels foreground wide-ranging changes in prevalent forms of life in the twenty-first century (see Jaeggi 2014; Basseler et al. 2015); they can be seen as articulations of significantly altered experiences in rapidly changing cultural contexts.

One of the most promising approaches to reframing concerns and contexts for the study of culture in the twenty-first century, therefore, seems to be to focus on the way in which cultures shape narratives and vice versa (see Nünning 2012). Renowned cultural psychologists go so far as to claim that "Storytelling becomes entwined with, even at times constitutive of, cultural life" (Bruner 2002, 31) and that, in our daily autobiographical practice of narrating our lives, "we draw some of our best material from master cultural narratives" (McAdams 2013, 84). Dan P. McAdams even goes so far as to maintain that we are still relatively ignorant about the key concepts and processes that shape cultures: "Beyond making vague references to things like 'my religious heritage' or 'the American Dream,' we tend to have remarkably little insight into the ways our lives are framed by cultural categories, values, and norms" (McAdams 2013, 271). His wide-ranging and brilliant account of the stories Americans live by could well serve as a model to be emulated in the study of other cultures:

> I would submit that life stories are more reflective of and shaped by culture than any other aspect of personality. Stories are at the centre of culture. More than favored goals and values, I believe, stories differentiate one culture from the next. I have argued throughout this book that the stories people live by say as much about culture as they do about the people who live and tell them. Our own life stories draw on the stories we learn as active participants in culture – stories about childhood, adolescence, adulthood, and aging. Stories capture and elaborate metaphors and images that are especially resonant in a given culture. Stories distinguish between what culture glorifies as good characters and vilifies as bad characters.
> (McAdams 2013, 284)

Exploring the ways in which cultures shape narratives just as much as narratives shape cultures could thus be an especially promising approach in order to reframe concerns and contexts for the study of culture in the twenty-first century. Since narratives can be conceptualized as "culturally mediated practices of (re)interpreting experience" (Meretoja 2018, 2), scholars engaged in both narrative studies and the study of culture need to be more "sensitive to the ways in which narratives as practices of sense-making are embedded in social, cultural, and historical worlds" (2). In order to come to grips with the ways in which cultures and narratives mutually constitute one another, we need to know much more about different communities' "stored narrative resources" (Bruner 1990, 67–68), what Bruner calls "culture's narrative resources" (2002, 93) and the cultural categories, values, and norms that frame and shape our lives.

3 Exploring Cultural Ways of Worldmaking as a Paradigm for the Study of Culture in the Twenty-first Century

If we accept the view delineated above that nobody can really tell what is happening right now due to an overload of information and a proliferation of competing accounts, narratives, and fictions (sensu Harari 2016, 2018), then challenging the hegemonic master narratives and reframing concerns and conjunctures for the study of culture in the twenty-first century needs to be supplemented with an exploration of how the worlds that we live in are fabricated or made. The news, for instance, purports merely to report what has happened but it really does much more than that. As Alain de Botton shrewdly observes in his book *The News*: "The news [...] fails to disclose that it does not merely *report* on the world, but is instead constantly at work crafting a new planet in our minds in line with its own often highly distinctive priorities" (de Botton 2014, 11). Media and especially the news thrive on catastrophes and crises, for instance, but what is often forgotten is that focussing on natural cataclysms and disaster and operating in a 24/7 crisis mode tell us more about the highly distinctive priorities of the media in question and their ways of worldmaking than about what is really going on in the world. As Susan Faludi has shown in her Pulitzer-Prize-winning and ground-breaking book *Backlash* (1991), the same holds true for the ways in which media campaigns often resort to dubious means such as skewed reporting with little or no evidence, wilfully fictitious news stories about alleged trends and the reinforcement of cultural myths and stereotypes that discredit feminist aspirations and obstruct women's equality.

In order to figure out what the main concerns and issues are in our present contexts and conjunctures, we need to know much more about how specific media and narratives establish and disseminate agendas, priorities, and worlds. Two related research questions concern the extent to which ways of worldmaking are not only imbued with cultural and ethical values but also implicated in power relations. The study of culture should thus pay much more attention to the ways in which media, metaphors, and narratives shape the cultural life of catastrophes and crises (see Meiner and Veel 2012; Nünning 2012), making highly biased media-worlds that strongly distort people's prevailing views. The main point here is not just that the degree of what the late Hans Rosling (2018) has felicitously called "factfulness" is often dubious or questionable in the crisis-prone worldview that the media tend to project but that we tend to lose sight of both the ways in which "mediashock" (Grusin 2015) and the worlds of news and popular culture are made, as well as long-term developments that are arguably much more important.

Although there is broad consensus by now that narratives are of great importance for the ways in which we make sense of our experiences and the world, neither narrative theory nor the study of culture have been much concerned with the ways in which events, stories, and fictional or real worlds are made, or with the functions that various forms of cultural worldmaking can fulfil (see Nünning 2010). Narratives are at work in processes such as identity formation, the forging of communities and nations, the negotiation and dissemination of norms and values, and the fabrication of storied versions of 'the world.' As Jerome Bruner aptly observed, "narrative, including fictional narrative, gives shape to things in the real world and often bestows on them a title to reality" (2002, 8).

The constructivist notions that provide the epistemological underpinnings of Nelson Goodman's approach pertain to a wide range of different domains of worldmaking and to many cultural ways of worldmaking that we find in the media. They range all the way from *Making Selves*, to borrow the subtitle of a seminal book by Paul John Eakin (1999), to worldmaking in such domains as politics, law, and economics. The question of ways of worldmaking is particularly important in the case of narratives (both factual and fictional), as well as other literary genres and artistic media, in that stories are endowed with performative power, functioning as world-building media that project alternatives to the world-models that we generally regard as 'reality.' In addition, narratives often self-reflexively foreground and explore many of the epistemological and ontological questions involved in worldmaking. It thus does not come as a surprise that Jerome Bruner observed that "Nelson Goodman's constructivism arms one well to appreciate the complexities of self- and life-making" (Bruner 1991, 17), although it is equally clear that his analytical toolbox needs to be supplemented by other concepts. W. J. T. Mitchell was the first to explicitly address the question of "exactly what Goodman is excluding under the rubric of value" (Mitchell 1991, 23), exploring "the scope of Goodman's project, what lies inside and outside the domain of his inquiry" (Mitchell 1991, 24). According to Mitchell, "there are three basic subject areas that Goodman routinely excludes from his system: values, knowledge, and history" (24), but there are also other domains, forms, and functions of worldmaking that did not fall within Goodman's philosophical purview (see Nünning and Nünning 2010, 12–16).

While there is quite a lot of research and debate on the ways in which narratives serve as one of the most important means of self-making, neither narrative theory nor the study of culture have been much concerned with the performative power that storytelling exerts in many domains beyond narrative fiction and autobiography. Narratives, for instance, also contribute to what may be called 'community-making,' with genres and culturally available plots serving as the main interfaces between the making of selves and the making of communities. Narratives can be endowed with performative power, actively moulding, constructing, or even creating the cultural and ideological conflicts that they purport to merely reflect or represent. The stories disseminated by George W. Bush and his administration are a case in point. As an examination of former President Bush's speeches shows, his narratives of crisis not only offer paradigm examples of how storytelling can serve to make worlds and generate conflicts, but they also serve to illustrate that even wilfully fictitious stories can become 'weapons of mass destruction' (see Nünning and Nünning 2017). Several notable exceptions notwithstanding, narrative theory has yet to fully grasp such influential and ubiquitous narrative ways of worldmaking as "The News" (see de Botton 2014), the so-called 'social' networking services like Facebook, and the forms and functions of storytelling in organizations, politics, law, economics, and many other fields. The study of culture would certainly stand to gain if it paid more attention to the multiple functions that narratives perform (see the articles in Erll and Sommer 2019).

As I have argued elsewhere (Nünning and Nünning 2010), the complex processes involved in cultural and, particularly, narrative ways of worldmaking could well serve as a paradigm for the study of culture in the twenty-first century. Since “no one area of study can come to terms with the multidimensional complexity of narrative worldmaking” (Herman 2011, ix), Goodman’s constructivist premises and his general approach tally well with the interdisciplinary nature of research in the study of culture. Although Goodman’s analytical theory of worldmaking provides a highly flexible framework for comparative inquiry, it needs to be further developed and supplemented so that it becomes applicable to a broad range of cultural activities and processes, including journalism, literature, film, music, computer games, and other media. An alliance between theories of symbol systems (as delineated by Ernst Cassirer and Goodman), which pertain equally well to verbal, non-verbal, and performative ways of worldmaking (see Mitchell 1991, 25), and approaches developed by cultural studies could be an important force in the current reconceptualization of the study of culture: Such an alliance can open productive possibilities for the analysis of both the relationship between cultural ways of worldmaking and their changing contexts, and the epistemological, historical, and cultural implications of symbolic systems and signifying practices involved in worldmaking. In addition, such a move could throw new light on both the diachronic development of cultural ways of worldmaking and their changing functions. Another reason why the approach dubbed ‘cultural ways of worldmaking’ could serve as a model for the study of culture in the twenty-first century is thus its broad scope, which opens up a wide range of possible applications to diverse disciplines and fields of inquiry. The latter include the ways in which not just the news, media and politics but also the arts, humanities, and sciences and their academic ways of worldmaking constitute our everyday worlds.

The main reason why the issues involved in cultural ways of worldmaking could serve as a paradigm for the study of culture in the twenty-first century is that “it shifts attention from ‘culture’ or ‘cultural objects,’ assumed to exist, ready to be examined, to the level of the concepts that we deploy to construct the objects of inquiry in the first place” (Nünning and Nünning 2010, 19). If we do not have access to the real world as such, and if the fictions, metaphors, and narratives we live by shape our mental worlds, then the most crucial issue is the question of how such worlds or world models are constructed and conceptualized in the first place. This question pertains to all the different actors, institutions, levels, and media involved in worldmaking, ranging all the way from first-order observations by participants in the cultural field to various second-order or even higher-level observations by which old and new media make their worlds. The paradigm of cultural ways of worldmaking could thus be productively aligned with other approaches in cultural, media, and social theory, such as Bruno Latour’s actor-network theory.

4 Re-Aligning the Study of Culture with Emerging Challenges, Concerns, and Problematics of the Twenty-first Century: New Conjunctures and Directions for the Study of Culture

The suggestions made in the previous section were not meant to imply that the topic of cultural ways of worldmaking should be regarded as *the* paradigm of the study of culture, although it would arguably merit much greater attention than it has hitherto been granted. It was rather meant to serve as a paradigm example of how the research questions, concepts, and theoretical frameworks of the study of culture could be reframed to better re-align them with emerging challenges and current concerns. Although the notion of conjunctures (see Grossberg 2010, 40–53, 57–101) as one of the key concepts of cultural studies has not gained much traction in European versions of the study of culture, I completely agree with Grossberg's claims that exploring cultural phenomena and processes should be conceptualized as an inherently dynamic, flexible, and open-ended project rather than as a fixed and static discipline, and that it should also strive to construct the most relevant conjunctures. What Grossberg dubs "conjunctural analysis" or simply "conjuncturalism" "is a description of change, articulation, and contradiction; it describes a mobile multiplicity, the unity of which is always temporary and fractured" (Grossberg 2010, 41). Rather than examining a clearly delimited set of events or objects, the study of culture should thus respond to the ever-changing concerns, contexts, and problems that emerge in particular social formations and their ongoing debates about cultural issues. What Grossberg claims about cultural studies arguably pertains just as well to the study of culture: "Too many have forgotten that cultural studies is about conjunctures, and that to do it successfully, it has to reinvent itself – its theories, politics, and questions – in response to conjunctural conditions and demands" (65).

With a view to the future of the study of culture, it seems of paramount importance to take the inherently dynamic nature of the cultural phenomena and processes that constitute the objects of inquiry into account, defining research priorities in response to changing challenges and contexts and constructing conjunctures accordingly. Rather than think in terms of a limited number of fixed research areas, it seems much more productive to attempt to identify emerging concerns, issues, and topics that cut across disciplines and research fields. Such key recent issues include, for instance, the proliferation of crises in finance and the economy, the challenges involved in climate change, global warming, and demographic change, and the far-reaching consequences of such wide-ranging transformations as digitalization, globalization, and global migration. Cutting across the dividing lines between culture, technology, science, and society, these and other processes all constitute transdisciplinary challenges that should change the research priorities of the study of culture. The study of culture has yet to come to terms with such challenges as those posed by ethics in digital cultures (Spiekermann 2019; see also Erll et al. 2008) and the equally challenging questions surrounding the impact of digital media on how we think (see Hayles 2012; S. Greenfield 2014), or *How to be Human in the Age of the Machine*, to quote the subtitle of Hannah Fry's (2018) recent book about the benefits and dangers of an increasingly algorithm- and data-driven world.

I would now like to single out at least some of the most important issues and trends to which the study of culture could devote much more attention. These include, for instance, the unprecedented degree of commercialization of culture (both high-brow and popular), the radically altered media ecology that constitutes the cultural environment, and the unprecedented rise of digital information technologies and networks (see e.g. Morozov 2013; A. Greenfield 2017). Among the catalysts that have recently enhanced changes in the cultural field are the acceleration of digital innovations, the unprecedented growth of digital monopolies (see Taplin 2016), and the pervasive "colonization" of both everyday life and the domestic sphere by networked devices, products, and services provided by digital information technologies (see A. Greenfield 2017, 36, 286 and passim). Although the ongoing boom of the radical technologies that Adam Greenfield has analyzed and critiqued in his seminal monograph *Radical Technologies* (2017) is such a complex topic that it cannot be delineated here, it is important for anyone who is trying to understand the ways in which contemporary cultures are evolving to remember just how comprehensively the digital information technologies and networks that so many people nowadays take for granted have changed everyday experience, completely reshaping prevalent forms of life in the twenty-first century:

> Networked digital information technology looms ever larger in all of our lives. It shapes our perceptions, conditions the choices available to us, and remakes our experiences of space and time. [...] It even inhibits our ability to think meaningfully about the future, tending to reframe any conversation about the reality we want to live in as a choice between varying shades of technical development. (A. Greenfield 2017, 8)

The dominant forms of life and everyday experience in what Roberto Simanowski has christened *Facebook-Gesellschaft* (2016), i.e., Facebook society, and, even more so, in the more recent worlds dominated by Instagram, Snapchat, and Twitter, are largely shaped by digital information technologies (see A. Greenfield 2017) and their relentless rhythms of round-the-clock communication and consumption that the art historian Jonathan Crary (2014) has exposed and critiqued in his brilliant but somber exploration of nonstop neoliberal capitalism, *24/7*. Crary's analyses tally well with the observations that the late sociologist Zygmunt Bauman made in his book *Consuming Life* (2007). Bauman uses the brilliantly polyvalent term 'consuming life' to describe the large degree to which contemporary forms of life are centered around the paradigm of consumption, transforming a society of producers into a society of consumers who end up consuming their lives. Contributing to a transformation of the consumer into a commodity, digital and social media (see Lanier 2018) arguably play a central role in the process of reconfiguring cultures and forms of life, affecting the very basics of life. As Crary (2014) shows, we are beginning to sacrifice sleep to a marketplace that operates 24/7, resulting in a collective fatigue that increasingly characterizes our dominant forms of life. The rapid acceleration of all walks of life in late capitalism has generated a 24/7 lifestyle that provides hardly any space to breathe. We are supposed to consume and communicate electronically round the clock, being monitored with digital surveillance techniques that would have made George Orwell's Big Brother blush with envy. According to Crary, this process will eventually lead to the end of sleep (see Crary 2014) and, in his brilliant analysis of the contemporary cultural imaginary, he illustrates perceptively how the widely used expression 'digital age' can be regarded as a questionable act of historiographic construction. It does not constitute a neutral representation of contemporary culture at all but rather blocks out a large number of aspects and experiences that are just as constitutive for the culture(s) of today but incommensurable with the culture's fixation on technical progress and growth: "This pseudo-historical formulation of the present age as a digital age, supposedly homologous with a 'bronze age' or 'steam age,' perpetuates the illusion of a unifying and durable coherence to the many incommensurable constituents of contemporary experience" (Crary 2014, 36).

Therefore, the disruptions brought about by the rise of digital information technologies and the concomitant changes in dominant forms of life, as far-reaching as they have been, are by no means the only relevant contexts against which recent developments and trajectories of contemporary cultures should be gauged. On the contrary, it would be equally (in)accurate to claim that we are living in an age of crises (e.g., the debt crisis, other financial crises, or the refugee crisis), an age of terrorism and the so-called 'war on terror' (see Hodges 2011), an age of surveillance, an age of climate change, and the Anthropocene (see Harari 2016, ch. 2), or an age of world-wide migration and refugee crises.

Moreover, any attempt to construct the most relevant conjunctures of today's cultures has to acknowledge that these diverse contextual developments are not isolated but rather interlinked in various ways. The banking and debt crises, for instance, were not just cataclysmic events in the systems of finance and the economy but rather had devastating consequences for society as a whole, changing cultures, the mental climate, and the dominant hierarchy of values in significant ways. As the prolific journalist and novelist John Lanchester has shown in his highly readable account of the financial crisis entitled *Whoops! Why Everyone Owes Everyone and No One Can Pay*, the "hegemony of economic, or quasi-economic, thinking" (Lanchester 2010, 187) has been so damaging for Britain and the Western world at large because the "economic metaphor came to be applied to every aspect of modern life, especially the areas where it simply didn't belong" (187–188). He goes on to argue that instead of having discussions about values and principles, the emphasis has almost exclusively been on costs:

> In Britain in the last twenty to thirty years that has all been the wrong way round. There was a kind of reverse takeover, in which City values came to dominate the whole of British life. There needs to be a general acceptance that the model has failed. [...] the model which spread from the City to government and from there through the whole culture, in which the idea of value has gradually faded to be replaced by the idea of price. (Lanchester 2010, 188)

These examples may suffice to illustrate what should be taken into consideration in attempts at constructing conjunctures that would re-align the study of culture with key C21 concerns. Even the apparently arcane world of finance or the bank and debt crises of the economy, just like ubiquitous digital information technologies, have had, and continue to have, far-reaching consequences for culture and society at large, shaping not just the dominant hierarchy of values and ideologies, but also the design of everyday life and prevailing notions about living together in a multicultural world. Although the major difficulties in constructing a research agenda and defining priorities derive from the fact that there are so many different cultural, economic, political, social, and technological contexts that could, and should, be taken into consideration, it is of great importance that we begin to construct new conjunctures by describing and explaining the concomitant changes in the cultural spheres and relating them to one or several of these contexts. Many cultural practitioners have responded to the various crises that have occurred in the aftermath of 9/11 and the so-called 'war on terror,' the banking and debt crises since 2008, as well as many further cultural and social issues that have shaped the new millennium. These include, for instance, the 'costs' of modernization, acceleration and globalization, which manifest themselves in, e.g., performance indicators, evaluations, and 'burnout' in the brave new worlds of 'new public management' (see Bartmann 2012) and disruptions in the job market in many branches and industries. Other questions about contemporary cultures revolve around the fact that we are living in multicultural societies that are increasingly marked by worldwide migration, competing forms of life and values, and conflicts of integration and identity that result from these developments.

Moreover, processes of slow change and gradual transformation such as climate change and global warming, "mind change" (S. Greenfield 2014) and the disruptions caused by digital technologies (see A. Greenfield 2017) in a wide range of cultural forms of production, from journalism and music (see Taplin 2017) to the literary field, certainly deserve much more sustained attention within the study of culture. The same holds true for the wide-ranging and devastating effects that economic competition and the doctrine of growth has had, and continues to have, on the ecology and environment: "Even as economic disparity is increasing, competition is urged with fundamentalist fervor as the single solution to all problems. Ecological health continues to elude us – and perhaps indeed depends upon the reconstruction of patterns of thought" (Mary Catherine Bateson 2000, xii; in: Bateson 1972/2000).

In order to ensure that the study of culture will have a sustainable future, it is not enough to merely address and respond to such changing concerns, contexts, and their respective problematics. Rather, we must attempt to develop such emergent problematics into fully-fledged research fields. The French literary historian Yves Citton has done just that with regard to what he has felicitously called the recent shift "From Attention Economy to Attention Ecology" (Citton 2018, 1), serving as a model well worth emulating. Although he concedes that "hyper-attention fed by digital acceleration is not inevitably going to undermine the foundations of our capacity for deep concentration," he hastens to add that "something major is being reconfigured, in which the distribution of attention already plays a major role. [...] Attention is *the* crucial resource of our epoch" (Citton 2018, 10). Citton not only outlines a highly fruitful approach to the new research field of what the title of his book names "the ecology of attention," but he also develops a coherent conceptual and theoretical framework for exploring the wide-ranging changes that we have witnessed in the spheres of "Attention Regimes" (27), "Attentional Capitalism" (44) and "The Digitalization of Attention" (63), to quote some of the key concepts used as chapter titles in Part I one of his book, in which he also includes some words of advice and warning that we would ignore at our own peril:

> Knowing how to choose our alienations and our enthralments, knowing how to establish vacuoles of silence capable of protecting us from the incessant communication that overloads us with crushing information, knowing how to inhabit the switches between hyper-focusing and hypo-focusing – this is what aesthetic experiences (musical, cinematic, theatrical, literary or video-gaming) can help us do with our attention, since attention is always just as much something that we *do* (by ourselves) as something that we *pay* (to another). (Citton 2018, 19)

At the risk of repetition, I would like to reiterate what I said at the beginning of this section: The discussion of these new contexts and concerns is not meant to suggest that these developments should be regarded as *the* new objects or the most important topics of the study of culture. They are rather meant to serve as paradigm examples that could illustrate how the research profile and priorities of the study of culture could be better aligned with major concerns, challenges, and pressing problems of the twenty-first century.

5 Recalibrating Key Concepts for the Study of Culture in the Twenty-first Century: Stock-taking and Enriching our Conceptual and Theoretical Frameworks

The next proposal follows directly from the previous one in that the suggestion to re-align the study of culture with emerging challenges, concerns, and problematics of the twenty-first century necessitates taking stock of the concepts we have been working with, recalibrating them and enriching the extant conceptual and theoretical frameworks by developing additional concepts in order to account for new contexts, conjunctures, and problematics. The notion of 'travelling concepts' as developed by the Dutch cultural theorist Mieke Bal (2002) has had a remarkable impact on recent approaches in cultural analysis and the study of culture at large, opening up new avenues for interdisciplinary exchange, while also introducing an important self-reflexive dimension to the field. Bal's fruitful project proceeds from the assumption that concepts are indispensable for the study of culture because they are "the tools of intersubjectivity" and "key to intersubjective understanding" in that "they facilitate discussion on the basis of a common language" (Bal 2002, 22). Bal also observes that concepts "offer miniature theories" (22), also referred to as "shorthand theories" (23), a claim based on the influential concepts that metonymically represent, or evoke, more complex theories – such as, for example, cultural memory (see Erll and Nünning 2008).

Following in Bal's footsteps, several recent volumes have not only traced the dynamic travelling of concepts between academic disciplines and across research cultures (see, e.g., Baumbach et al. 2012), but they have also provided an overview of main concepts and cutting-edge research fields in the study of culture (see Neumann and Nünning 2012). The chapters in these volumes also show that the "meaning, reach, and operational value" (Bal 2002, 24) of concepts differ between disciplines, academic cultures, and historical periods. Concepts in the study of culture are usually not univocal, fixed, or firmly established. Rather they are dynamic and flexible, undergoing semantic changes as they travel back and forth "between disciplines, between individual scholars, between historical periods, and between geographically dispersed academic communities" (Bal 2002, 24), which are often shaped by different national research cultures and traditions.

Taking stock of the travelling concepts we have been working with in the study of culture and their respective journeys, however, is arguably not enough if we want to realign our conceptual and theoretical frameworks with current concerns and emerging fields. This involves much more than simply making an inventory of key concepts. Rather, it entails a sober exploration of the unspoken assumptions, implications, and ideological baggage that concepts in the field of cultural analysis typically entail. To a much greater extent than in the sciences, we do not deal with clearly defined concepts but rather conceptual metaphors and metaphorical concepts. Travelling across various axes, key concepts in the study of culture tend to gravitate into the force fields of metaphors and narratives (see Baumbach et al. 2012). Imbued with, and shaped by, their respective historical and national traditions, concepts typically come with an array of semantic implications, often including ideological freight, unconscious biases, and normative implications.

In order to decide whether established concepts are still adequate in addressing current concerns and research questions, an especially promising approach may well be to carefully examine their semantic implications and their respective affordances and constraints. While the notion of constraints refers to both the limitations of a concept, pattern or shape and its unacknowledged ideological implications or presuppositions, the term 'affordance' foregrounds the range of possibilities or potentialities a concept entails or opens. In her convincing and powerful proposal for a new formalism that serves to connect aesthetic, literary, and symbolic forms to historical, political, and social contexts, Caroline Levine introduces the concepts of affordances and constraints as follows:

> To capture the complex operations of social and literary forms, I borrow the concept of *affordance* from design theory. *Affordance* is a term used to describe the potential uses of actions latent in materials and designs. [...] Let's now use affordances to think about form. The advantage of this perspective is that it allows us to grasp both the specificity and the generality of forms – both the particular constraints and possibilities that different forms afford, and the fact that those patterns and arrangements carry their affordances with them as they move across time and space. (Levine 2015, 6)

This perspective can be applied to the key concepts we work with in the study of culture. We can interrogate the respective affordances and constraints of different concepts. In his analysis of the concept of globalization, Grossberg has, for instance, convincingly demonstrated that discourses of globalization tend to have very particular affordances and constraints in that the presuppositions of this concept inevitably "set up a particular structural logic – an inescapable binary logic of the global vs. the local, which is applied across every possible dimension" (Grossberg 2010, 60). The same arguably holds true for many of the key concepts of postcolonial theory and postcolonial discourse analysis, to mention just two additional cases in point. As Dipesh Chakrabarty has shown in his influential book *Provincializing Europe*, every case of transferring a cultural, economic or political concept, model or theory from one context to another is "a problem of translation" (Chakrabarty 2000/2008, 17) – a translation of existing worlds, their "conceptual horizons" and their thought-categories into the context, concepts and horizons of another life-world (see Chakrabarty 2000/2008, 71). He also draws attention to the important but often unacknowledged facts that any seemingly "abstract and universal idea" can "look utterly different in different historical contexts," no country is "a model to another country," "historical differences actually make a difference," and "no human society is a *tabula rasa*" (Chakrabarty 2000/2008, xii). What Chakrabarty observes about the "universal concepts of political modernity" also holds true for every approach and concept in the study of culture that is transferred from one academic context or discipline to another: such travelling concepts "encounter pre-existing concepts, categories, institutions, and practices through which they get translated and configured differently" (xii). This should be kept in mind when trying to gauge the challenges and possibilities offered by the notion of travelling concepts in general and when assessing the usefulness and limitations of particular key concepts for coming to terms with C21 concerns.

The need to recalibrate and update key concepts in the study of culture is not only a result of constantly changing contexts. It also arises from the fact that concepts are 'operative terms' (see Welsch 1997); they are never merely descriptive but "also programmatic and normative" (Bal 2002, 28). Concepts construct and change the very objects to which they are applied (see Welsch 1997, 20), "entailing new emphases and a new ordering of the phenomena within the complex objects constituting the cultural field" (Bal 2002, 33). It is thus vital for the development of the study of culture in an interdisciplinary and transnational framework that we maintain awareness of these epistemological implications and continue to develop new and more nuanced concepts for addressing the constantly changing cultural fields of the twenty-first century.

Relatively recent additions to our conceptual repertoire have been the concepts of affect, media events and "mediashock" (Grusin 2015), 'ritual dynamics' as developed by the science of ritual (see Michaels 2010–2011), transculturality, resilience, and the notion of cultural resources. The latter is the key concept around which the projects in the collaborative research centre "RessourceCultures" at the University of Tübingen revolve, which focuses on the socio-cultural dynamics of using resources. In order to gain a better understanding of how cultures and societies manage to cope with the kinds of crises that Europe has been faced with for more than a decade, the study of culture should, for instance, explore the cultural resources of resilience.

Recalibrating key concepts for the study of culture involves not only developing new categories and enriching extant theoretical frameworks but also reframing and retheorizing concepts that have been around for a while but were never fully explicated (see Berning et al. 2014). An obvious case in point is Raymond Williams' rich notion of "structure of feeling," which has only recently (re)gained the attention it deserves in a volume that explores the importance of affectivity in various research areas of the study of culture (see Sharma and Tygstrup 2015). Similarly, the essays in a volume edited by Doris Bachmann-Medick and Jens Kugele (2018) revisit established analytical tools in the study of migration, showing that we need to reframe migration if we approach it from a conceptual perspective and confront established terminologies with recent cultural and discursive frameworks as well as historical and political realities that are all too often referred to as 'the refugee crisis,' a biased and loaded term that already frames the events in an ideologically and politically charged manner. Changing concepts in this field involves, for instance, taking practices of visibility and visualization (including invisibility and making unwelcome people invisible) into account, reframing mobility and the structure of the collective unconscious, taking a fresh look at border regimes and borders as conflict zones, and reconceptualizing migration as translation (for detailed explorations, see the essays in Bachmann-Medick and Kugele 2018).

In order to avoid possible or even obvious misunderstandings, I would like to emphasize, however, that the plea to re-align the study of culture with emerging challenges, concerns, and C21 problematics is by no means meant to suggest that the concepts and issues surrounding, e.g., identity politics have lost any of their former relevance. On the contrary, both the categories of race, class, and gender, and ongoing debates about structural racism, blatant or latent misogyny, and other ingrained forms of inequality continue to be as relevant as they were in the twentieth century. One could even go so far as to say that there is a dire need for more research on the subtle forms that racism, misogyny, and anti-feminism have taken in the twenty-first century and that there has been a backlash against both anti-racism and feminism. As Reni Eddo-Lodge (2018, 99–100) has persuasively shown, in Britain there has been "a backlash against conversations about white privilege" (Eddo-Lodge 2018, 99) and "a backlash against any and all anti-racist organising" (100). Similarly, in the preface to the 15th anniversary edition of *Backlash*, originally published in 1991, Faludi shows that we have witnessed several developments in the twenty-first century that are arguably "worse than backlash"

(2006, xiii) in that "the very fundaments of feminism have been recast in commercial terms" (xiv): "The feminist ethic of economic independence has become the golden apple of buying power", the "feminist ethic of self-determination has turned into the golden apple of 'self-improvement'" (xv), and "the feminist ethic of public agency has shape-shifted into the golden apple of publicity" (xv).

The recent boom of so-called 'social media' and the concomitant emergence of a "Facebook society" (Simanowski 2015), a Twitter-filter bubble, Instagram, Snapchat, and YouTube cultures revolving almost entirely around people's physical appearance not only amply substantiate these prophetic statements, but they also underscore the need to rethink feminism and identity politics in new conceptual, contextual, and theoretical frameworks. Moreover, as recent research in intersectionality has demonstrated, we should redirect our attention from a focus on just one of the categories of difference that define identities to the various ways in which age, class, ethnicity, gender, sexuality, nationality, and others are entangled and interlinked with radically altered media environments, networked digital information technologies, and an equally changed economy and ecology of attention (see Citton 2018). In addition, such recent media phenomena as the worlds generated by Instagram, Twitter, and YouTube, or "WeTube" (see Greif 2016, 200–210), and the peculiar "Reality of Reality Television" (177–199) have yet to receive the degree of attention that they arguably deserve as new cultural ways of worldmaking.

6 Responding to the Epistemological Crisis and Proliferation of Fictions: Reclaiming Authority, Credibility, and Truth for the Study of Culture

The next proposal is a direct follow-up to the previous one in that the key concepts of objectivity and truth upon which research in the humanities and the sciences are based, just like the notion of scientific knowledge, have recently been challenged. Although the challenging of scientific truth is not a recent phenomenon but something that has quite a long and sorry history, it has gained new urgency in an age in which the distinctions between fact and fiction have become increasingly blurred and in which oxymora like 'alternative facts' or dubious notions like 'fake news' and 'post-truth' are gaining currency. As the historians of science Naomi Oreskes and Erik M. Conway have demonstrated in their seminal book *Merchants of Doubt* (2012), not only journalists and politicians but also groups of scientists have been involved in disseminating doubt about "the Truth on Issues from Tobacco Smoke to Global Warming," as the subtitle of their book succinctly states. In recent years, debates about what is really going on, about whether the consensus that science has established about global threats such as climate change and global warming can be trusted, has become even more ferocious, often revealing a blatant lack of respect for research and the very notion of scientific knowledge. If counter-narratives without a shred of scientific evidence can be constructed and disseminated by popular and so-called social media, gaining wide acceptance and even credibility, then genuine research will sooner or later be in dire straits.

One of the most pernicious effects of these alarming tendencies, which pertain to every academic discipline, is the proliferation of "ideological fictions" (Harari 2016, 151). Such fictions not only have the power to sow seeds of doubt (see Oreskes and Conway 2012, 66) and call scientific consensus into question, but they can even undermine the notions of objective reality and scientific truth: "As human fictions are translated into genetic and electronic codes, the intersubjective reality will swallow up the objective reality and biology will merge with history" (Harari 2016, 151). Although Harari has quipped that "humans have always lived in the age of post-truth" and that homo sapiens could even be defined as "a post-truth species, whose power depends on creating and believing fictions" (2018, 233), the proliferation of 'alternative facts,' fake news, and ideological fictions constitutes a real challenge to academic work, scholarly research, and the notion of scientific knowledge, threatening the very existence of universities as institutions in their own right.

On the other hand, the proliferation of various kinds of fictions is also a great challenge and opportunity for the study of culture in that it opens up an important new field of research. As experts in both literary fiction and cultural ways of self-, community-, and worldmaking (see section 2 above), scholars engaged in the study of culture should apply their conceptual expertise, methodological know-how, and analytical research techniques to understand how such ideological fictions are constructed and disseminated as well as the functions they fulfill. Simply dismissing ideological fictions as bullshit (sensu Harry Frankfurt 2005), fake news, or lies is arguably a serious sin of omission in that it would fail to acknowledge both the fact that "some fake news lasts forever" (Harari 2018, 231) and that such fictions often have pernicious effects. We might therefore be much better advised to heed Harari's clarion call and put the examination of fictions onto our research agenda, because fictions serve as important ways of meaning- and sense-making:

> In the twenty-first century fiction might therefore become the most potent force on earth, surpassing even wayward asteroids and natural selection. Hence, if we want to understand our future, cracking genomes and crunching numbers is hardly enough. We must also decipher the fictions that give meaning to the world. (Harari 2016, 151)

I would even go so far as to claim that the alarming proliferation of bullshit, fake-news and fictions not only tends to undermine the authority, credibility, and reliability of scholarly research and of scientific knowledge, but that these tendencies have also become so widespread that they arguably constitute nothing less than an epistemological crisis. A brief look at how the renowned philosopher Alasdair MacIntyre explains the connection between culture and narrative schemata will hopefully clarify this claim. According to MacIntyre, cultures can be understood as communities that share foundational schemata: "Consider what it is to share a culture. It is to share schemata which are at one and the same time constitutive of and normative for intelligible action by myself and are also means for my interpretations of the actions of others" (MacIntyre 1977, 453). When global processes become more complicated, and when it becomes more difficult to know the truth about the world because of competing accounts, then such a co-existence of incompatible schemata and irreconcilable world models can lead to an epistemological crisis: "it is also the case that the individual may come to recognise the possibility of systematically different possibilities of interpretation, of the existence of alternative and rival schemata which yield mutually incompatible accounts of what is going on around him. Just this is the form of epistemological crisis encountered by ordinary agents and it is striking that there is not a single account of it anywhere in the literature of academic philosophy" (MacIntyre 1977, 454). Although the publication of MacIntyre's pioneering essay on the subject dates back more than forty years, no publication that I have come across has managed to offer a better account of how such an epistemological crisis can be resolved:

> When an epistemological crisis is resolved, it is by the construction of a new narrative which enables the agent to understand both how he or she could intelligibly have held his or her original beliefs and how he or she could have been so drastically misled by them. The narrative in terms of which he or she at first understood and ordered experiences is itself made into the subject of an enlarged narrative. The agent has come to understand how the criteria of truth and understanding must be reformulated. He has to become epistemologically self-conscious. (MacIntyre 1977, 455)

Three of the most challenging and important tasks for the study of culture, at least from my point of view, would thus be to devote much more attention to the proliferation of ideological fictions and to an exploration of the resulting epistemological crisis, to reclaim authority, credibility, and truth for the study of culture as well as scientific knowledge in general, and to attempt to tell better narratives and stories than those that currently occupy the headlines and shape the agendas of universities (see section 10 below). Even though the majority of people may well continue to prefer fictions to the truth, it is up to us to remind the public that "the scientific community has been our most reliable source of knowledge for centuries" (Harari 2018, 244).

Movements such as the series of international demonstrations and rallies held under the banner of 'March for Science' serve to show that an increasing number of scholars and scientists from around the world are beginning to understand how urgent it has become to stand up for academic freedom, the indispensable value of research, and scientific knowledge. It is high time that the study of culture not only joined forces with movements such as 'March for Science' and 'Fridays for Future' but also started to employ its expertise and research methodologies in an exploration of the proliferation of fictions, the concomitant epistemological crisis, and their far-reaching detrimental effects.

7 Fostering Internationalization and Pluralism: Transnationalizing Approaches for the Study of Culture in the Twenty-first Century

Although the next proposal might amount to forcing an open door in an age of globalization and worldwide mobility, I would still like to suggest that we should continue to foster more sustained international collaboration and transnational approaches to the study of culture. The study of culture is, after all, itself very much a cultural practice characterized by local traditions and national specificity. As I have discussed elsewhere (see Nünning 2014), approaches to the study of culture as practised in different countries still display considerable differences due to factors such as language, intellectual style, respective cultural contexts, historical developments of disciplines and approaches, and institutional differences between national research cultures and traditions. German *Kulturwissenschaften* and British cultural studies, for instance, are two national traditions with significant differences. The development of genuinely transnational, or even trans-European, approaches to the study of culture is still a desideratum for future research.

Some recent contributions to research have, however, begun to fill the void. These include approaches that either cut across national traditions or have successfully travelled from one research culture to others. A number of influential 'cultural turns' (Bachmann-Medick 2016) in the humanities or 'cultural sciences' (*Kulturwissenschaften*) as well as the notions of 'travelling concepts' (Bal 2002) and 'translation' offer promising ways of overcoming boundaries between research cultures and national traditions. During the last two decades, there have been sustained attempts at *Internationalizing Cultural Studies*, to borrow the title of an anthology edited by Ackbar Abbas and John Nguyet Erni (2005), as the study of culture has been one of the most rapidly developing fields in European and American universities and has also emerged in Asian and Australian scholarship (see Bachmann-Medick 2014, 1–22).

We deliberately chose the more neutral and open term 'study of culture' over the Anglo-American term cultural studies or the German notion of *Kulturwissenschaften* for this volume and the entire book series. It signals that our project of developing transnational approaches to the study of culture does not refer to any narrow understanding of the object of study, a particular theoretical approach, national research tradition or school of thought, as is the case with, for example, 'cultural studies,' 'cultural analysis' (Bal 2002, 6–8), 'cultural materialism,' and 'cultural criticism' (Belsey 2003). The goal is, rather, to enhance the dialogue among these and other approaches, disciplines, and cultures of research to foster self-reflexive, interdisciplinary, international, and potentially even transnational approaches.

The development of transnational approaches to the study of culture does not privilege any one approach but should rather display a commitment to theoretical and methodological pluralism. An approach resulting from an anthropological, semiotic, and constructivist understanding of culture that characterizes many recent approaches is a prerequisite for the rich exchange that takes place in transdisciplinary and international research undertakings. Approaches that have cut across disciplinary and national research traditions include, e.g., cultural semiotics, cultural anthropology, historical anthropology, literary anthropology, the new cultural history, cultural ecology, and area studies (for an overview, see Nünning and Nünning 2008). Although the traditions, research foci, and methodologies of these different ways of studying culture differ substantially, these approaches all embrace inter- or transdisciplinary collaboration and an international, or even global, orientation.

In an article entitled "Cultural Studies and the Transnational," the Canadian cultural theorist Imre Szeman has demonstrated that the notion of the transnational (referred to as a "concept-metaphor"; Szeman 2007, 200) "forces us to consider seriously that the very object of cultural studies – culture – has been radically changed in ways that require the activity of the field to shift from what has remained its basic orientation: the study of cultural objects and practices of everyday life in relation to power" (202). Szeman also carefully delineates three levels on which the transnational functions within cultural studies: cultural studies as a transnational discipline; the field's examination of transnational contexts, issues, and sites; and the political and epistemological challenges involved in the transfer of British and American cultural studies to other contexts.

While it may well be "apparent that cultural studies as a professional practice is now truly transnational" (203), what is much less clear is whether there have been any sustained attempts at inter- or transnationalizing research traditions and practices themselves.

In addition, the notion of transnationalization challenges established Eurocentric and American concepts used to denote modern cultures and collectives such as the nation-state and the polity. It draws attention to the interconnections between polity, nation, and culture in their various manifestations within language, media, memory, and identity. Itself a major mode of the diffusion, transfer, and problematization of key concepts, a transnational perspective epitomizes the emergent character of concepts and the necessity for greater self-reflexivity. It also illuminates transnational cultural phenomena (e.g., Hollywood and Bollywood movies, popular music and MTV, a new understanding of world literature) that have proliferated in the age of globalization and the historical and heuristic reconfigurations of culture, society, and the polity that these phenomena have demanded.

Szeman states that "an interrogation of the potential cultural parochialisms and conceptual blind spots of cultural studies constitutes [...] one of the most important and compelling 'theoretical' projects in the field today" (2007, 206). Transnational approaches to the study of culture could serve as important contributors to such a project. The same holds true for my plea to recalibrate key concepts and explore the processes of appropriation, reframing (see Berning et al. 2014), and translation that are involved in the travelling of concepts. To the extent that the meaning of such concepts must be constantly renegotiated, a sustained enquiry into the dynamics of such travelling, including the "'translational' processes" (Chakrabarty 2000/2008, 19) and politics involved and the genealogies of the concepts in question (see section 5 above), is a prerequisite for the development of transnational approaches to C21 challenges in the study of culture.

8 Reinventing the Practices and Institutional Frameworks of Knowledge Production: The Study of Culture as a Collaborative, Evidence-Based, and Interdisciplinary Practice

Another daunting but important task for anyone interested in securing viable futures for the study of culture is to get involved in the attempt to shape and remake the institutional sites and practices of knowledge production. Since the study of culture is an interdisciplinary project, we need institutional frameworks that foster and support collaborative work and interdisciplinary conversation across different cultures of research. It is certainly not a coincidence that cultural studies only began to rise to prominence after the Centre for Contemporary Cultural Studies had been established at the University of Birmingham in 1964 and that, even decades later, international luminaries such as Lawrence Grossberg continue to sing their praises for the stimulating research environment that the CCCS generated (see Grossberg 2010).

Moreover, research cultures are themselves subject to historical change, and both cultural studies and German *Kulturwissenschaften* have undergone far-reaching developments and important innovations in recent years. In comparison to the programmatic mission encapsulated in the name of the Birmingham Centre for Contemporary Cultural Studies, which suggests that cultural studies "is a field devoted entirely to the immediate present" (Felski 2003, 501), for instance, from today's point of view the "rationale for isolating the study of popular, contemporary culture from high culture and the culture of the past now seems purely historical" (Belsey 2003, 91). In an article entitled "Beyond Literature and Cultural Studies," Catherine Belsey made a programmatic proposal that calls for "a new discipline [...], beyond literature and Cultural Studies, that would explicitly treat all culture as its province, and would take full advantage of the attention French theory pays to the signifier" (Belsey 2003, 99).

Taking my cue from Cathy Davidson's call for what she called "a Project Workplace Makeover" (2011, 167), I would even go so far as to make a similar plea for 'a Project Institution Makeover' for the study of culture. Davidson is certainly right when she observes that it is high time that we began to rethink and reinvent our institutions for research and teaching: "What we haven't done yet is rethink how we need to be organizing our institutions – our schools, our offices – to maximize the opportunities of our digital era" (Davidson 2011, 12). There has also been relatively little debate about how we should organize our universities, faculties, and departments to make the best of both the scholarly expertise that we find in different departments and faculties that pertain to the study of culture. We also need to overcome the disparity between disciplinary specialization dating back to the nineteenth and twentieth centuries and the concerns and challenges we are facing in the new millennium, almost all of which demand interdisciplinary collaboration. The standardized ways in which our institutions have traditionally been organized "may have worked for the twentieth century, but do they always and necessarily make sense for the twenty-first" (Davidson 2011, 220–221)? Moreover, the limits imposed by disciplinary boundaries and bureaucracy are often "hostile to creative thinking" (Graeber 2016, 146), posing unnecessary administrative obstacles rather than fostering collaboration across disciplines and impending conceptual breakthroughs. Such an institutional makeover would also have to address, and critique, such detrimental developments as the ongoing bureaucratization and corporatization of many universities, the increasing reliance on extramural funding, research assessment exercises, and grant agencies. This has involved an "extraordinary squandering of human creativity" (Graeber 2019, 188), not least due to the concomitant proliferation of administrative tasks, paperwork, and increase in what the anthropologist David Graeber, who is fortunately never one to mince words, has designated "bullshit jobs" (2019).

Since the predominant division of labor between faculties, departments, and disciplines is anything but conducive to fostering interdisciplinary projects, we arguably need institutions and research centres that are specifically designed for the study of culture. Re-aligning the study of culture with emerging concerns presupposes that we continue to challenge disciplinary boundaries and develop the study of culture as an interdisciplinary practice. None of the daunting developments that threaten us most in the twenty-first century seem to do us the favor of falling into the scope of just one academic discipline. No matter whether we are dealing with phenomena or processes such as climate change, "mind change" (Greenfield 2014) and other effects of digitalization, or ways of worldmaking, for that matter, coming to terms with any of these or other current challenges involves interdisciplinary collaboration.

As the exemplary discussion of some of the emerging concerns in section 4 above has already indicated, an institutional makeover of the humanities and social sciences is arguably not enough. Coming to terms with such transdisciplinary challenges necessitates forging new collaborative interfaces between the study of culture and disciplines such as economics, law, medicine, psychology, and theology, to name but a few in which cultural aspects are particularly prominent. Initiating dialogues across disciplinary borders between the humanities, social sciences, and life sciences is an important first step, but in order to explore new cross-disciplinary research fields like those briefly discussed above, we need to establish new institutions for important emerging fields like environmental humanities, medical humanities, and the cultures of the economy and law.

Recent international developments have considerably broadened the aims and scope of what falls under the purview of cultural studies or the study of culture, both historically and as far as the synchronic range of forms of art and culture are concerned. Cases in point include the Amsterdam School for Cultural Analysis (ASCA), the European Summer School in Cultural Studies (ESSCS), the Lisbon Summer School for the Study of Culture, and the International Graduate Centre for the Study of Culture (GCSC). These international research networks and institutions have managed to foster new models of collaboration and bring the modes of reasoning and research methodologies from a broad range of disciplines (ranging from history, sociology, and political science to literary and theatre studies) into the study of culture. They have also firmly established and institutionalized the study of culture as a collaborative, evidence-based, and interdisciplinary practice, thus making important contributions to a better understanding of current issues that no discipline can get to grips with on its own: "Collaboration by difference is an antidote to attention blindness. It signifies that the complex and interconnected problems of our time cannot be solved by anyone alone" (Davidson 2011, 100). Such a method is extremely well suited for the kinds of inter- and transdisciplinary projects needed in the study of culture for the twenty-first century, as Davidson's comments illustrate: "Collaboration by difference respects and rewards different forms and levels of expertise, perspective, culture, age, ability, and insight, treating difference not as a deficit but as a point of distinction" (100). Seen in this light, collaboration by difference might be one of the keys for updating dominant methods in such a way as to re-align them for the cross-disciplinary challenges we face in the new millennium:

> If the twentieth century was all about training experts and then not paying attention to certain things because the experts would take care of the matter for you, the twenty-first is about crowdsourcing that expertise, contributing to one another's fund of knowledge, and learning how to work together towards solutions to problems.
> (Davidson 2011, 258)

Such international research networks and institutions as those mentioned above could well serve as models to be emulated if we want to ensure that the study of culture will be able to make important contributions towards solving problems and thus really have a sustainable future for the simple reason that people will begin to understand that such a problem-solving paradigm is indispensable for society. These models have not only demonstrated how the transfer of approaches, concepts, and methodologies between diverse disciplines and academic cultures can be organized and advanced, but they have also managed to create sustained structures for research organization and research training in the field of the study of culture. Since such structures provide the necessary institutional basis for initiating research in innovative, interdisciplinary research areas, investing creative energy, resources, and time in the development and building of such institutions and initiating new projects might be one of the most promising ways of ensuring sustainable futures for the study of culture.

9 Moving beyond Critique, or: A Vision for the Study of Culture as a Problem-Solving Practice for the Twenty-first Century

The second but last proposal can be very brief, because it refers to an important plea that Rita Felski developed in great detail concerning literary studies (see Felski 2015). My point here is simply that her observations about the preponderance of critique over other approaches and ways of knowledge production pertain just as much to the study of culture, and arguably even more so to cultural studies, as to literary criticism and literary studies. Anglo-American forms of cultural studies have traditionally been conceived of as politically engaged projects that put a heavy emphasis on critique. British cultural studies were developed as a response to concrete social and political challenges of the British class system and as a politically motivated project aimed at producing changes in society and strategies of resistance. While culture and politics have always been inextricably intertwined in this research tradition, the German tradition of *Kulturwissenschaften*, which can be traced back to the late nineteenth and early twentieth centuries, has quite a different genealogy, lineage, and non-political agenda, is largely an academic enterprise that explores cultural phenomena as objects of academic research without an eye towards engendering political change.

Notwithstanding such differences, however, Felski's observations about what she calls "the malaise of critique" (Felski 2015, 119) and her plea for the need to move beyond the dominant model of critique applies equally well to cultural studies: "Yet the malaise of critique could also free us up to reassess our current ways of reading and reasoning: to experiment with modes of argument less tightly bound to exposure, demystification, and the lure of the negative" (Felski 2015, 119–120). Similarly, the range of theoretical and methodological approaches she suggests as fruitful alternatives to critique could also open up new horizons for research in the study of culture, in which, e.g., phenomenology and other non-political forms of reading (e.g. "actor-network theory, post-historicist criticism, affective hermeneutics"; Felski 2015, 182) have enjoyed greater prominence than in American or British cultural studies.

I would like to go even further, however, in that I conceive of the study of culture not as a dominantly critical project geared at debunking, demystifying or exposing forms of discrimination or ideologies, but as a paradigm actively engaged in problem-solving. Putting the emphasis on problem-solving would not only involve challenging ingrained assumptions and practices, but it would also arguably be very productive in initiating conversations across disciplinary borders between the humanities, social sciences, and life sciences (see section

8 above). It is in such cross- or trans-disciplinary research fields like environmental humanities, medical humanities, and the cultures of the economy and law that scholars involved in the study of culture could really make worthwhile contributions to solving, rather than merely critiquing, real problems – even though critique has and always will also have an important role to play. The questions of whether the study of culture should be about critique or problem-solving is not an either/or but rather a both/and issue, but we would arguably be well advised if we moved beyond critique and managed to develop better stories, both for the study of culture and the future(s) of the world at large.

10 Developing Positive Visions for the Future(s) of the Study of Culture: Imagining Alternative Positive Futures and Telling Better Future Narratives

Taken together, the previous proposals could be summed up in the overall suggestion to develop positive visions for the future(s) of the study of culture by telling better narratives of the futures of both the world we live in and the interdisciplinary field in which we work. We could thus turn recent insights into the reality-constituting function of narratives to our own advantage and bear in mind that "narratives both expand and diminish our sense of the possible" (Meretoja 2018, 2). Instead of accepting the stories disseminated either in the realms of politics and the media or the largely apocalyptic and dystopian visions of the future in popular culture, we should bear in mind Grossberg's wise and witty reminder that "*Bad Stories Make Bad Politics!*" (Grossberg 2010, 64). My final and overarching proposal is thus that we should scrutinize hegemonic master narratives that no longer make sense of the world as it is and try to invent much better stories.

To propose narrative constructions for the future of the study of culture, we could follow the constructivist advice of Brian McHale:

> it is important to distinguish among better and less good stories – "better" not in the sense of objectively *truer* (a criterion discredited by the constructivist approach), but in terms of such criteria as rightness of fit, validity of inference, internal consistency, appropriateness of scope, and above all *productivity*. (McHale 1992, 9)

If we look at the dominant master narratives of modernity and capitalism in terms of McHale's criteria, we find that they no longer display rightness of fit, have ceased to be appropriate in today's world and are no longer sustainable. Much more work needs to be done to gauge the complex challenges and concerns with which we are faced, and to conceive and develop fully-fledged transnational approaches and concepts for the study of culture that meet such criteria.

It therefore seems apt to conclude by once again stressing the need for developing and debating the trans/national dimension of the study of culture. Anyone interested in transnational approaches to the study of culture that are re-aligned with C21 problematics can profit considerably from comparing different national approaches and considering the ways in which such influential traditions are discursively constructed and institutionally implemented. What is needed for the development of such approaches is an enhanced degree of self-reflection about different national traditions in 'doing' the study of culture, the promotion of greater "transnational literacy" (Bal 2002, 291), an openness to interdisciplinarity, and a questioning of one's own academic routines. Several promising new departures have served to internationalize and even transnationalize approaches to the study of culture, such as the volumes and anthologies *Internationalizing Cultural Studies: An Anthology* (Abbas and Erni 2005), *The Worlding Project: Doing Cultural Studies in the Era of Globalization* (Connery and Wilson 2007), *New Cultural Studies: Adventures in Theory* (Hall and Birchall 2007), and *The Trans/National Study of Culture: A Translational Perspective* (Bachmann-Medick 2014), all of which delineate promising trajectories for developing a transnational study of culture.

We have also witnessed sustained attempts to develop new forms of 'global' cultural studies. Although an Institute for Global Cultural Studies (IGCS) was founded as early as 1991 at Binghampton University, with other universities offering programmes on global cultural studies following suit (e.g., Point Park University and most recently Université Jean Moulin Lyon 3), there have also been critical voices expressing scepticism against the very notion of 'a global cultural studies.' Jon Stratton and Ien Ang's warning that the "'internationalization' of cultural studies cannot mean the formation of a global, universally generalisable set of theories and objects of study" (1996, 363) may still serve as a timely reminder that the field of the study of culture may not lend itself particularly well to universalizing or transnationalizing gestures, and even less to attempts to develop universal theories of any cultural object, phenomenon, or process. In a particularly stimulating and thought-provoking review essay, Imre Szeman assesses the notion of "*Global* Cultural Studies" (Szeman 2011), carefully gauging both the risks and promises of such an ambitious project, and the political and epistemological problems that it entails. We face both "national-cultural situations, events, and circumstances" and a more or less "shared global critico-theoretical discourse" (Szeman 2011, 148), though the latter indeed tends to be "heavily weighted towards ideas emerging from Anglo-American and European traditions" (148). I not only agree but would add that we should redress the balance by provincializing (sensu Chakrabarty 2000) hitherto dominant approaches and following up on the ten proposals made here.

While cultural studies may well have gone international or even transnational in some instances, there is still a great need to further develop innovative and truly transnational approaches to the study of culture that are equipped to come to terms with such global challenges and transnational questions of the twenty-first century as highlighted by Harari:

> What will happen to the job market once artificial intelligence outperforms humans in most cognitive tasks? What will be the political impact of a massive new class of economically useless people? What will happen to relationships, families and pension funds when nanotechnology and regenerative medicine turn eighty into the new fifty? What will happen to human society when biotechnology enables us to have designer babies, and to open unprecedented gaps between rich and poor? (Harari 2016, 269)

Although it may be a tall order to try to change disciplinary traditions, institutional contexts, and theoretical frameworks dating back to the twentieth century (if not before) in such a way as to align them with the concerns and challenges of the twenty-first century, there is definitely a real need for the kind of conceptual and institutional project makeover outlined above (see section 8).

Moreover, we should neither give up hope nor forget that everything could be different, to translate the title of Harald Welzer's (2019) recent book. Even ingrained disciplinary traditions and institutional arrangements can be changed if there are enough dedicated and enthusiastic people willing to collaborate in order to challenge and change them: "Don't give up hope. Departmental cultures can change. And there can be subcultures of support within departments creating pockets of resistance to the effects of the corporate university" (Berg and Seeber 2016, 84). The study of culture will only have a future if the scholars working in the field are prepared to take full responsibility for it and take Max Tegmark's wise words that "we need more mindful optimists" to heart (Tegmark 2017, 334). It is certainly high time that we **"re-orientate ourselves from an exclusive preoccupation with retrospectively making meaning(s) to the creative activity of making future(s), prospectively"** (Bode and Dietrich 2013, 107; bold-print emphasis in the original).

The ten proposals made above are an attempt to sketch some of the tasks that are involved in such an endeavor, outlining what the most promising trajectories may be. I am perfectly aware, however, that these proposals are relatively general and that they need to be fleshed out in a much more detailed fashion. Therefore I would like to leave the last words to the late psychologist Christopher Peterson, whose ambivalent self-assessment of his own vision for the future of psychology also expresses my own sense of striving to strike the right balance between aiming high while being perfectly aware of the limitations of what we can achieve:

> As a research psychologist, my goal is to do studies that are important and interesting, that answer questions about weighty matters, and that suggest ways to enhance the psychological good life. This is a vision with which I am quite happy, although it is of course vague. When I start to flesh out the vision, the process too often gets mired down by my worries about the means. (Peterson 2013, 321–322)

References

Abbas, Ackbar, and John Nguyet Erni, eds. *Internationalizing Cultural Studies: An Anthology*. Malden, MA/Oxford/Carlton: Blackwell, 2005.

Allatson, Paul. *Key Terms in Latino/a Cultural and Literary Studies*. Malden, MA/Oxford/Carlton: Blackwell, 2007.

Bachmann-Medick, Doris, ed. *The Trans/National Study of Culture: A Translational Perspective*. Berlin/Boston: De Gruyter, 2014.

Bachmann-Medick, Doris. *Cultural Turns: New Orientations in the Study of Culture*. Berlin/Boston: De Gruyter, 2016.

Bachmann-Medick, Doris, and Jens Kugele, eds. *Migration: Changing Concepts, Critical Approaches*. Berlin/Boston: De Gruyter, 2018.

Bal, Mieke. *Travelling Concepts in the Humanities: A Rough Guide*. Toronto: University of Toronto Press, 2002.

Bal, Mieke. "Working with Concepts." *Travelling Concepts in English Studies. EJES: European Journal of English Studies* 13.1 (2009): 13–24.

Bartmann, Christoph. *Die Rückkehr der Diener. Das neue Bürgertum und sein Personal*. Munich: Hanser, 2016.

Basseler, Michael, Daniel Hartley, and Ansgar Nünning, eds. *Emergent Forms of Life in Anglophone Literature: Conceptual Frameworks and Critical Analyses*. Trier: WVT, 2015.

Bateson, Gregory. *Steps to an Ecology of Mind*. With a new Foreword by Mary Catherine Bateson. Chicago/London: University of Chicago Press, 2000 [1972].

Bauman, Zygmunt. *Leben als Konsum*. Hamburg: Hamburger Edition, 2009 (orig.: *Consuming Life*. London: Polity, 2007).

Baumbach, Sibylle, Beatrice Michaelis, and Ansgar Nünning, eds. *Travelling Concepts, Metaphors, and Narratives: Literary and Cultural Studies in an Age of Interdisciplinary Research*. Trier: WVT, 2012.

Becker, Gaylene. *Disrupted Lives: How People Create Meaning in a Chaotic World*. Berkeley/Los Angeles/London: University of California Press, 1997.

Belsey, Catherine. "Beyond Literature and Cultural Studies: The Case for Cultural Criticism." *Belgian Journal of English Language and Literatures* New Series 1. Thematic Issue (2003): 91–100.

Berg, Maggie, and Barbara K. Seeber. *The Slow Professor: Challenging the Culture of Speed in the Academy*. Toronto/Buffalo/London: University of Toronto Press, 2016.

Berning, Nora, Ansgar Nünning, and Christine Schwanecke, eds. *Reframing Concepts in Literary and Cultural Studies: Theorizing and Analyzing Conceptual Transfers*. Trier: WVT, 2014.
Bode, Christoph, and Rainer Dietrich. *Future Narratives: Theory, Poetics, and Media-Historical Moment*. Berlin/Boston: De Gruyter, 2013.
Bruner, Jerome. *Acts of Meaning*. Cambridge, MA/London: Harvard University Press, 1990.
Bruner, Jerome. "Self-Making and World-Making." *Journal of Aesthetic Education* 25.1 (1991): 67–78.
Bruner, Jerome. *Making Stories: Law, Literature, Life*. Cambridge, MA/London: Harvard University Press, 2002.
Chakrabarty, Dipesh. *Provincializing Europe: Postcolonial Thought and Historical Difference*. Reissue, with a new preface by the author. Princeton/Oxford: Princeton University Press, 2008 [2000].
Citton, Yves. *The Ecology of Attention*. Cambridge/Malden, MA: Polity, 2018.
Connery, Christopher Leigh, and Rob Wilson, eds. *The Worlding Project: Doing Cultural Studies in the Era of Globalization*. Santa Cruz: New Pacific, 2007.
Crary, Jonathan. *24/7: Late Capitalism and the Ends of Sleep*. London/New York: Verso, 2014.
Cunningham, Guy Patrick. "Fragmentary: Writing in a Digital Age." 2016. http://www.themillions.com/2012/01/fragmentary-writing-in-a-digital-age.html [accessed: 22 November 2019].
Davidson, Cathy C. *Now You See It: How Technology and Brain Science will Transform Schools and Business for the 21st Century*. London: Penguin, 2011.
De Botton, Alain. *The News: A User's Manual*. London: Pantheon, 2014.
Eakin, Paul John. *How Our Lives Become Stories: Making Selves*. Ithaca: Cornell University Press, 1999.
Eddo-Lodge, Reni. *Why I'm no Longer Talking to White People about Race*. London et al.: Bloomsbury, 2018.
Erll, Astrid, and Ansgar Nünning, eds. *Cultural Memory Studies: An International and Interdisciplinary Handbook*. Berlin/New York: De Gruyter, 2008.
Erll, Astrid, Herbert Grabes, and Ansgar Nünning, eds. *Ethics in Culture: The Dissemination of Values Through Literature and Other Media*. Berlin/New York: De Gruyter, 2008.
Erll, Astrid, and Ansgar Nünning, eds. *A Companion to Cultural Memory Studies*. Berlin/New York: De Gruyter, 2010.
Erll, Astrid, and Roy Sommer, eds. *Narrative in Culture*. Berlin/Boston: De Gruyter, 2019.
Faludi, Susan. *Backlash: The Undeclared War Against American Women*. New York: Three Rivers Press, 2006 [1991].
Felski, Rita. "Modernist Studies and Cultural Studies: Reflections on Method." *Modernism/Modernity* 10.3 (2003): 501–517.
Felski, Rita. *The Limits of Critique*. Chicago/London: University of Chicago Press, 2015.
Fluck, Winfried, Donald Pease, and John Carlos Rowe, eds. *Re-framing the Transnational Turn in American Studies*. Hanover: Dartmouth College Press, 2011.
Franck, Georg. *Ökonomie der Aufmerksamkeit. Ein Entwurf*. Munich: Hanser, 1998.
Frankfurt, Harry. *On Bullshit*. Princeton/Oxford: Princeton University Press, 2005.
Fry, Hannah. *Hello World: How to be Human in the Age of the Machine*. London: Transworld/Black Swann, 2018.
Goodman, Nelson. *Ways of Worldmaking*. 1978. Indianapolis: Hackett, 1992.
Grabes, Herbert, Ansgar Nünning, and Sibylle Baumbach, eds. *Metaphors Shaping Culture and Theory. REAL: Yearbook of Research in English and American Literature* 25. Tübingen: Narr, 2009.

Graeber, David. *The Utopia of Rules: On Technology, Stupidity, and the Secret Joys of Bureaucracy*. Brooklyn/London: Melville House, 2016.
Graeber, David. *Bullshit Jobs: A Theory*. London/New York: Penguin, 2019.
Greenfield, Adam. *Radical Technologies: The Design of Everyday Life*. London/New York: Verso, 2017.
Greenfield, Susan. *Mind Change: How Digital Technologies are Leaving their Mark on our Brains*. London: Penguin, 2014.
Greif, Mark. *Against Everything: On Dishonest Times*. London/New York: Verso, 2016.
Grossberg, Lawrence. "Does Cultural Studies Have Futures? Should It? (Or What's the Matter with New York?)." *Cultural Studies* 20.1 (2006): 1–32.
Grossberg, Lawrence. *Cultural Studies in the Future Tense*. Durham, NC/London: Duke University Press, 2010.
Grusin, Richard. "Mediashock." *Structures of Feeling: Affectivity and the Study of Culture*. Eds. Devika Sharma and Frederik Tygstrup. Berlin/Boston: De Gruyter, 2015. 29–39.
Hall, Gary, and Clare Birchall, eds. *New Cultural Studies: Adventures in Theory*. Athens: The University of Georgia Press, 2007.
Hänggi, Marcel. *Fortschrittsgeschichten. Für einen guten Umgang mit Technik*, Frankfurt a.M.: Fischer, 2015.
Harari, Yuval Noah. *Homo Deus: A Brief History of Tomorrow*. London: Harvill Secker, 2016.
Harari, Yuval Noah. *21 Lessons for the 21st Century*. London: Jonathan Cape, 2018.
Hayles, N. Katherine. *How We Think: Digital Media and Contemporary Technogenesis*. Chicago/London: University of Chicago Press, 2012.
Herman, David. "Editor's Column: Principles and Practices of Narrative Worldmaking." *Storyworlds* 3 (2011): vii–x.
Hodges, Adam. *The "War on Terror" Narrative: Discourse and Intertextuality in the Construction and Contestation of Sociopolitical Reality*. Oxford: Oxford University Press, 2011.
Jaeggi, Rahel. *Kritik von Lebensformen*. Berlin: Suhrkamp, 2014.
Kovach, Elizabeth. *Novel Ontologies after 9/11: The Politics of Being in Contemporary Theory and U.S.- American Narrative Fiction*. Trier: WVT, 2016.
Lanchester, John. *Whoops! Why Everyone Owes Everyone and No One Can Pay*. London: Penguin, 2010.
Lanier, Jaron. *Ten Arguments for Deleting Your Social Media Accounts Right Now*. London: Bodley Heade, 2018.
Levine, Caroline. *Forms: Whole, Rhythm, Hierarchy, Network*. Princeton/Oxford: Princeton University Press, 2015.
Lynch, Michael Patrick. *The Internet of Us: Knowing More and Understanding Less in the Age of Big Data*. New York/London: Liveright Publishing Corporation, 2016.
MacIntyre, Alasdair. "Epistemological Crises, Dramatic Narrative, and the Philosophy of Science." *The Monist* 60.4 (1977): 453–471 (reprinted in: MacIntyre, Alasdair. *The Tasks of Philosophy: Selected Essays*. Vol.1. Cambridge: Cambridge University Press, 2006. 3–23.)
MacIntyre, Alasdair. *After Virtue: A Study in Moral Theory*. London: Bloomsbury, 2013 [1981].
McAdams, Dan P. *The Redemptive Self: Stories Americans Live By*. Oxford: Oxford University Press, 2013.
McHale, Brian. *Constructing Postmodernism*. London: Routledge, 1992.
Meiner, Carsten, and Kristin Veel, eds. *The Cultural Life of Catastrophes and Crises: Facts, Forms, Fantasies*. Berlin/New York: De Gruyter, 2012.

Meretoja, Hanna. *The Ethics of Storytelling: Narrative Hermeneutics, History, and the Possible*. Oxford: Oxford University Press, 2018.
Mersmann, Birgit, and Hans G. Kippenberg, eds. *The Humanities between Global Integration and Cultural Diversity*. Berlin/Boston: De Gruyter, 2016.
Michaels, Axel, ed. *Ritual Dynamics and the Science of Ritual*. Vol. 1–5. Wiesbaden: Harrasowitz, 2010–2011.
Mitchell, W.J.T. "Realism, Irrealism, and Ideology: A Critique of Nelson Goodman." *The Journal of Aesthetic Education* 25.1 (1991): 23–35.
Morozov, Evgeny. *To Save Everything, Click Here: Technology, Solutionism and the Urge to Fix Problems That Don't Exist*. London/New York: Allen Lane, 2013.
Neumann, Birgit, and Ansgar Nünning, eds. *Travelling Concepts for the Study of Culture*. Berlin/New York: De Gruyter, 2012.
Nünning, Ansgar. "Making Events – Making Stories – Making Worlds: Ways of Worldmaking from a Narratological Point of View." *Cultural Ways of Worldmaking: Media and Narratives*. Eds. Vera Nünning, Ansgar Nünning, and Birgit Neumann. Berlin/New York: De Gruyter, 2010. 191–214.
Nünning, Ansgar. "Narrativist Approaches and Narratological Concepts for the Study of Culture." *Travelling Concepts for the Study of Culture*. Eds. Birgit Neumann and Ansgar Nünning. Berlin/New York: De Gruyter, 2012. 145–183.
Nünning, Ansgar, ed. *Metzler Lexikon Literatur- und Kulturtheorie. Ansätze – Personen – Grundbegriffe*. 5th ed. Stuttgart/Weimar: Metzler, 2013 [1998].
Nünning, Ansgar. "Towards Transnational Approaches to the Study of Culture: From Cultural Studies and *Kulturwissenschaften* to a Transnational Study of Culture." *The Trans/National Study of Culture: A Translational Perspective*. Ed. Doris Bachmann-Medick. Berlin/Boston: De Gruyter, 2014. 23–49.
Nünning, Ansgar, and Vera Nünning. "Ways of Worldmaking as a Model for the Study of Culture: Theoretical Frameworks, Epistemological Underpinnings, New Horizons." *Cultural Ways of Worldmaking: Media and Narratives*. Eds. Vera Nünning, Ansgar Nünning, and Birgit Neumann. Berlin/New York: De Gruyter, 2010. 1–25.
Nünning, Ansgar, and Vera Nünning. "Conceptualizing 'Broken Narratives' from a Narratological Perspective: Domains, Concepts, Features, Functions and Suggestions for Research." *Narrative im Bruch. Theoretische Positionen und Anwendungen*. Eds. Anna Babka, Marlen Bidwell-Steiner, and Wolfgang Müller-Funk. Göttingen: Vienna University Press, 2016. 37–86.
Nünning, Ansgar, and Kai Sicks, eds. *Turning Points: Concepts and Narratives of Change in Literature and Other Media*. Berlin/New York: De Gruyter, 2012.
Nünning, Vera, and Ansgar Nünning. "Stories as 'Weapons of Mass Destruction', or: George W. Bush's Narratives of Crisis as Paradigm Examples of Ways of World- and Conflict-Making (and Conflict-Solving?)." *Narrative(s) in Conflict*. Eds. Wolfgang Müller-Funk and Clemens Ruthner. Berlin/Boston: De Gruyter, 2017. 189–229.
Nünning, Vera, and Ansgar Nünning. "Cultural Concerns, Literary Developments, Critical Debates: Contextualizing the Dynamics of Generic Change and Trajectories of the British Novel in the Twenty-first Century." *The British Novel in the Twenty-First Century: Cultural Concerns – Literary Developments – Model Interpretations*. Eds. Vera Nünning and Ansgar Nünning. Trier: WVT, 2018. 21–52.
Nünning, Vera, Ansgar Nünning, and Birgit Neumann, eds. *Cultural Ways of Worldmaking: Media and Narratives*. Berlin/New York: De Gruyter, 2010.

Oreskes, Naomi, and Erik M. Conway. *Merchants of Doubt: How a Handful of Scientists Obscured the Truth on Issues from Tobacco Smoke to Global Warming*. London/New York: Bloomsbury, 2012.
Peterson, Christopher. *Pursuing the Good Life: 100 Reflections on Positive Psychology*. New York/Oxford: Oxford University Press, 2013.
Phelan, James. *Living to Tell About It: A Rhetoric and Ethics of Character Narration*. Ithaca/London: Cornell University Press, 2005.
Rosling, Hans. *Factfulness: Ten Reasons Why We're Wrong About the World – and Why Things Are Better than You Think*. London: Hoddert and Stoughton, 2018.
Sharma, Devika, and Frederik Tygstrup, eds. *Structures of Feeling: Affectivity and the Study of Culture*. Berlin/Boston: De Gruyter, 2015.
Simanowski, Roberto. *Facebook-Gesellschaft*. Berlin: Matthes & Seitz, 2016.
Spiekermann, Sarah. *Digitale Ethik. Ein Wertesystem für das 21. Jahrhundert*. Munich: Droemer, 2019.
Stratton, Jon, and Ien Ang. "On the Impossibility of a Global Cultural Studies: 'British' Cultural Studies in an 'International' Frame." *Stuart Hall: Critical Dialogues in Cultural Studies*. Eds. David Morley and Kuan-Hsing Chen. New York: Routledge, 1996. 361–391.
Szeman, Imre. "Cultural Studies and the Transnational." *New Cultural Studies: Adventures in Theory*. Eds. Gary Hall and Clare Birchall. Athens: The University of Georgia Press, 2007. 200–218.
Szeman, Imre. "*Global* Cultural Studies?" *The Minnesota Review* 76 (2011): 147–161.
Taplin, Jonathan. *Move Fast and Break Things: How Facebook, Google and Amazon Have Cornered Culture and what it Means for All of Us*. London: Macmillan, 2017.
Tegmark, Max. *Life 3.0: Being Human in the Age of Artificial Intelligence*. London/New York: Penguin, 2017.
Turkle, Sherry. *Reclaiming Conversation: The Power of Talk in the Digital Age*. New York: Penguin, 2015.
Twenge, Jean M. *iGen: Why Today's Super-Connected Kids are Growing Up Less Rebellious, More Tolerant, Less Happy – and Completely Unprepared for Adulthood*. New York/London: Atria, 2017.
Welsch, Wolfgang. "Transkulturalität." *Universitas* 52. 607 (1997): 16–24.
Welzer, Harald. *Selbst Denken. Eine Anleitung zum Widerstand*. Frankfurt a.M.: Fischer, 2013.
Welzer, Harald. *Die smarte Diktatur. Ein Angriff auf unsere Freiheit*. Frankfurt a.M.: Fischer, 2016.
Welzer, Harald. *Alles könnte anders sein. Eine Gesellschaftsutopie für freie Menschen*. Frankfurt a.M.: Fischer, 2019.
Woolf, Maryanne. *Tales of Literacy for the 21st Century*. Oxford: Oxford University Press, 2016.
Woolf, Maryanne. *Reader, Come Home: The Reading Brain in a Digital World*. New York: HarperCollins, 2018.

Andreas Langenohl

The "Future Sense" and the Future of the Study of Culture

1 Introduction: The Future Unravelling the Present

The title of this collection of essays might, at first glance, sound rather speculative. How are we to know what future awaits the study of culture? After all, the future seems to be, by modern definition, that which cannot be known and is not accessible in the present, following such diverse scholars as Frank Knight, Reinhart Koselleck, or Niklas Luhmann. What is more, this fundamental 'uncertainty' regarding the future may be celebrated as indicating the potentially open horizon of contemporary societies – an openness that has to be defended against attempts to foreclose and to predefine it, as seems to be happening in such diverse arenas of permanent crisis as accelerating global climate change, the mushrooming of security apparatuses, or out-of-bounds financial speculation. Should the study of culture not join the struggle for openness, and against predefinition?

However, one might interpret the title of this collection also in a grammatically different sense, such that the study of culture might 'have' an own mode of future, or futurity. A brief look at language, still one of the most important and paradigmatic areas of research in the study of culture, already shows that different languages have different modalities of referencing the future. For instance, in English we can differentiate between something that 'will' happen, something that 'will have' happened, and something that is imminent, something that 'is going to' happen. Seen from this angle, 'the future of the study of culture' might refer less to a prediction or anticipation, and instead might carve out a research field in its own right. Its question is: How are we to understand things and events with an index of futurity, and what might differentiations in that understanding be?

Recently, two prominent scholars in the study of culture have been attempting to formulate an understanding of futurity that critically reflects on concepts of the future that derive future openness from its supposed unknowability and uncertainty. Literary studies scholar Leslie A. Adelson and cultural anthropologist Arjun Appadurai have both argued that the future might be understood as a function of the present – not in the sense that the future depends on and follows from present events and decisions, but in the sense that it informs, or may inform, the present precisely inasmuch as the present seems to be caught in a violent and hopeless synchronicity. This "future sense," as Adelson calls it (2017, 29), is thus invoked in a bid to wrest the present away from its seemingly total interlocking in

https://doi.org/10.1515/9783110669398-004

a horizon of, in Theodor W. Adorno's terms, "permanent catastrophe" (Adelson 2017, 78). For Adelson, the recent work of Alexander Kluge provides a vista for identifying a futurity in present conditions that is bound to unravel these conditions as they unfold in catastrophe, where "world-making [...] is constantly undone" (Adelson 2017, 5). For Appadurai, it is the solidarity practices of marginalized social groups all over the world that form globe-girdling alliances whose cooperative ethos is informed by experiences that become salient in the present, but transgress and break open its violent synchronicity and fatedness for disaster. Insisting on a mode of futurity that becomes activated in the present but denies its synchronicity, both Appadurai (2013, 1) and Adelson (2017, 64) speak of the future as a "fact," or more precisely, of a "cultural" and a "social fact," respectively.

In this article, I will discuss these suggestions of a 'future sense' that take issue with notions of futurity that reduce the future's openness to its intangibility in the present. As one shared concern in both suggestions is social cooperation, I will then contextualize them with a precursor in the political history of the twentieth century, namely, 'prefigurative politics' at whose core, I suspect, one can find a similar concern with the problem that a future that is radically shut off from the present loses its transformative impulse for present cooperation. Prefigurative politics has been experimented with since the 1960s, when it emerged out of frustration both with the violent synchronicity of instrumental mainstream politics and with instrumentalist understandings of revolutionary action. Today, prefigurative politics is related back to particular strands of anarchist thought (see Day 2005, 91–128; see also Graeber 2013, 89, 186–195 *et passim*) that sought to carve out social and political spaces within the existing society that would envision future modes of cooperation without waiting for a revolution to happen. Without claiming the existence of a genealogy between Adelson's and Appadurai's understanding of futurity with prefigurative politics or anarchist thought, I will concentrate on shared contextual features that, to my mind, interrelate these approaches: an urge to decipher the future as impacting the present; a characterization of that present that insists on its own normality while being caught in functional imperatives that constantly engender disaster and steer toward greater disaster; and a focus on a field of social cooperation that emerges against the odds.

In the last section, I will conclude on some points that I regard as significant for this volume's dedication, specifically on the question of what the 'future sense' might harbor for the future of the study of culture. Emerging at different sites and settings, it seems that the 'future sense' is historically circumscribed by a double dismay at modernity's catastrophism and its tendency to turn the future into a mere externality of the present. In this context, the question arises what 'utopia' can mean for a present that learns to adapt its sensorium to futurity.

2 Futurity in Reading and Interrelating

Adelson's current book on Alexander Kluge's "cosmic miniatures" (Adelson 2017, 1) can be read as the continuation of an interest in the literary articulation of discontinuities. In an earlier phase of her work, Adelson analyzed Turkish-German literature with a view to how this literary production in the German language might be understood not as a mingling of different 'cultures' – that of the 'classical' German canon and that of the Turkish 'origin' of writers such as Zafer Şenocak and Feridun Zaimoğlu – but as a site of literary representations whose gist cannot be reduced to a cultural genealogy (see Adelson 2005). Her manifesto "Against Between" (Adelson 2001) challenged the view that these writers and their works cover a space 'between' cultures, instead arguing that this production has to be understood as *German* literature on its way of shedding off not only classical canonical references but the very idea of cultural and literary tradition in terms of continuity. This approach had a methodological side as well. Adelson's analyses do not so much refer to the plot level or the use of metaphors and tropes, which all too easily lend themselves for the 'representation' of 'cultures,' but rather to the textual creation of polyvalent narrative positionalities and spatial markers that shuttle opaquely between geographical referentiality and spatialized phantasm. These narrative devices make culturalist and genealogical interpretations of those texts more and more implausible. Developing this research strategy further, Adelson compared Turkish-German literature with writings of Alexander Kluge. In "Experiment Mars" (Adelson 2008), she argues that, in these writers, the invocation of memories of the past as informing the present gives way to the invocation of the future. These analyses do not only diagnose particular uses of the *referential* category of future in literary writing. More importantly, they converge on a shared insistence on making the future *perceptible* in narration, thus opening a vista on what Adelson terms the "protean abstraction" of futurity.[1]

In *Cosmic Miniatures and the Future Sense*, Adelson focuses on Alexander Kluge against the background of the engagement of his writings with critical theory. While the memory boom in the study of culture has led many scholars to depict the main contributions of Walter Benjamin or Theodor W. Adorno in their invoking the past as a potential horizon for thinking redemption amidst a ruinous world, Adelson argues that another reading is possible, and more accurate, that dissects the potentially enabling and emancipating presence of a sense of the future in the catastrophic present. Kluge's work, according to

1 See Adelson 2013, 215. See also the literary analyses in the special issue of *The Germanic Review* to which this text by Adelson is the introduction.

Adelson, vouches for such reading and develops it further; his current miniature writing represents a critical-interpretative elaboration on critical theory, which, as Adorno's *Minima moralia* demonstrates, at significant points took the form of the miniature itself. Again, the main methodological device in the reconstruction of this argument is a formal-narratological one, as Adelson is interested in the production of complications concerning the narrative positionality that create "experiential portals in time" (Adelson 2017, 44, 69, 127) for the future to enter the present.

Here I can refer only to a few examples of Adelson's narratological readings. Analyzing Adorno's miniature "Heliotrope" in a preparatory step in laying out the notion of 'future sense' (Adelson 2017, 95–110), Adelson demonstrates that the narrator in this miniature shifts from a voice that would be possible to identify with the little boy anticipating a house guest as a person who for him represents a "radiant other life" (Adelson 2017, 108), to a voice that actualizes that radiant future not as a possibility but as a certainty, speaking, as it were, from the position of the *fulfilled* dreams and hopes of the boy. The mode of futurity crystallizing around this shifting narratorial voice is thus not one of anticipation and hope under the proviso of uncertainty, but one that installs that future as a referential certainty in the present. The point of Adelson's interpretation of narrative futurity thus relates both to "longing" (Adelson 2017, 107), as a vision of a radiant future life represented by the house guest, and to a "sense," that is, an imaginary and sensual capacity – "a long-distance sense organ of temporal perception" (Adelson 2017, 200) – that is conveyed in the practice of reading as it follows the shifts and turns in narrative perspective. Through these turns, the reference object of that longing acquires a "utopian dimension" (Adelson 2017, 106) that has a locale in the present. The 'future sense' is a mode of perception "that becomes, however incrementally, phenomenologically accessible to social experience through reading" (Adelson 2017, 196).

Thereby, it has to be borne in mind that the 'future sense' manifests itself against the background of experiences of macro-violence in the twentieth century that instill a sense of 'permanent catastrophe,' and are bound to leave hope only in the quality as "counterfactual hope" (Adelson 2017, 218). Comparing Adorno's miniatures with those of Kluge, Adelson finds that, in Adorno, the 'future sense' as conveyed in the narrative structure of "Heliotrope" still bears the mark of anticipation, even if one that somehow stubbornly and against the odds of a crushed world embraces certainty. In contrast to this, according to Adelson, Kluge's 'future sense' is "more robust" (Adelson 2017, 77), as it assembles the future on par with the present, thus challenging the latter's catastrophic reality directly. Her reading of one of Kluge's miniatures on Fritz Bauer, the legal and political architect of the 1960s Auschwitz trials in Frankfurt, shows how this is achieved: The miniature posits a first-person narrator witnessing Bauer's funeral and at the same time seeing the dead Bauer addressing inmates at a prison as 'comrades' (an episode verified in Bauer's biography). For Adelson, this represents "a perlocutionary speech act in the sense that Bauer is not addressing an existing collective but a future collective he would like to call into being" (Adelson 2017, 244–245). Moreover, all collectives alluded to in the text are almost never circumscribed through a first-person plural pronoun but through its conspicuous absence: "The historical voices of Kluge's conjunctive cultivation of differential temporalities in narrative perspective are not simply plural, collective, or even collaborative, but cooperative instead" (Adelson 2017, 245). Adelson's interpretation thus stipulates the 'future sense' as a mode of building collectivity not through the invocation of a 'we' that always threatens with essentialization, but through envisioning a cooperation between different human beings that is vouched for in the narrative experience enabled by the text's formal narrative structure. That structure becomes the point of entry for a modality of hope that, while unavoidably being 'counterfactual' in the face of the mass atrocities of the twentieth century, is nevertheless a "real counterfactual force" (Adelson 2017, 246).

While Leslie A. Adelson reconstructs political sociality, as based on future cooperation that is operative in the form of narrative address already in the present, through the practice of narratological reading, Arjun Appadurai grounds his understanding of the future as a "cultural fact" (2013, 1) on his fieldwork among marginalized, oppressed, and exploited groups in Mumbai. According to his analysis, the forms of sociality and cooperation found among the members of these groups can be regarded a counter-project to contemporary capitalism, which he characterizes, with Naomi Klein, as "disaster capitalism" (Appadurai 2013, 295–296). The reference to futurity is established by Appadurai, in a way that is comparable to Adelson's pronouncement of the 'future sense' as an antidote to 'permanent catastrophe,' through a juxtaposition with the temporal colonization of the present and the future through out-of-bounds financial markets. The diagnosis that financial capitalism imposes its extractive and exploitative imperatives on societies through a particular commodification of the anticipated future refers back to Klein's notion of "disaster capitalism," as evident from financial instruments that bet on the probability of future natural and political disasters to happen; but it has also been observed by other scholars (see Lee and LiPuma 2002, 203–207). As financial capitalism draws out a certain scenario of the future as punctuated by more or less probable and more or less profitable catastrophes, Appadurai finds in the mundane practices of subaltern populations and in their efforts to create networks of solidarity a different register of futurity, that would be circumscribed by concepts like 'hope' or 'ethics of possibility.' This mode of futurity can be regarded a 'cultural fact' inasmuch as it emerges from everyday capacities and at the same time creates another capacity, which Appadurai terms the "capacity to aspire" (Appadurai 2013, 179).

Here, I will take a closer look at Appadurai's (2013, 197) interpretation of "cosmopolitanism from below" on the side of the urban poor that, according to this analysis, is both the result of forced adaptive strategies to survive and a platform from which to embark on cooperation in the direction of emancipation. In this analysis, the author refers in particular to female sex workers in Mumbai who were urged to acquire language competencies demanded by their interactions with their male clients, who come from different cultural and linguistic backgrounds in India (Appadurai 2013, 206–208). Their multilingual and multicultural proficiency has thus formed not by dint of an intellectual decision in the horizon of a philanthropic ethics, but as the consequence of highly exploitative and commodified social relationships. Appadurai writes: "Such cosmopolitanism is hard won, unsupported as it is by the apparatus of literacy and cultural privilege or by the practices of leisure and self-cultivation" (Appadurai 2013, 208). Yet at the same time, as a side effect this has led women to greater ability and social resourcefulness with respect to the forging of social links not preordained by exploitation, as they use their communicative proficiencies to transform their neighborhoods[2] and to link up with other marginalized groups and communities in India and elsewhere.

This is an example of how Appadurai conceptualizes cosmopolitanism as having, despite its highly problematic genesis, the potential to open up horizons of change, which he terms the "capacity to aspire" (Appadurai 2013, 179). The capacity to aspire can be regarded as a "navigational capacity" (Appadurai 2013, 188) that – where it is situated within social structures, as it is an unevenly distributed resource – encourages, and rests on, cooperation with others, and hence is more than the sheer aspiration toward individual social upward mobility. Rather, the capacity to aspire forms connections and enables cooperation in the present by dint of a future as part of a web of sociality whose present saliency resides in that very cooperation.

To conclude this section, I want to focus on conceptual commonalities between the two ways that the future as a 'fact' is introduced by the two bodies of scholarly work discussed here. The concept of the future as a social or cultural fact is invoked in a refusal of a catastrophic present, which threatens to prolong itself into the future. Futurity is thus outspokenly anti-utopian, as it claims its

2 The term 'neighborhood' has been used by Appadurai 1996, 181–183, in terminological function in order to characterize a given form of social organizations that forms in interaction with its context (for instance, its natural environment).

topos in the present; we might say that it is only through the present that futurity can actually turn into a future that would be different from a mere prolongation of the disastrous present. This conceptualization thus necessarily entails a radical deconstruction of the modernist conception of the future as principally 'open,' as that future is preordained by the self-reproductive proneness to disaster of the present. That modernist conception of the future as that which is open and thus cannot be known gives way to a much darker vision of the future as that which *will* happen in an undeniably catastrophic way if the present continues as it does. In Adelson, this gloomy picture of the future is rendered in critical-theoretical terms such as 'counterfactual hope' and 'permanent catastrophe,' developed in the wake of the Holocaust as a macro-crime that put an end to any indolent trust in the future as necessarily better than the present. In Appadurai, we find a comparable diagnosis that focuses on financial capitalism's push to commodify the future, a commodification that rides on a drive toward inescapable, if unpredictable and thus potentially profitable, catastrophe.

In the horizon of these circumstances, the notion of the future as a 'fact' suggests an activation of futurity *beyond* 'mere' hope, that is, of a future that has a greater robustness than an expectation or an anticipation as it plays itself out already in the present – as a "bridge in time to the utopian dimension of a 'future without life's miseries,' as Adorno put it" (Adelson 2017, 245). While the notion that hope cannot be but counterfactual remains salient as a necessary reminder of the seemingly insurmountable misery of the present 'damaged life,' both the 'future sense' and the 'capacity to aspire' refer to practices – practices of reading and practices of interrelating – that challenge the threatening facticity of a *bad* future by the facticity of a present futurity that is capable of tearing the present out of its violent and catastrophic synchronism.

Finally, we find in both bodies of work a focus on cooperation as a site where futurity can manifest as an imagination-orienting practice. The stunning improbability of this invocation of cooperation becomes graspable if one contemplates the circumstance that modernity has itself compromised the notion of cooperation to ethical incomprehensibility. Adelson (2017, 239) alludes in her work to the fact that the Holocaust was brought about by the expansion of the principles of the industrial and bureaucratic division of labor into the organization of the extermination of the Jewish population. Appadurai (2013, 244–245) builds his critique of present-day capitalism on insights that emphasize the perversion of the means-ends relationship in finance, as when instruments designed to insure against financial risks become the object of financial bets, which comes along with a reduction of any understanding of cooperation to 'corporate ethics.' And yet, cooperation figures in both theorists as a futurist challenge to these advanced stages of decline of modernist and capitalist modes of reproduction. For Adelson, envisioning cooperation posits an alternative to the invocation of a collective identity (Adelson 2017, 244–246), which historically served as an ideological antidote to radical functional differentiation and division of labor. For Appadurai, cooperation is the unforeseen consequence of a radical commodification and exploitation of human beings who, in order to make their living, must adapt to their circumstances and thereby acquire the capacity to transform these circumstances. This kind of cooperation, however socially marginal or narratologically presuppositional, signifies not so much an alternative future but an alternative present: It does not anticipate future cooperation, but instills cooperation in the present so as to decouple the present from its functional entanglement with an imminent catastrophic future that prolongs the present.

To envisage the future as a social or cultural 'fact,' thus, has implications first and foremost for the way that the relationship between present and future is conceptualized. This conceptualization is necessarily historical inasmuch it gains its momentum from the contemporary diagnosis of a present that threatens to 'roll over' into the future. The 'fact' of the future, by way of contrast, intervenes into the synchronist functionality of the future for the present's cataclysms. In the next section, I will turn to what I believe is a similar conceptualization of futurity as informing the present that might help to even better understand the intervening effect of futurity in the present so as to break open the hold of the present on the future, and the role that cooperation might play in this.

3 Prefigurative Politics

The term 'prefigurative politics' is currently used in social movements that form against the societal, political, economic, and environmental consequences of a contemporary drive in capitalism that is often termed 'neoliberal.' In the wake of Michel Foucault's (2008) studies on governmentality, which introduced the term 'neoliberal' in order to characterize a post-classical discursive order of capitalism, scholars have distinguished neoliberalism from classical nineteenth-century industrial capitalism along the following features. Rather than being a form of deregulated and market-centered *economic* activity that has historically been characterized as 'laissez-faire,' neoliberalism envisions a *societal* order that aims at installing competitive markets, which are held to be the superior mechanism of social coordination, as the core principle of the whole social order (Harvey 2005; Gertenbach 2007). The point is thus not only to 'deregulate' markets from all norms and regulations alien to the principle of competition, but to *create* and *expand* competitive markets and legal institutions to safeguard the principle of competition in them. Unlike Foucault, who referred mainly to German postwar ordoliberalism as the site where neoliberalism became articulated, newer studies depict the takeoff of neoliberalism in the expansion and the deepening of the financial system after the end of the Bretton Woods agreement. Recent social movements like Occupy Wall Street, which oppose the consequences of this globalizing mode of societal rearrangement, address the complicity of state institutions in the setup, maintenance, and deepening of neoliberalism. 'Prefigurative politics,' for them, heralds a practice of the political that creates a maximum distance between those institutions and the organization of protest and resistance against them: the point is not to buy into the legitimacy of state institutions through, for instance, addressing them with demands that would provide legitimacy to those institutions' claims to 'represent' the political forces of society (Graeber 2013, 87–99). In this respect, prefigurative politics breaks with the reduction of the political to 'representation' in the sense of representative democracy (Sitrin and Azzelini 2014). The political system is denied the function of being a point of crystallization for political action and empowerment. Therefore, prefigurative politics also takes issue with the politics of recognition as it implies that demands made by social forces be 'recognized' by the political system, thus again installing that system as the major hinge of the political (see Day 2005, 66–90; Nail 2013).

While prefigurative politics is invoked in a bid to articulate alternative political projects that bypass the political system as an addressee, it also has a particular index of futurity. The term was first used in order to describe practices of radical political dissent in the historical context of the civil rights movements in the US, and more precisely, of the protests of student and staff at campuses such as Berkeley in the 1960s. According to Wini Breines's (1989) historical reconstruction, what was particular about these protests were the ways that alternative political goals and agendas were articulated in tandem with community-building practices that envisioned modes of sociality directly corresponding to the political aims of those groups and networks. These practices were 'prefigurative' in the sense that they heralded modes of social encounter that would also characterize an alternative social and political commons of the future (see Day 2005, 19–45).

The first focus of those protests' critiques was, thus, an instrumental understanding of the political that relied on a distinction between political means and political ends. Prefigurative politics, in contrast, targeted a correspondence and non-instrumental coherence between political means and ends. This privileged community-building over the formation of political organizations, as the latter – like many communist-leaning student organizations – were often seen as embodying and prolonging the dictate of political instrumentalism even if their aims were to overthrow the present political system (Breines 1989, 18–66). In this respect, prefigurative politics was envisaged as a "questioning of instrumental rationality"

(Breines 1989, 50). Community-building, in contrast, was embraced as a political strategy that built possible future communal structures into present ones, thereby redeeming the political viability of the former while probing and developing alternative ways of decision-making in the latter. Crucially, this establishment of a continuity between present community and future commune challenged the main instrument of democratic decision-making, which is the majority vote, as it appeared to be itself complicit in the instrumentalist narrowing of the political in established democratic political systems, relying on a distinction between those represented and those representing them (see Poletta 2002; Nail 2013). Up to the present day, a critique of the majority vote, as the key embodiment of representative politics, is therefore at the heart of prefigurative politics, which keeps experimenting with different practices of consensus-based decision making (see Sitrin 2006; Sitrin and Azzelini 2014; Graeber 2013, 196–207, 210–227).

Critiques have been launched against prefigurative politics as a mode of practicing the political that is highly vulnerable to irritations (see Breines 1989, 48–49). Compared to political organizations with a strict means-ends instrumentalism, prefigurative politics might seem to be somewhat ineffective and always threatened by a relapse into communal romanticism (see Breines 1989, 67–95). Its principle of consensus-based decision-making has been confronted with critiques since antiquity, as Egon Flaig, a political anthropologist focusing on collective decision rules, has demonstrated (Flaig 2013). However, the instrumental success of prefigurative politics is not the concern of this article, as neither Adelson's "future sense" nor Appadurai's "future as a cultural fact" is concerned with how these two modes of futurity 'deliver.' Instead, prefigurative politics, I would argue, is another instantiation of a 'future sense,' this time decidedly applied to rearticulating notions of the political as such. At the same time, it shares with Adelson's and Appadurai's suggestions a concern with a present that is seen as catastrophic, and with a future that threatens to be reduced to a mere continuation, and ultimately verification, of the present's proneness to catastrophe. In other words, prefigurative politics is historically circumscribed and motivated by a present that seems to have always already captured the future within an ideological horizon of a future that is allegedly unknown, unknowable, and hence irrelevant for the present.

Richard Day (2005, 91–128) has genealogically connected prefigurative politics with off-mainstream traditions of anarchist political thought. His conceptual concern is the rejection of the critique that anarchism is utopian, which was put forth often by Marxist thinkers. Focusing on the works of Gustav Landauer, Day dissects a peculiar motif in anarchist thought to conceptualize revolutionary political agency without recourse to a revolutionary political collectivity – typically, class – and to the instrumental character that present political action must maintain to such future collectivity. Instead, Day argues, Landauer proposed the concept of 'structural renewal' that anticipated some key components of the later prefigurative politics. Its main idea was that, in order to transform the present capitalist and oppressive system, one has to start out with collaborative potentials already partially established in everyday interactions and affinities. 'Structural renewal' therefore did not have to wait for the advent of a revolutionary subject, constructed by the instrumental action of political organizations, but could proceed from these mundane alternatives to an oppressive system in order to corrode it from within. The temporality articulated in the concept of 'structural renewal' anticipated that of 'prefigurative politics' in the sense that existing communal structures were deemed to bear the kernel of an imminent end of the oppressive societal system, and at the same time foreshadowed such communal structures on a much larger scale.

4 Conclusion: What the 'Future Sense' Might Harbor for the Future of the Study of Culture

If the notion of "future sense" as proposed by Adelson has been chosen to start and to end this article, it is because that notion is fundamental for the issues raised here. For the transdisciplinary field of the study of culture, it seems to me that Adelson's conception of the "future sense" is both cautious and foundational at the same time. It is cautious because it circumscribes the future sense not as heralding any radically new political or societal project (Adelson 2017, 62), but rather as an entry point – a "portal," as Adelson says – for reconsidering the ways that present and future may relate to each other. As a new way of perception that can be added to our senses, it requires "cultivation" (Adelson 2017, 196, 246) through the practice of "reading" (Adelson 2017, 196) that does not necessarily re-constellate fields of social practice, but first of all the ways that we relate to the world as one that is always already interpreted (see Weber 1904). Yet the concept of the 'future sense' is foundational too, referring to the capacity of localizing the future in the present, with the effect that the future turns from a utopian into a 'topian' point of reference, that is, to something "that is not consistently deferred" (Adelson 2017, 78), but *there* to be perceived.

The contributions discussed in this article do not share all aspects with respect to how they attempt to recalibrate understandings of how future and present relate to one another. However, for the question organizing the present volume – what are the futures of the study of culture? – they together bear important inspirations as well as questions.

First, they urge us to rethink the inherited modernist juxtaposition of the 'present future' and 'future present,' in Niklas Luhmann's terms (Luhmann 1976;

see also Luhmann 1998). The gist of Luhmann's distinction is that the future can be present only ever as an expectation formed in and by the present, which has conceptually nothing to do with how the present will present itself at any point in the future. In other words, the future 'as such' is an externality of the present. This distinction writes forth a tradition to conceive of modernity as an episteme and a phenomenology that radically cuts the bond between 'experience' and 'expectation,' as in Reinhart Koselleck's distinction between '*Erfahrungsraum*' and '*Erwartungshorizont*' (Koselleck 2004; see also Adelson 2017, 126). While both experiences and expectations undoubtedly inform present social action, expectations are often associated with a quality of 'ficticiousness' regarding their reference objects, as those objects are held to belong to the future as being 'open' or 'contingent.'[3] Yet, modernity's conceptual decoupling of the present from the 'open' or 'contingent' future might, only seemingly paradoxically, be complicit in 'rolling over' the present's tendencies into the future, because many present problematic or even catastrophic social, political, and economic tendencies operate on such notions as future contingency. This celebration of contingency can be most clearly seen in financial market practices, whose condition of possibility is the claim that the future is open and therefore a projection screen for expectations, manifested as bets.

Second, we need to account for the fact that the "future sense" is a "sense organ" (in Adelson's terms) whose "cultivation" is historically embedded and circumscribed. This historical circumscription, as can be seen from the above discussion, is informed by a sense of present, ongoing, and imminent catastrophe as a paradigmatic experience since the twentieth century at least. Adelson's reflections take up the critical theoretical thread to contemplate modernity as a "permanent catastrophe." Appadurai posits the "future as a cultural fact" against disaster capitalism and systematic exploitation on a mass scale. The radically democratic project of 'prefigurative politics' seems to be returned to whenever the suspicion arises that the institutions of representative democracy might either not be able to cope with ongoing disaster or, worse, be complicit with it.

Third, these studies invite scholars of culture to turn their attention to social cooperation and collective decision-making as phenomena that might have a presence apart from modernity's 'division of labor' and apart from invocations of collective identity. In other words, conceptualizing cooperation must resist the temptation of sociological functionalism as well as that of culturalist identitarianism. Instead, cooperation might be envisioned as a lateral – or, in an

---

3 For instance, see Beckert 2013 who diagnoses "fictional expectations" as core components of financial capitalism.

activist idiom, "horizontal" (Sitrin 2006; Nail 2013) – activity that actualizes forms of address in the present that not only anticipate their broadening in the future but make the potentiality of such broadening the basis of their present experiencing.

Fourth, the following question arises: What becomes of utopia when an alternative, more 'topian' sense of futurity is in the present's reach? Is the revolutionary political inclination of utopia lost for the 'future sense'? While all the contributions discussed in this article insist on a 'future sense' that rejects the severing of present from future that, in the modernist account, is often associated with the genre of utopia, it might still be possible to engage, as Adelson (2017, 126) argues that Kluge does, in an approximation of utopia as a "paradigm for imagining social perfection" whose futurity is not an index of inaccessibility but a "social fact" (Adelson 2017, 87) that structures the present. 'Utopia' might be revisited as a potential conceptual kernel of a future temporality whose significance resides in its quality of being, as Adelson defines the 'future sense,' "an anti-realist and non-empirical but nonetheless real dimensional phenomenon in time" (Adelson 2017, 29). The conceptual labor that such revisiting requires might start with a deconstruction of an array of understandings that have been instrumental in rejecting utopia as a modality of social and political perception. This concerns not only the notion of the 'empirical' but also its permutations like probability and risk that, as Appadurai reminds us, foster a belief in the empirical precisely as they play with, and bet on, the possibility of its unreality. It also concerns notions of political 'effectiveness' that dismiss the significance as social and cultural facts of however fleeting and ephemeral experiments with prefigurative politics, at the double cost of taking political instrumentalism for granted without interrogating the latter's 'effectiveness.' For the study of culture, the 'future sense' might thus be a call not only to robustly withstand the pressure to collapse the real into the empirical, but also to dissect the permutations of the empirical as they reach out into the unreal.

References

Adelson, Leslie A. "Against Between: A Manifesto." *Unpacking Europe: Towards a Critical Reading*. Eds. Salah Hassan and Iftikhar Dadi. Rotterdam: Nai Publishers, 2001. 244–255.

Adelson, Leslie A. *The Turkish Turn in Contemporary German Literature: Toward a New Critical Grammar of Migration*. New York: Palgrave Macmillan, 2005.

Adelson, Leslie A. "Experiment Mars: Contemporary German Literature, Imaginative Ethnoscapes, and the New Futurism." *Über Gegenwartsliteratur. Interpretationen und Interventionen*. Ed. Mark W. Rectanus. Bielefeld: Aisthesis, 2008. 23–49.

Adelson, Leslie A. "Futurity Now: An Introduction." *The Germanic Review* 88.3 (2013) (special issue *Futurity Now)*: 213–218.

Adelson, Leslie A. *Cosmic Miniatures and the Future Sense: Alexander Kluge's 21st-Century Literary Experiments in German Culture and Narrative Form*. Berlin/Boston: De Gruyter, 2017.

Appadurai, Arjun. *Modernity at Large*. Minneapolis/London: Minnesota University Press, 1996.

Appadurai, Arjun. *The Future as Cultural Fact: Essays on the Global Condition*. London/New York: Verso, 2013.

Beckert, Jens. "Imagined Futures: Fictional Expectations in the Economy." *Theory and Society* 42.3 (2013): 219–240.

Breines, Wini. *Community and Organization in the New Left, 1962–1969: The Great Refusal*. New Brunswick/London: Rutgers University Press, 1989 [1982].

Day, Richard J.F. *Gramsci is Dead: Anarchist Currents in the Newest Social Movements*. London: Pluto Press, 2005.

Flaig, Egon. "To Act with Good Advice: Greek Tragedy and the Democratic Political Sphere." *The Greek Polis and the Invention of Democracy: A Politico-cultural Transformation and Its Interpretations*. Eds. Johann P. Arnason, Kurt A. Raaflaub, and Peter Wagner. 1st ed. Oxford: Wiley, 2013. 71–98.

Foucault, Michel. *The Birth of Biopolitics: Lectures at the Collège de France 1978–1979*. New York: Picador/Palgrave, 2008.

Gertenbach, Lars. *Die Kultivierung des Marktes. Foucault und die Gouvernementalität des Neoliberalismus*. Berlin: Parodos, 2007.

Graeber, David. *The Democracy Project: A History, a Crisis, a Movement*. London/New York: Penguin, 2013.

Harvey, David. *A Short History of Neoliberalism*. Oxford: Oxford University Press, 2005.

Koselleck, Reinhart. *Futures Past: On the Semantics of Historical Time*. New York: Columbia University Press, 2004.

Lee, Benjamin, and Edward LiPuma. "Cultures of Circulation: The Imaginations of Modernity." *Public Culture* 14.1 (2002): 191–213.

Luhmann, Niklas. "The Future Cannot Begin: Temporal Structures in Modern Society." *Social Research* 43 (1976): 130–152.

Luhmann, Niklas. "Describing the Future." *Observations of Modernity*. Stanford, CA: Stanford University Press, 1998. 63–74.

Nail, Thomas. "Zapatismo and the Global Origins of Occupy." *Journal for Cultural and Religious Theory* 12.3 (2013): 20–35.

Polletta, Francesca. *Freedom Is an Endless Meeting: Democracy in American Social Movements*. Chicago/London: University of Chicago Press, 2002.

Sitrin, Marina. "Introduction." *Horizontalism: Voices of Popular Power in Argentina*. Ed. Marina Sitrin. Oakland/Edinburgh: AK Press, 2006. 1–19.

Sitrin, Marina, and Dario Azzellini. "Introduction." *They Can't Represent Us! Reinventing Democracy from Greece to Occupy*. Eds. Marina Sitrin and Dario Azzellini. Oakland/Edinburgh: AK Press, 2014.

Weber, Max. "Die 'Objektivität' sozialwissenschaftlicher und sozialpolitischer Erkenntnis." *Gesammelte Aufsätze zur Wissenschaftslehre*. Tübingen: Mohr (Siebeck), 1988 [1904]. 146–214.

II No Future? Politics and Concerns

Nicole Anderson

Pre-Post-Apocalyptic Culture: The Future(s) of the Humanities

> She walks through her sunken dream
> To the seat with the clearest view
> And she's hooked to the silver screen
> But the film is a saddening bore!
> For she's lived it ten times or more
> She could spit in the eyes of fools
> As they ask her to focus on
> Sailor's fighting in the dance hall
> Oh man! Look at those cavemen go
> It's the freakiest show,
> Take a look at the law-man
> Beating up the wrong guy
> Oh man! Wonder if he'll ever know
> He's in the best-selling show,
> Is there life on Mars?

These words are from David Bowie's popular song *(Is there) Life on Mars?*, which was released in 1971. I want to suggest that the reason why the song has, and continues, to be so popular is because it speaks strongly to our cultural imaginations, that is, it speaks to where we are going as societies, cultures, and as a species (or not), and it projects, and therefore anticipates, a future time and place that, like all good sci-fi, captures our imaginations, and projects our hopes, fears, horrors, dreams, and our wonder onto something somewhere else, if not better. Apart from being a cultural phenomenon in and of itself that has consequently, to some extent, had a huge influence on the way people feel about space, time, and the future, Bowie's song, while making reference to great musicians, filmmakers, and philosophers, is also a political commentary on our times. The song therefore serves as a good sci-fi trope, to which I will keep referring throughout this paper, for thinking about the "futures" of studying culture in the political climate of today and tomorrow.

Let me start by giving you Bowie's description of the meaning of his song: "A sensitive young girl's reaction to the media [or to the film she is watching on the silver screen] [...] she finds herself disappointed with reality [...] that although she's living in the doldrums of reality, she's being told there's a far greater life somewhere, and she's bitterly disappointed that she doesn't have access to it" (Pegg 2011, 144). The question "Is there life on Mars?" then could be interpreted as a plea, a call, to that something "other" as Jacques Derrida would say; to a

https://doi.org/10.1515/9783110669398-005

future, to another world that may *perhaps* come or save us, as the girl in the song seems to crave, from our "caveman"-like behaviors that we paradoxically exhibit in our modern times. We could argue the question of the song's title is a call to and for a *new* Enlightenment, not one solely determined by privileging reason and autonomy or individual monadism (in the Kantian sense), but rather one that also includes a more contextual connection between people, and the environments (both local and global) in which we live.

However, while Bowie's song has shaped the public imagination about space and the future, I would also argue that he does not necessarily present a utopic vision. This is because the question: "Is there life on Mars?" is also dystopic precisely because the question remains only a question. It is therefore perhaps a cynical political and cultural commentary on the times (the 1970s), but one just as relevant and pertinent today. In other words, the question becomes a commentary on the way we culturally represent ourselves to each other. For example, the girl watching a film about "sailor's fighting" and a "law-man beating up the wrong guy" that she has experienced in reality "ten times or more", suggests a repetitive and circular representation of ourselves, one that endorses and perpetuates the simplistic "caveman"-like behavior in the media and on the silver screens, of our politicians, lawyers, and war-machines: "Oh man! Look at those cavemen go." At the very end of the song there is a drumbeat that is an intertextual musical reference to Stanley Kubrick's film *2001: A Space Odyssey* (which was released in 1968, 3 years before Bowie's song). The drumbeat in the movie's theme song, referenced by Bowie, is taken from German composer Richard Strauss's composition, *Also sprach Zarathustra* (1896). One of the scenes in the movie accompanying Strauss's music is the ape (or early "caveman") throwing a bone into the sky. The bone, a symbol of the tools and weapons humans learn to use, brings us right back – and thus repeats the circularity of human behavior – to the line in Bowie's song: "Oh man! Look at those cavemen go." But the title of Strauss's music, is in turn, as we know, a reference to Friedrich Nietzsche's famous book, *Thus Spoke Zarathustra* (1883–1891), in which Nietzsche opposes Immanuel Kant's form of Enlightenment (or ascetic ideal). Rather, as Paul Patton argues, Nietzsche believes in the "possibility of a different type of human being, the 'overman' represented by the coming of a child without *ressentiment* at the natural conditions of life, including death, and with no need of a belief in super-sensible worlds," that is, belief in worlds we cannot experience with our senses and are therefore transcendent (including alien worlds) (Patton 2012, 139, 142–143). The child in Bowie's song that references Strauss's *Also sprach Zarathustra* and in turn *2001: A Space Odyssey* (1968) could be reminiscent of the coming of the child that represents Nietzsche's overman. And while *Space Odyssey* also projected a "future human," perhaps an Nietzschean "overman" in space, on the ground, on earth, in 1968, we saw various anti-authoritarian revolutions of varying intensity throughout Europe: in Poland, Prague, Germany, and arguably most famously in Paris with the student riots that brought the nation to a standstill in just over a couple of weeks. Much of the revolts at this time, as Angela Merkel put it, was a response to the political repressive socialism of the day on the one hand, and the free-market economy that in turn was blamed for appalling conditions for workers and increased fees for students on the other (Dülffer 2008).

In summing up the '68 movement on its fortieth anniversary in 2008, Daniel Cohn-Bendit claimed that "1968 was a European movement," one that "gave rise to a new form of society all over Europe. Today we are on the path to a common identity" (Cohn-Bendit 2008). Compare that to what is happening in Europe now, and reflect on the fact that only nine years after Cohn-Bendit's claim, on 23 June, 2016, the UK voted to leave the EU (what has been termed "Brexit"). In the immediate aftermath of the vote to leave the EU, there were threats from Ireland and Scotland for a new referendum to leave the UK, and talk of Nexit (the "next" country touted to leave was the Netherlands). If there was a common identity or a potential for one, it was shattered on 23 June, 2016. The line in Bowie's song in which he sings "Rule Britannia is out of bounds" is uncannily prophetic, evidenced by a display of sovereignty when Boris Johnson made clear in his speech on the morning of the outcome of the vote that "we can now take back democratic [meaning sovereign] control ... we can control our own borders" (Johnson 2016). And as Philip Stephens in his column in the UK *Financial Times* on 23 June argued: "The 2008 financial crash, austerity and the grossly uneven distribution of the rewards of globalisation have all taken a toll" (Stephens 2016). This is indeed a far cry from a united EU on a path to a "common identity," as Stephens goes on: "Not so long ago British politicians of almost all shades were proud of Europe's role as a catalyst for the spread of freedom and democracy beyond its borders. Governments of right and left championed the EU accession of formerly communist states" (Stephens 2016). But now, and quoting this time politician and Brexit supporter Michael Gove, "people in this country have had enough of experts" (Gove 2016). Indeed, those experts are not just "organisations" initially suggested by Gove, but one could argue they also include academics and academic institutions. I'll come back to this disillusionment with academic experts shortly, but for the moment while it is possible to see the Brexit revolt as a repetition of political uprisings of the 1960s, as Karl Marx argues such repetitions are often in form only because history repeats itself "once as tragedy and again as farce" (Marx 2013 [1852], 9), or as Bowie sings: It's an old story, seen "ten times or more."

This discussion of Brexit is an example, a case study, or a way to think about the statement, "the futures of studying culture," because the vote to Leave was much more than just about economics, politics, fear of terrorism, and dislike of immigration. It is, I would argue, also about culture and 'identity.' Indeed for some, such as David Morris, it represents the new culture wars between classes and between the left and right: "The most overwhelming Leave constituencies are a social milieu that is remote, both literally and figuratively, from higher education [...] Class, education and geography dominate above all else, far more so than the policy debates about the economy and immigration" (Morris 2016). Moreover, no longer wanting a cosmopolitan identity, Britons have been lamenting a long-passed idea of "Britishness" (and thus culture), which they believe immigration has diluted (McQuillan 2016a). My aim here is not to side or comment on whether or not Britain was right or wrong to decide to leave the EU. Rather what is more interesting is that Brexit not only exposed the fault lines in class relations, but that in the lead-up to and fall-out from the vote, while there were many academics commenting about the economic and political implications of Brexit, there were very few academic voices linking the debate to ethics, responsibility, and social justice.

Jacques Derrida would definitely have had something to say, and in fact, still does, when we consider how and why his work has ongoing relevance. For instance, writing on the mass demonstration in Paris in 1996 over the issue of refugees and immigration, in his article "On Cosmopolitanism" (2001), which is relevant to the Brexit issue today, Derrida rigorously unpacks the issues by focusing on the word "cosmopolitanism." Derrida argues that "[s]ince the Revolution," France has portrayed its cosmopolitan identity by "being more open to political refugees in contradistinction to other European countries" (Derrida 2001, 10). And yet, despite this, Derrida locates a double bind or contradictory imperative within the concept of cosmopolitanism. As summarized by Simon Critchley and Richard Kearney in the preface to Derrida's article:

> [O]n the one hand, there is an unconditional hospitality which should offer the right of refuge to all immigrants and newcomers. But on the other hand, hospitality has to be conditional: there has to be some limitation on rights of residence. All the political difficulty of immigration consists in negotiating between these two imperatives. (Derrida 2001, x)

I would argue that any negotiation of this nature involves ethics because it has to account for the needs of both hosts and refugees. Of course Derrida has a lot more to say about this negotiation and the nature of hospitality, and while Derrida wrote "On Cosmopolitanism" in 1997, twenty years later we could say that the Brexit vote to Leave was perhaps a failure by our politicians to take on this negotiation and a further failure of both academics and politicians to even communicate the issues involved to the general population. Still I can't help wondering what Derrida would have to say about this situation. I recall a conversation

Derrida and I were having about 1968, in which he was telling me that intellectuals (the professors) en masse were supporting, and were often found on the front line with, the students and workers. Foucault was one of them, and Derrida harbored in his home some of the student rioters wanted by authorities. And while the situation is different today, and while there are many academic commentators on Brexit, the questions remain: where are the professors, the intellectuals, the academics today, not just as commentators on an event like Brexit, but shapers of the future of culture and politics and our responsibilities and ethics towards others, to all communities and classes in our societies? Or, to put the question another way: What should be the role of higher education in a situation like Brexit? To some British academic commentators there is an argument that the lack of an active role in Brexit by higher education institutions has created a legitimation crisis. Max Weber and Jürgen Habermas broadly define a legitimation crisis as a situation in which the citizens of a political regime lose belief in the system, such as, public institutions (Weber 2012, 382; Habermas 1987, 140; see also Habermas 1975). Now some British academic commentators have been arguing that a legitimation crisis was played out in and through Brexit, with the vote to Leave as much about an economic as it was an anti-intellectual or a higher education legitimation crisis. The crisis can be summed up by Martin McQuillan when he argues:

> The widening of the graduate population and the expansion of immigration to the United Kingdom are perhaps the two most significant factors that have changed the cultural composition of Britain in the last two decades. The EU vote is a direct challenge to the latter; it also feels like a warning shot to the former. As the so-called "post-truth" politics of the campaign demonstrated, it is not an exaggeration to say that the referendum was an anti-intellectual experience and that the outcome is a rejection of the values espoused by those in universities who argued for Remain. (McQuillan 2016b)

Indeed, 103 universities in the UK wrote an open letter to the British voters appealing to them to vote Remain. But admirable as this was, and while it was absolutely needed, in some ways the letter missed its mark and failed to take account of this culture war and this anti-intellectualism that McQuillan comments on. Beginning the open letter with this line, "we are gravely concerned about the impact of a UK exit from the EU on our universities and students" (*Independent* 2016), we can perhaps see why it missed its mark. The letter did not address at all what universities could do for those working classes who do not share the same values, who are simply struggling to survive, and often have no chance or opportunity of ever getting to university. After all, to use Bowie's interpretation of his song: "the girl is being told there's a far greater life somewhere, and she's bitterly disappointed that she doesn't have access to it" (Pegg 2011, 144).

Of course in this climate of legitimation crisis the university and research communities are easy targets. Because while universities argued that leaving the EU would be a mistake, could not be supported by any reasonable evidence, and would be bad for national institutions such as universities, as David Morris argues, "[t]he sceptical listener only hears self-interest and a wish to preserve the status quo above all else," and they hear this precisely because "[t]he highly educated and their educators have done relatively well in *Britain* since the recession and austerity, even with the diversity of experience within these groups" (Morris 2016). In other words, within some western societies, large portions of the population, not just Britain, are tilting in terms of their cultural values to the extreme right, and what this means, as Philip Stephens puts it, is that increasingly "[f]ear counts above reason, anger above evidence, lies and prejudice claim equal status with facts, and political popularism wins the day, and this is a consequence of trust in democratic and educational institutions being at a very low point in the history of Western democracy" (Stephens 2016). But when academics start to associate political popularism with the ignorance of the masses, this ironically just widens the cultural and class divide and further perpetuates mistrust in academic institutions. One could just as easily argue that, like the working class response, the educated position of being for "Remain" is equally ideological and/or a knee-jerk reaction because, as I mentioned earlier, it is perhaps based on thinking of the working class as uneducated bigots, and thus not really paying attention to the desperation coming from austerity measures and the failure of neoliberalism to distribute wealth fairly. But perhaps this situation may have been averted if the role of higher education, and here I particularly include the humanities and cultural studies, had contributed (say through more effective public discourse and through extensive and consistent influence on the secondary school system) to developing a citizenry that would be capable of making informed decisions: that is, decisions based on being able to critically choose among various options articulated in one form or another between the left and right and necessary for understanding not only the complexity but the ethical implications of a given situation or event like Brexit. Therefore, perhaps it is ethical education and decision-making that will enable us to go beyond the ideologies of both left and right so that we can negotiate and make informed decisions with the others' interests in mind. That is, ethical education enables us to focus on the interests and needs of the other, no matter how different they may be to "us."

Meanwhile, if Brexit has revealed a legitimation crisis in Higher Education, then we can conclude that academics are seen as part of an elite, as experts in control (economically and intellectually) who have let the people down. In this Brexit climate that *will* effect Europe, as well as in the larger world context of potential ecological and hence political crisis with increasing refugees and immigration born out of war, fundamentalism and climate change (Syrian refugees are partly a result of climate forces as well), with the forthcoming struggle for raw materials, such as food and water, and with the closing or policing of borders, the explosions of social divisions and exclusions, that is, increased tribalism and forms of apartheid (Žižek 2016), it is no wonder that a discourse of post-apocalyptic culture is on everyone's lips, and why Slavoj Žižek's (2016) claim that we are "living in end times" and approaching an apocalyptic state taps into the current *Zeitgeist*. In fact, as Claire Colebrook puts it, this explains why "Fredric Jameson suggested that it was easier to imagine the end of the world than the end of capitalism" (Colebrook 2016, 21; see Jameson 2003, 76). In fact, this could be called a pre-post-apocalyptic culture precisely because it is, and it will be climate change and fundamentalism, coupled with the rise of the Internet and the supposed democratization of knowledge, that differentiates us now from 1968 and other political upheavals of the past.

Given all of this, given the talk of living on the verge, or in a pre-post-apocalyptic culture, given the alleged demise in trust in academic institutions, there seems to be an increasing dominant way of thinking about the future of studying culture and the humanities, and that is, that the humanities will *either* slide into non-existence, in other words, there is no future at all, because, with the mistrust in 'experts' such as academics, universities and particularly the study of the humanities, including culture, will be seen as increasingly indulgent by the general populace, *or* we will nonetheless muddle along catering only to a small elite cut off from the mainstream and increasingly less relevant and a tiny voice that makes no difference among the dominant political issues of the day. Unless academics have given up entirely, this view looks bleak.

But why care? Should we care? When Žižek and other philosophers tell us we are living in the "end times," like good cultural theoreticians, the question we need to be raising here, as Claire Colebrook argues, is "whose 'end times' and whose 'world' [is] deemed to be ending?" (Colebrook 2016, 21). Put more concretely: The "end of the world" that is being pre-emptively mourned in post-apocalyptic culture is a highly specific world of hyper-consuming, personalized, neoliberal, and narrowly post-human "man." It is only possible to say that "we" are easily imagining the end of the world when what is presented as "the end" in films, pop-songs, and philosophy "is the end of Western affluence and white privilege" (Colebrook 2016, 22). We could put the question to higher education institutions as well. It could well be argued and indeed has been that implicit in statements like "the futures of studying culture" or the "value of the humanities" (and you could replace humanities with universities here), is a mournful narcissism of our own values, beliefs, ideologies, and privileged positions.

Having said this, whether or not we are on the verge of a post-apocalyptic culture is actually not the issue. If there is some kind of apocalypse (whether it is nuclear, viral, terrorist, etc.), when and if it happens, everything, let alone the question of the future of studying culture will be redundant. As Derrida argues, the absolute future is unpredictable, incalculable, absolutely unknowable, and therefore unpreventable, and this is what Derrida calls the "future to come" (Derrida 1994, xix). However, there is also another future, what Derrida calls the "future-present" (Derrida 1976, 67). This is a near future, one that can be calculated and anticipated based on the present and the past, and through reason and imagination this future can be shaped (shaped enough to perhaps prevent our own man-made apocalypse). Therefore, this for me is the future of which we can take hold. A future in which we can assert the value of the humanities, or the futures of studying culture, as I do, as a more positive and hopeful attempt to step out of this narcissism and make ourselves relevant to the people that need us. After all, helping politicians, journalists, and the general public understand and solve the social and ethical issues of our age is the university's, and especially the humanities', most pressing mission. For Morris, the "challenge is to make the complexities and caveats of research resonate with the everyday experience of citizens" (Morris 2016). The alternative is to waste away in crumbling ivory towers, relics, or ruins of an age past.

The study of culture, and the humanities more broadly, offers the perfect position to help others understand our social and cultural problems and provide answers to them; with the theoretical and empirical tools in our toolbox we have the ability to respond to the most important issues of our times with flexibility and with speed. The study of culture, and the humanities, provides not only knowledge and understandings of human behavior, society, and culture, but also what it means to be human. Moreover, the study of culture and the humanities knows how to tell a story about who we are and where we are going, much like Bowie's song; the humanities need to tell that story more strongly, and through that story start shaping the cultural, social, political, and ethical imagination in positive and socially just ways.

Furthermore, unlike the current break-up of the EU, what is needed now more than ever is that cultural studies and the humanities collaborate with the sciences (in research and in teaching), which will involve translating the ways in which various disciplines see the world, and use various theoretical and empirical frameworks to provide a new vision of the world, as well as new innovative implementations of those visions. Unlike the EU breakup, we need to come together in unity (not cynically for our own economic survivals as C.P. Snow (1959) writing in the 1950s analyzed the reason for *the divide* between the humanities and sciences); we need to come together around a bigger picture so that we can start shaping or at least influencing the way others will ethically see the future history of the world. The more relevance we have the more we will be in a position to grow the humanities (which has been on the decline since the 1970s) by influencing public perception. As Leonard Cassuto, professor of English and American Studies at Fordham University, stating the obvious, argues:

> Administrators are looking outside the liberal arts because they feel financial pressure. The numbers show that liberal arts graduates actually do very well after graduation, but there is no denying that prospective college students – and their tuition-paying parents – do not believe that. The percentage of liberal arts majors is trending downward, and shrinking enrolments result in college teaching jobs that don't get replaced. (Cassuto 2016)

If we care about studying culture and the humanities, if we care about the future of these things, then we should probably be asking ourselves the hard questions, if we aren't already. And I will pose some of those questions in a moment. I do believe there will be the humanities and the study of culture in the future, although it might not look like what it does today. So leaving aside the current discourse about the end of times (which we may very well be in), if we assume we are going to be around as a species, then the question is: When we think about the future/s of the humanities and the study of culture, how far into those future/s should our projections go? Should we think only of the next 10 years? In fact let's project into the far distant future, and let's assume that the answer to the question "Is there life on Mars?" in Bowie's song, is *us*: the human species.

When we teach culture and the humanities on Mars, what will it look like? On Mars the environment will obviously be different, and our responses to the "world" or planet around us will be different, therefore the students, teachers, and researchers will be different, and so too what we learn, teach, and research. In this Martian environment, which we might be able to populate thanks to our scientists and economists, what contribution will the humanities and cultural studies have made? What will the humanities and cultural studies equip us with to not only survive, but also live and become, in that environment? What do the humanities give us today to enable the potential for that possible future? How do we prepare our students today for the many possible futures that we can only imagine? How do we make ourselves relevant today to enable possible futures?

Here are a few ideas or answers to these questions (obviously they are not the only ones): The humanities and cultural studies are what help us to communicate our understanding of the world around us (whether it be on Mars or Earth). Or to put it another way, the humanities and cultural studies are the study of how we represent ourselves to each other (currently through philosophy or art; literature or film; sociology, languages, cultures, history, anthropology or social media, and so on). The humanities, then, give us the critical thinking and communication skills as well as the problem-solving tools in which to think laterally about, and respond flexibly and with speed to a fast-evolving world and the challenges of the times (now and in the future). Most importantly, the humanities teach us not only to communicate our understanding of the world around us to others, but enables us to transform, shape, or create the worlds we live in, as it provides us with the ability and the tools to describe visions for a better (ethical, moral, and socially just) future for all. It is through these critical thinking tools that we (not only students but also teachers and researchers) acquire the skills to learn and critically reflect. After all, if we as educators don't learn, and don't adapt by envisioning the future, then how can we teach and research? The future of the humanities in large part, then, comes down to what happens in the classroom and how we collaborate with each other as well as the sciences and industry here, now, today, *and* also the way we connect for, and convey or communicate to students and the general public, the relevance of the details and particularity of what is learnt in the classroom and in our research to the larger world in general.

Acknowledgement: This paper is based on my lecture at GCSC/GGK Anniversary Symposium "Futures of the Study of Culture", July 7–8, 2016, Alte Gießerei (Justus Liebig University, Giessen, Germany). I thank Jens Kugele, Doris Bachmann-Medick, and Michael Basseler, and the GCSC board members, for inviting me to speak at this important symposium.

References

Cassuto, Leonard. "This Picture Tells a Story." *Inside Higher Ed* (6 June 2016). <https://www.insidehighered.com/views/2016/06/06/responding-state-humanities-today-essay> [accessed: 28 February 2018].

Cohn-Bendit, Daniel. "Stop the Comparisons with 1968." *Café Babel* (23 January 2008). <http://www.cafebabel.co.uk/society/article/daniel-cohn-bendit-stop-the-comparisons-with-1968.html> [accessed: 28 February 2018].

Colebrook, Claire. "The Play of the World: The End, the Great Outdoors, the Outside, Alterity, and the Real." *Derrida Today* 9.1 (2016): 21–35.

Derrida, Jacques. *Of Grammatology*. Transl. Gayatri Spivak. Baltimore, MD: The Johns Hopkins University Press, 1976.

Derrida, Jacques. *Specters of Marx*. Transl. Peggy Kamuf. New York/London: Routledge, 1994.

Derrida, Jacques. *On Cosmopolitanism and Forgiveness*. Transl. Mark Dooley and Michael Hughes. London/New York: Routledge, 2001.

Dülffer, Meike. "1968 – A European Movement?" *eurotopics*. European Press Review (March 2008). <http://archiv.eurotopics.net/en/home/presseschau/archiv/magazin/geschichte-verteilerseite-neu/achtundsechzig-2008-03/debatte-1968-2008-03/> [accessed: 28 February 2018].

"EU Referendum: An Open Letter to UK Voters from Leaders of 103 British Universities." *Independent* (20 June 2016). <http://www.independent.co.uk/news/uk/politics/eu-referendum-an-open-letter-to-uk-voters-from-leaders-of-96-british-universities-a7092511.html> [accessed: 28 February 2018].

Gove, Michael. "Interview with Faisal Islam." *Sky News* (3 June 2016). <https://www.youtube.com/watch?v=GGgiGtJk7MA> [accessed: 29 November 2019].

Habermas, Jürgen. *Legitimation Crisis*. Transl. Thomas McCarthy. Boston: Beacon Press, 1975.

Habermas, Jürgen. *The Theory of Communicative Action*. Vol. 2. Boston: Beacon Press, 1987.

Jameson, Fredric. "Future City." *New Left Review* 21 (2003): 65–79.

Johnson, Boris. "Boris Johnson Hails Brexit Victory – Full Statement." *The Telegraph* (24 June 2016). <http://www.telegraph.co.uk/news/2016/06/24/boris-johnson-hails-brexit-victory---full-statement/> [accessed: 28 February 2018].

Marx, Karl. *The Eighteenth Brumaire of Louis Napoleon*. Transl. Daniel De Leon. Belmont, NC: Wiseblood Classics, 2013 [1852].

McQuillan, Martin. "Taking a Leave of our Senses in the EU Referendum." *Wonkhe Report* (23 June 2016a). <http://wonkhe.com/blogs/comment-taking-a-leave-of-our-senses/> [accessed: 29 November 2019].

McQuillan, Martin. "The Morning After the Referendum the Night Before." *Wonkhe Report* (24 June 2016b). <http://wonkhe.com/blogs/comment-the-morning-after-the-referendum-the-night-before/> [accessed: 29 November 2019].

Morris, David. "What Next for Universities in Our Disunited Kingdom?" *Wonkhe Report* (23 June 2016). <http://wonkhe.com/blogs/comment-universities-dis-united-kingdom/> [accessed: 29 November 2019].

Patton, Paul. "McCarthy's Fire." *Styles of Extinction: Cormac McCarthy's* The Road. Eds. Julian Murphet and Mark Steven. London/New York: Continuum, 2012. 131–144.

Pegg, Nicholas. *The Complete David Bowie*. London: Titan Books, 2011.

Snow, Charles P. *The Two Cultures*. Cambridge: Cambridge University Press, 1959.

Stephens, Phillip. "The Perils of a Populist Paean to Ignorance." *Financial Times* (23 June 2016). <https://www.ft.com/content/bfb5f3d4-379d-11e6-a780-b48ed7b6126f> [accessed: 1 March 2018].

Weber, Max. *The Theory of Social and Economic Organization*. Eastford, CT: Martino Fine Books, 2012.

Žižek, Slavoj. *Living in the End Times*. London/New York: Verso, 2016.

Isabel Capeloa Gil

The Global Eye or Foucault Rewired: Security, Control, and Scholarship in the Twenty-first Century

The great antagonist in J.R.R. Tolkien's *Lord of the Rings*, Sauron, is metonymically described in the novel and visualized in Peter Jackson's cinematic trilogy as the 'great eye.' The 'Eye of Sauron,' the 'red eye,' and the 'great eye' are epithets that arguably connote an embodied feeling to the penultimate villain in Tolkien's trilogy. This is reported in a letter sent by Tolkien to his friend Mrs. Eileen Elgar on October 3, 1963: "[...] in a tale which allows the incarnation of great spirits in a physical and destructible form their power must be far greater when actually physically present. [...] Sauron should be thought of as very terrible" (Carpenter 1981, 246). Throughout the saga, the thought of Sauron trumps the character's materiality. Sauron is less an active driver of antagonistic action than he is a sensation of danger and fear. He is less a character than an ambiance conveyed through the terror of pervasive, continuous, absolute, and totalitarian observation.

The *Lord of the Rings* trilogy, written roughly between 1937 and 1949, substantiates, in Sauron's eye, a particular twentieth-century panopticophobia: the fear of universal control via sight, at a time when the tools to expand the capacity for control over populations were starting to grow exponentially. As the century unfurled, visual control widened and became pervasive, from the improvement of weapon target accuracy via optics, to the introduction of visual technologies in the public sphere.[1] In fact, the very project of modernity in its dual dimension of progress on the one side, and violent exploitation on the other, is a by-product of the Enlightenment project equating the progress of reason to the widening of a politically controlled system of images. If anything is common to the projects of late and early modernity, it is the organization around plans of total visuality. They encompass simultaneously utter control and utter sight, defined by Nicolas Mirzoeff as the mandate to see and control everything, everywhere, all the time (Mirzoeff 2015, 20). This plan of total visuality hence becomes a strategic driver in the organization of the social and in the partition of the sensible, as well as an overhaul in the wider production of meaning. As such, a twenty-first-century agenda for the study of culture will unavoidably deal with visuality beyond modes of mediation and representation to ask how and under which conditions the pro-

1 On modernity and vision, see Paul Virilio 2009 as well as Jonathan Crary 1999.

https://doi.org/10.1515/9783110669398-006

duction of meaning is shaped by technologically induced visual surveillance. The gathering of visual data, the visual control, taps into what Yuval Noah Harari has termed the algorithmic hacking of humanity (Harari 2018, 308). It changes the way in which cultural subjects look at the world and make sense of it, going all the way deep into the very sense of social and political subjectivity, and rebooting the wider understanding of culture as the practices, discourses, and products that aggregate the way in which humans interact with each other and the world around them.

And yet, Sauron and even Orwell's 'Big Brother' speak to a different world, a world of centralized and unified surveillance, of vertical linear power structures, a pre-modern panoptical world in fact, clearly identified by Michel Foucault in his reading of Jeremy Bentham's panopticon. Under twenty-first century conditions, the control system of technical vision is multipolar, diffuse, overarching, and overwhelming, cutting across all human action from biometric identification to smart phones, personal computers to CCCR cameras at airports, playgrounds, schools, churches, restaurants, and in the public space. From the practice of facial recognition effective now in China (Fig. 1) to body scanners at airports, the project of visual domination acts to produce a transparent body, stripped of its opaque singularity.[2]

Despite a semblance of democratic empowerment, because all that are seen are apparently also agents of seeing, the system of images that structures late modernity, and the resulting explosion of control centers across multiple, diffuse, and diverse sight points, is consistently built upon a few centers of dominance. These range from national security agencies to private surveillance providers, from media conglomerates to the entertainment industry, and even from the influencer's Instagram page to the structured streaming of images by social media platforms, from Facebook to Instagram, Twitter, and YouTube. Beyond the containment of the dangerous in the correction institutions (prisons, reformatories, hospitals, and asylums), the control of the stream of produced images has become a privileged tool: to manage those who see, but also limiting access and thereby influence; regulating, monitoring, and restraining. The control of visual regimes, whether made visible or invisible, has shaped modernity's "structure of feeling" (Williams 1961).

As such, it has unavoidably impacted the production of scholarly research particularly in the humanities, as it articulates questions of democratic access and inclusion (Appadurai 2006), and of citizenship (Azoulay 2008) crafted by an

2 See on body scanners, the forthcoming work by Ilios Willemars. *The Body as Placeholder: Incorporated Subjects in Digital Art*.

understanding that the right to see is melded with the right to be seen and heard by the State. But it also connotes the anxiety over technological transformation (Turkle 2009), over the future of the human subject, and the right to privacy and embodiment (Scott 1998; Chamayou 2011). The transparent body under surveillance equates to the subsumption of flesh to numerical abstraction. The utopia of vision as empowered access to knowledge that inspired the metaphor of sight for the European Enlightenment oozes into A.I.-powered smartglasses for facial recognition (Fig. 2) materializing the dystopia of global snooping and algorithmic control via visualization in countries like China (Mozur 2018).

While surveillance, following the rising interest in security and terrorism, has given vent to prolific scholarly production, not least following the studies inaugurated by David Lyon (Lyon 1994) and William Staples (Staples 1997), the hermeneutical toolbox of cultural studies reveals ample possibilities in the exploration of the crossover between the regimes of the visual and technologies that structure them, among them the visual arts, film, and literature. And then, as Lawrence Grossberg argues, it is our task to question the context (Grossberg 2010), widening the field to explore the (in)visible conditions that structure the project of total visibility and its impact in shaping late modernity's forms of subjectivity, ways of belonging, communicating, and living.

1 The Terror of Sight

In Western iconology, the sense of sight[3] has traditionally symbolized knowledge and/as control, be it over the body (as in Aristotle),[4] the city (the sovereign eye), or the world (the eye of God). Pre-modern approaches to sight connote both a creative dimension – as in Nicholas of Cusa, *De Visione Dei* – and a phobic, terrifying side, epitomized in the frightening stare depicted, for instance, in Caravaggio's famous rendering of the Gorgo or Medusa (1597).

3 Sight is understood as a multimodal category in the arguments presented by Jonathan Crary in *Suspensions of Perception*, which define it as a historically situated and culturally constructed category that is irreducibly a mixed multisensorial modality, encompassing touch, hearing, taste, and smell, Crary 1999, 3.

4 In *Metaphysics*, Aristotle writes: "All men by nature desire to know. An indication of this is the delight we take in our senses; for even apart from their usefulness they are loved for themselves; and above all others the sense of sight. For not only with a view to action, but even when we are not going to do anything, we prefer seeing (one might say) to everything else. The reason is that this, most of all the senses, makes us know and brings to light many differences between things" (Aristotle 1999, I, 1, 980a 21–26).

In 1453, Nicholas of Cusa anthropocized the idea of God's vision and represented divine sight as embodied and culturally situated perception. In chapter two of his *De Visione Dei*, titled "Absolute Sight encompasses all modes of seeing," he wrote that:

> Notice that in those who have sight, sight varies as a result of the variety of its contractedness. For our sight is conditioned by the affections of the organ and of the mind. [...] But sight that is free from all contractedness [...] encompasses at one and the same time each and every mode of seeing. For without Absolute Sight there cannot be contracted sight. But Absolute Sight encompasses all forms of seeing – encompasses all modes in such a way that it encompasses each mode. And it remains all together free from all variation. For in Absolute Sight every contracted mode of seeing is present uncontractedly. (Cusa 1999, 683)

For Nicholas of Cusa, seeing is not a natural given, but a situated multisensory endeavor. He distinguishes between individual sight that is both varied and limited, and the Absolute Sight that overcomes "contractedness" as it encompasses all forms of seeing. The above excerpt suggests the difference between finite vision and the infinite spherical vision of God, which is nonetheless modeled after human experience. Written to accompany a visual experiment, the text instructs the monks of Tegernsee to circumambulate an all-seeing icon of God. Developing the concept of infinite space that had found a geometrical representation in the vanishing point,[5] Cusa places it within God's eye, which retains its medieval and religious function by gazing out of the painting at the monk according to inverse perspective. As Arianne Conty writes, Cusa constructs a veritable sociology of belief, in which believing the word of the other, of the community, is necessary for each individual's understanding of God (Conty 2012, 480–481). In this sense, there can be no unmediated experience of the divine, for God, as infinite and absolute, cannot be reduced to a single perspective. Cusa's experiment suggests that individual seeing is an unreliable source for communal belonging, because what we see cannot be shared. However, one can share the trust in the vision of God conveyed by the Scripture, which thus overcomes singularity and beacons to universal meaning.

God's vision, hence, organizes a system of dependence of the diverse and limited upon the unlimited central and omnipresent Eye of God as it presents a system of power that though unseen is clearly located at the idealized heart/center of the universe. The Eye of God is the vanishing point of a universe continuously observed and controlled by the transcendent.

A different strand is invoked by the trope of the Gorgo. The Medusa, here depicted in Caravaggio's famous painting, the Gorgon slain by Perseus, is the

5 The geometrical perspective was developed by Filippo Brunelleschi in Florence in 1413.

epitome of seeing as trauma and panic (Fig. 3). Let us consider for a moment the story of the Gorgon.[6] Medusa shares with her sisters Stheno and Euryale a terrifying appearance. These monsters with snake hair were said to possess huge teeth and a beard, in addition to a powerful shrieking voice, and a gaze that turned all those who beheld them into stone. Medusa (also Médusa, the 'Lady,' Gil 2007, 329), unlike her kin, was mortal. That is why Perseus, in order to save the young Andromeda, and aided by the goddess Athena, could resort to trickery to approach the beast and capture her head. Cloaked in a cape that rendered him invisible and armed with a mirror, Perseus approached the monster in arrears, guiding his moves with the aid of the mirrored reflection. When the Gorgo's gaze hit the mirror, she saw her own reflection, hence turning herself to stone. The Medusa presents the phobia of the gaze, the horror of perception in a world marked by catastrophe and frozen to stone by the panic of seeing. As a metaphor of the observer, the monstrous body of the Medusa presents the dislocation of the spectator from the order of things. Medusa's glaring eyes and guttural cry present in Italian philosopher Adriana Cavarero's words the "human appalled by his very being and contemplating the unspeakable act of his own annihilation" (Cavarero 2005, 25).

The double logic of creation and terror, of producing a subjectivity within a logic of terrifying control, precisely to limit the possibilities of the controlled subject to unleash that very same terror (be it the plague or crime), is what lies at the node of the Foucauldian surveillance model, inspired by Jeremy Bentham's model of the panopticon. Foucault's work has inspired social theory along the past 40 years, interested as it was in understanding the complex modulations of power in its articulation with disciplinary technologies and the very institutions of modernity (the prison, the factory, the hospital, the school, the church, the court) and how they impacted the production of subjectivity. Bentham's panopticon is an enlightening trope in this strategy of control that brings about a radical paradigmatic shift in the production of modern subjectivity.

> Hence the major effect of the Panopticon: to induce in the inmate a state of conscious and **permanent visibility** that assures the **automatic functioning of power**. So to arrange things that the surveillance is permanent in its effects, even if it is discontinuous in its action; that the perfection of power should tend to render its actual exercise unnecessary; that this architectural apparatus should be a machine for creating and sustaining a power relation independent of the person who exercises it; in short, that the inmates should be caught up in a power situation of which they are themselves the bearers. To achieve this, it is at once too much and too little that the prisoner should be constantly observed by an inspector: too little, for what matters is that he knows himself to be observed; too much,

6 On the duplicity of the Medusa, see Vernant 1985; Gil 2007.

> because he has no need in fact of being so. In view of this, Bentham laid down the principle that **power should be visible and unverifiable. Visible: the inmate will constantly have before his eyes the tall outline of the central tower from which he is spied upon. Unverifiable: the inmate must never know whether he is being looked at at any one moment; but he must be sure that he may always be so.** [...]
>
> The Panopticon is a / machine for dissociating the see/being seen dyad: in the peripheric ring, one is totally seen, without ever seeing; in the central tower, one sees everything without ever being seen. (Foucault 1975, 201–202; my emphasis)

The importance of the panopticon is to produce a permanent visibility-inducing machine that becomes the hallmark of modern power, structured upon the ability to dissociate this ever present and all-seeing entity from the seen. That is, hiding agency, the voyeuristic machine of vision produces subjectivities structured on utter control and transparency, because in the Foucauldian conditions of high modernity, though power is unverifiable, it is utterly visible. Foucault's reading of Bentham's pedagogical panopticon has effectively not only given vent to a wide array of studies on the trials and mechanisms of what David Lyon in 1994 named "surveillance society" (Lyon 1994), but arguably has fielded a new culture as well, a culture of surveillance, creatively producing surveyed subjectivities (Staples 1997). This culture has impacted scholarship and created a new sub-field of surveillance studies, situated on the cusp of politics and interpretation, philosophy and sociology, technology and the humanities.

2 The Culture of Visual Surveillance and Scholarship in the Humanities

The conditions under which the technologies of surveillance operate in the post 9/11 world have changed. Not only does the center no longer hold, but the directions, the flows, and the locations from which the panoptic control is exercised have multiplied and changed. Whereas the panopticon, in Bentham's assertion and Foucault's reading of Bentham, rests on the potential visibility of the observer and on the sheer and radical stability of the position from which power is exercised, in the post 9/11 world, power is still unverifiable and permanent, but it is utterly invisible, exercised from a multiplicity of even contradictory positions. The post 9/11 culture of surveillance is best epitomized in a global eye, made up of multiple smaller eyes, that exert power in what Arjun Appadurai calls a "volatile morphing" (Appadurai 2006, 83), of reciprocal and contradictory observation, a control spreading rapidly over the globally networked 4.0 industry. The volatility, and the multiple and multidirectional seeing positions, conflate a social technological drive that radical critics, such as William Staples in *The Culture of Surveillance*, have called the visual "pornography" of total self-display (Staples 1997). Display and technologically mediated self-presentation become extraordinarily effective modes of self-surveillance.

James C. Scott's remarkable *Seeing Like a State* (1998) insightfully places the operation of the state at the center of the transformations in the conflation of knowledge, vision, and control. Scott's research shows how state power and the modern project is not simply a project of absolute visibility, but of restricting access to vision and of constraining the right to look, exerted through forms of rationalization and simplification.

> Certain forms of knowledge and control require a narrowing of vision. The great advantage of such tunnel vision is that it brings into sharp focus certain limited aspects of an otherwise far more complex and unwieldy reality. This very simplification, in turn, makes the phenomenon at the center of the field of vision more legible and hence more susceptible to careful measurement and calculation. Combined with similar observations, an overall, aggregate, synoptic view of a selective reality is achieved, making possible a high degree of schematic knowledge, control, and manipulation. (Scott 1998, 11)

As the new vanishing point of secular modernity, the state thus enacts a model of control that rests on the expansion of technological capacity – to see everything, everywhere, everytime – but also on the concomitant narrow selection of visual data, that leads in turn to a simplification ultimately aimed at making the phenomenon more legible, but also more schematic and prone to manipulation. Suggestively, Scott's diagnosis of simplified legibility opens up the field for a scholarly questioning of the project from the perspective of a counter-legibility, enacted on the intersection of the visible and the visual with the textual and the speakable, inviting complexity, where simplification is becoming the rule.

In liquid modernity, the panopticon that organized the populations with the ultimate end of keeping in, in gated areas, the undesirable, damaged, sick and dysfunctional parts of society, has given way to a different logic. As Zygmunt Bauman argued, in the new post-panoptic society, the surveillance *dispositif* is creative of subjectivities that are not simply produced from the standpoint of a general dominating entity, but it is fragmented and heterogeneous.[7] It operates through state and corporate entities and it reflects

7 Post-panoptic developments have been studied by scholars such as David Lyon, proposing the *synopticon* as a technology mediated device of permanent observation, Lyon 2006, 35, or Didier Bigo 2011, proposing the term *ban-opticon* to define the way in which profiling technologies are used to create surveilled and excluded identities; see Bauman and Lyon 2013.

> [...] a more general phenomenon of surveillance philosophy and surveilling equipment wrapped around the task of 'keeping away' instead of 'keeping in', as the panopticon did, and drawing its life juices and developmental energy from the currently unstoppable rise of *securitarian* preoccupations, not from the *disciplining* urge as in the case of the panopticon.
> (Bauman and Lyon 2013, 58)

If the age of the (post) panopticon in its steady and heavy institutional sense is over, as Jean Baudrillard (1994) announced in *Simulacra and Simulation*, and has been substituted by pure representation, a world of simulacra meaning nothing but themselves, what societies have experienced is the effective tightening of social control through volatile abstraction (refugees as swarms, the migrant caravans, etc.), eliciting a radical transformation in surveillance modes. They have changed to become more pervasive, permanent, lighter, faster, and increasingly supporting preemptive action. Having dissociated, in Foucault's terms, the seeing from the seen, the global surveillance eye enacts a seeing **ex-ante**, even before there is anything to be seen. One of the most remarkable examples of this evolution is the drone, be it the civilian gadget form that hovers over malls and parks, and that is consistently pushed as the new trendy object of consumption in airport tax-free shops all over the globe, or in its military form as the UAV or UCAV,[8] potentiating the pervasiveness of an unseen repressive and murderous power machine.

In its military form, the drone captures the imagination, just as it is a product of a heritage of cultural representations of phobia. Naming is an act of power. Named the Reaper, the Predator, or the Gorgo, the UCAVs capture the imagination of fear and project the drone controller's imagination of power and control over a vision of society and culture. The wrong attire in the wrong place, a children's game that falls outside the scope of the cultural understanding of the observer, a gesture out of hand, a social gathering outside the pattern of Western sociality, may trigger a click with deadly impact. What goes on in the command room at Creech Air Force Base in Nevada, where the US Air Force's drone squadron operating in Afghanistan is located, and in other military sites engaged in the 'war from a distance' and in cyberwar, is not simply a matter for politics, military sciences, and social studies. This is a matter of culture, a second wave of surveillance studies in which we who partake in meaning-making activities must engage.

One of the most remarkable studies of the cultural implications of the age of the drone, Grégoire Chamayou's *Drone Theory* (2011), argues that in the new stage of the war on terror, precisely the business of detection and preemptive annihilation trains military personnel as cultural analysts. The stakes are about

8 UAV – Unmanned aerial vehicle; UCAV – Unmanned combat aerial vehicle.

schematizing patterns of life, detecting narratives as they appear in visual form and the deviations from those patterns. As one analyst, quoted by Chamayou says, "You're now getting into a culture study. [...] You're looking at people's lives. [...] Essentially, the task consists in distinguishing between 'normal' and 'abnormal' activity in a kind of militarized rhythm-analysis, that takes on increasingly automatized forms" (Chamayou 2011, 43). According to Chamayou, the world has entered into a new stage where surveillance is now a tool not simply for strategic but also for effective annihilation, a process that folds the cultural with the technological, creating an assemblage that radically shifts the cultural organization of the social. The combat-ready drone epitomizes a global eye that has now become a tool to advance preemptive mechanisms of branding, excluding, and annihilating unlivable life. Chamayou considers six major trends in the new surveillance and annihilation culture:

1. The principle or permanent surveillance or permanent watch: freed from the constraints of a human body (that of the pilot) the drone is a resilient and resistant body of iron that enlarges the possibilities of constant 'geo-spatial' overwatch, continuously sending back data to supercomputers, that work to analyze, simplify, and read patterns into it;
2. The principle of the totalization of perspective or synoptic viewing: to see everything, everywhere all the time;
3. The principle of creating an archive of everyone's life: the drone feed is marked by 'archive fever,' as optical surveillance is not limited to present time, but its strength lies precisely in the ability of recording and archiving a myriad of sights that are later built into stories of people's lives. In the drone archive life stories are continuously created and recreated;[9]
4. The principle of data fusion: the feed is multisensory, drones also capture voice communication and data from various devices (computers, mobile phones, etc.). The archival intent is to fuse all of these elements in a common story;
5. The principle of the schematization of forms of life: the data feed provides for a 'cartography of lives,' which has become the epistemic basis of armed surveillance. It is no longer simply about identifying individuals, but of schematically co-creating them;
6. The principle of the detection of anomalies and preemptive anticipation: this is a principle that draws on the cultural competences, on the visual literacy of the analyst and allows for a reading of the present to become a prediction of the future.

---

9 This happens when the system is perfect: that is, because there is no ability to store, retrieve, and analyze all the data fed by these unmanned vehicles.

Conflating the post-panopticon's dispersive power, and its total presentness and subsumption into total representation, drone theory speaks to the metamorphic transformations of contemporary surveillance culture. To question the doctrine is, I contend, to question the nature of representation itself. But although the case has been made as to the urgency of understanding how post-panopticon surveillance appropriates cultural literacy for counterinsurgency, the question must be asked: is it simply in diagnosis that the task of the study of culture rests? Arguably, we are not problem-solvers, but no less important or impactful is the task of critique, of understanding, and questioning the perturbations in the normative. Under the spectacularly spectacled conditions of our late modernity, the counter-visuality of sorts we may be able to induce, may come from areas of activity that are not necessarily scholarly, but are certainly objects of scholarly inquiry. Art forms materialize a type of criticality that is increasingly intervening in the public debate, bringing to bear new languages that speak to supplementary modes of knowledge production. Speaking to visually trained audiences, they suggest a mode of representation situated beyond the textual abstraction, on the one hand conveying a new materialism but on the other requiring a sensitive awareness to modes of representation that collapse the traditional device between fact and fiction, authenticity and simulacrum. My final example resorts to art to ask for fault lines in the representation of the drone narrative in what is a widening opportunity to discuss the new and intricate entanglements between real politics, facts, and the imaginary work of artistic creation.

Journalist and director Laura Poitras took surveillance as the node of concern of her exhibition, *Astro Noise*, which opened at the Whitney Museum in New York City in February 2016. *Astro Noise* is an installation of the affective assemblages that rule the communal in the post 9/11 world. The visitor is guided through a transformed curatorial space that organizes the experience of living under surveillance around representational modes. In doing so, the viewer is plunged into the landscape of permanent war we live in, and is strategically transformed from detached observer into participant, from being an object of surveillance into an artistic subject.

In the first room of the exhibition space, located on the top floor of the Whitney, two large double screens hang in the middle of the space with video projections to each side of the screen. On the one side, Poitras' short film *O'Say Can you See (*2001/2011) is projected, featuring onlookers viewing the wreckage of the World Trade Center. Made up of slow-motion reaction shots of by-standers, unwilling observers to the spectacle of destruction, the film creates a sense of separation from reality. As Poitras claims, reaction shots reveal the limits of representation (2016, 36); they question the frame and dislocate the hegemony of the seen *vis-a-vis* the indeterminacy of the seeing. Then, in the background of that same screen, the viewer is confronted with footage released by the American military under the Freedom of Information Act, depicting an interrogation of would-be Taliban suspects in Afghanistan in February 2002. American pain and awe are glued to the torture of others, a mutual implication whereby the viewer is instructed to read the visual embedding of what Achille Mbembe calls a necropolitics, a definition of life based on the exercise of control over mortality and as simple deployment and manifestation of power (Mbembe 2003). This is also an invitation to read the impossible representation of the torture of others. Just as the reality of the fallen towers remains outside the frame of the awe-stricken observers, so too does the reality of torture, lie outside the scope of re-presentation.

The visitor then moves to Room 2 and the installation *Bed Down Location*, where she is invited to lie down in a huge bed looking up at the skies of multiple Middle Eastern cities and New York, just as drone activity is perceived overhead. The skies of Baghdad or Mosul, New York or Kabul appear strikingly alike and yet radically different, pervaded by the fleeting overhead sound of a buzzing engine, a drone, almost imperceptible to viewers. Lying in bed, the visitors are unaware of being recorded by an infrared camera, the feed of which is being streamed live in Room 4. In fact, the whole space of the exhibition is a space of permanent observation, by Poitras' camera, by the military recording, the infrared, and the museum.

Next, the visitor is then guided through a dark corridor with small openings inviting the voyeuristic gaze. Beyond the opening, there are cases with surveillance records, redacted data, pictures, video logs of past Guantánamo prisoners, and writing material. The viewer takes up the position of the gatekeeper in the surveillance system. The objects inside the illuminated cases are visible all together and all the time to the intrusive viewer who remains protected in the dark hall, confronted with the real pain of others. The hall is a transition space, a liminal area that leads up to the transformation of the viewer into object, or rather to the conflation of the viewer-as-object, which takes place in Room 3, as the display of the infrared camera in Room 2 is presented on screen. In the hallway leading out of the installation, this data is equated with a different surveillance product, as viewers visualize the data garnered by UK military satellite surveillance over the Gaza strip. This material is transformed by dint of the aesthetic gesture, blurring the boundaries between art and life, that also speaks to the musealization of surveillance culture.

Poitras' project is about reciprocity, about mutual implication in the project of total war, but it is also a project about the language of the visible and the ways in which citizens, exhibition visitors, artists, scholars, and activists are led to operate within this "surveillant assemblage" (Weizman 2017). Ultimately it is also a project about storytelling, about creating narratives that model social and political constellations. In the age of the post-panopticon, Poitras invites the surveyed to understand emotionally, disbanding the categories between viewer and the viewed and stressing the mutual vulnerability that is the trademark of this new stage of the social.

In an interview, Poitras claims this is an exhibition that turns the museum viewer into the protagonist and forces her to make choices. Insomuch as it reproduces the conditions that place individuals under the global eye of surveillance, art acts out a paradoxical role that replicates the circumstances of control just as it produces critique, or an artistic criticality in Irit Rogoff's sense (Rogoff 2003),[10] that is, a critique embedded in the position of the artist as a situated body in a constellation of mutually engaged relations. Poitras states: "I am interested not just in the experience of the viewer as protagonist moving through a narrative journey but in how other bodies create that narrative experience" (Poitras 2016, 33). This is a project for a new communality, a call for recognition and engagement enacted through art. It is a project conflating the educated position of critique with the artistic edge of creation, a model of collaboration in knowledge, and a critical act to traverse the complexity of our present.

3 Coda

While the metaphor of sight as a model of cultural engagement has prevailed in Western aesthetics from Aristotle to Nicolas of Cusa, and from myth to science and art, the liquid condition of the twenty-first century has brought on radical changes. A top-down hierarchical *dispositif* structured upon the exercise of discipline over docile bodies, the panoptical model devised by Jeremy Bentham and revised by Michel Foucault turned vision into a strategy of domination and power, a simultaneously creative and repressive mechanism producing docile subjectivities held together by the institutional power of the disciplinary apparatus. In the post-panoptical world of pervasive seeing-inducing technology, where everyone is watching and being watched, the visible has become the locus of a new cultural model that is no longer set on discipline but on surveillance; on organizing

10 Rogoff defines the move from critique to criticality as a motion of productive engagement, moving from blame to power, that is: "[...]from finding fault, to examining the underlying assumptions that might allow something to appear as a convincing logic, to operating from an uncertain ground which while building on critique wants nevertheless to inhabit culture in a relation other than one of critical analysis; other than one of illuminating flaws, locating elisions, allocating blames," Rogoff 2003, xx.

areas of observed participation, of shaping questions and monitoring answers; and on devising the problems, the bodies, and the institutions that organize the social. A metadiscipline that works across contextual deployment by analyzing phenomena and, as Siegfried Kracauer put it, its "surface-level expressions" (Kracauer 1988, 75), the study of culture has consistently addressed the multiple ways in which the changes of the post-panoptical world infiltrate and shape cultural experience, as well as how art and culture can work to resist these oblique powers and produce new modes of seeing the world. Surveillance studies and the meaning-inducing strategies that are at work in the world of total visuality shape a field of inquiry that requires a transdisciplinary approach, an understanding of contextual hermeneutics and mediation as well as the narratological tools to unpack the identitarian storytelling secretly conducted in and through surveillance. Questions most likely to inform the wider undertaking ask: what does the new surveillance do to the understanding of the human? How does it affect subjectivity and artistic creation? What are the tools used to infiltrate the organization of the sensible? In the world of total visuality, can there be an outside to cultural models produced and enacted within the limits of surveillance? And what are the new languages that produce this outside? Because, as always, the work of culture will likely be the strategic opening at the beginning of resistance.

References

Appadurai, Arjun. *Fear of Small Numbers.* Durham, NC: Duke University Press, 2006.
Aristotle. *The Metaphysics.* Trans. Hugh Lawson-Tancred. Harmondsworth: Penguin, 1999.
Azoulay, Ariella. *The Civil Contract of Photography.* Cambridge, MA: Zone Books, 2008.
Baudrillard, Jean. *Simulacra and Simulation.* Ann Arbor: University of Michigan Press, 1994.
Bauman, Zygmunt, and David Lyon. *Liquid Surveillance: A Conversation.* Cambridge/Malden, MA: Polity, 2013.
Bigo, Didier. "Security: A Field Left Fallow." *Foucault on Politics, Security and War.* Eds. Michael Dillon and Andrew W. Neal. Basingstoke/New York: Palgrave Macmillan, 2011. 100–110.
Carpenter, Humphrey, ed. *The Letters of JRR Tolkien.* New York: Houghton Milffin, 1981.
Cavarero, Adriana. *For More Than One Voice: Toward a Philosophy of Vocal Expression.* Stanford: Stanford University Press, 2005.
Chamayou, Grégoire. *Drone Theory.* New York: Penguin, 2011.
Conty, Arianne. "Absolute Art: Nicholas of Cusa's *De Visione Dei.*" *Religion and the Arts* 16 (2012): 461–487.
Crary, Jonathan. *Suspensions of Perception: Attention, Spectacle and Modern Culture.* Cambridge, MA: MIT Press, 1999.
Cusa, Nicholas of. *Visions of God.* Boston: Book Tree, 1999.
Foucault, Michel. *Discipline and Punish: The Birth of the Prison.* New York: Vintage Books, 1995.

Gil, Isabel Capeloa. *Mitografias: Figurações de Antígona, Cassandra e Medeia no Drama de Expressão Alemã do Século XX*. Lisboa: INCM, 2007.
Grossberg, Lawrence. *Cultural Studies in the Future Tense*. Durham, NC/London: Duke University Press, 2010.
Grusin, Richard. *Premediation: Affect and Mediality After 9/11*. Basingstoke/New York: Palgrave Macmillan, 2010.
Harari, Yuval Noah. *21 Lessons for the 21st Century*. New York: Random House, 2018.
Kracauer, Siegfried. *The Mass Ornament*. Cambridge, MA: Harvard University Press, 1988.
Lyon, David. *The Electronic Eye: The Rise of Surveillance Society*. Cambridge: Polity, 1994.
Lyon, David. *Surveillance Society: Monitoring Everyday Life*. Buckingham/Philadelphia: Open University Press, 2001.
Lyon, David. "9/11, Synopticon, and Scopophilia: Watching and Being Watched." *The New Politics of Surveillance and Visibility*. Eds. Kevin D. Haggerty and Richard V. Ericson. Toronto: University of Toronto Press, 2006. 35–54.
Mbembe, Achille. "Necropolitics." *Public Culture* 15.1 (2003): 11–40.
Mirzoeff, Nicholas. *How to See the World*. London: Pelican, 2015.
Mozur, Paul. "Inside China's Dystopian Dreams: A.I., Shame and Lots of Cameras." *The New York Times* (8 July 2018). https://www.nytimes.com/2018/07/08/business/china-surveillance-technology.html [accessed: 25 January 2019].
Poitras, Laura. *Astro Noise: A Survival Guide for Living Under Total Surveillance*. New York: Whitney Museum, 2016.
Rogoff, Irit. "What is a Theorist." *Was ist ein Künstler?* Eds. Katharina Sykora et al. Berlin: Transvia, 2003.
Scott, James C. *Seeing Like a State: How Certain Schemes to Improve the Human Condition Have Failed*. New Haven: Yale University Press, 1998.
Staples, William G. *The Culture of Surveillance: Discipline and Social Control in the United States*. New York: St. Martin's Press, 1997.
Turkle, Sherry. *Simulation and Its Discontents*. Cambridge, MA: MIT Press, 2009.
Vernant, Jean-Pierre. "Die religiöse Erfahrung der Andersheit. Das Gorgogesicht." *Faszination des Mythos. Studien zu antiken und modernen Interpretationen*. Ed. Renate Schlesier. Basel: Stroemfeld/Roter Stern, 1985. 399–420.
Virilio, Paul. *War and Cinema: The Logistics of Perception*. London/New York: Verso, 2009.
Weizman, Eyal. *Forensic Architecture: Violence at the Threshold of Detectability*. Cambridge, MA: Zone Books, 2017.
Williams, Raymond. *The Long Revolution*. Harmondsworth: Penguin, 1961.

ANNEX

Fig. 1: A video showing facial recognition at Megvii intelligence company in Beijing. Credits: Gilles Sabrie *New York Times* (published July 8, 2018).

Fig. 2: Chinese Police wearing A.I.-powered smartglasses. Credits: Reuters.

Fig. 3: Michelangelo Merisi da Caravaggio, *Medusa* (1597). Le Gallerie degli Uffizi, Florence.

Richard Grusin

No Future: The Study of Culture in the Twenty-first Century

Preface

This lecture was written for the occasion of a celebration in July 2016, marking the 10th and 15th anniversaries respectively of the University of Giessen's GCSC and GGK. It was, in retrospect, a more innocent time, when my colleagues from outside the US were asking curiously if Donald Trump could really be elected president of the United States, which I assured them was extremely unlikely to happen. If I were asked to write a similar lecture today my premediation of "no future" for the study of culture in the twenty-first century would seem even more on point than it might have then.

My own involvement with these two graduate programs began in March 2010 when I delivered the opening plenary address at "The Arts of Mediation," a summer conference at the Catholic University of Lisbon, organized as part of the PhDnet, of which the University of Giessen is a founding member. At that time I was in the process of finalizing the details of an offer to direct the Center for 21st Century Studies (C21) at the University of Wisconsin-Milwaukee. Initial discussions with Ansgar Nünning about the GCSC persuaded us both that it would be good to work together. Within a year, we had written and signed off on a memorandum of agreement between the two centers that we would collaborate in the future. Beginning with master classes, lectures, and meetings in Giessen, we developed plans for collaboration. Working primarily with Martin Zierold and Beatrice Michaelis, I helped to organize the May 2013 GCSC conference on "The Re/turn of the Nonhuman in the Study of Culture," which was co-sponsored by C21. Our most substantial collaboration was a $1.5 million Andrew Mellon Foundation-funded program in Interdisciplinary Graduate Humanities Education Research and Training (IGHERT), a program that also involved humanities centers at University of California-Santa Cruz and Australian National University, Canberra.

In 2011 while working out the details of our letters of intent to collaborate, I sent Ansgar Nünning a copy of the inaugural lecture I had delivered at University of Wisconsin-Milwaukee in October 2010, entitled "The Future of 21st Century Studies." I like to think that the Anniversary Symposium on "Futures of the Study of Culture" was at least partly indebted to that inaugural lecture, which argued that the future of the study of culture lay in the interdisciplinary academic field

https://doi.org/10.1515/9783110669398-007

called twenty-first-century studies. I know that my argument in this lecture shares some points of contact with that earlier piece, which addressed the future of twenty-first-century studies, particularly as it manifests itself in the arts, humanities, and social sciences, the traditional areas of focus among centers like C21 and GCSC.

1 Framework for the Study of Culture in the Twenty-first Century

What do I mean by claiming that there is no future to the study of culture in the twenty-first century? My thesis is that we are (or should be) nearing the end of the study of culture, and that if we as academics continue to study it as we have since at least Kant and Herder we will run the risk of irrelevance, or worse. In this chapter I maintain that there is no future for the study of culture if it does not include the study of key concerns of the twenty-first century, including especially those ecological, geopolitical, and economic issues that threaten the existence of culture as we know it, but which have historically been defined in opposition to cultural issues.

I intend by the phrase "no future" to make two allusions. The first is to the powerful refrain to the Sex Pistols' 1977 anthem, "God Save the Queen," whose sympathy for the working class and resentment of the monarchy have recently refigured themselves very differently in the wake of the 2016 Brexit referendum (Sex Pistols 1977). In alluding to the Sex Pistols song I mean to call attention to the way in which the study of culture in the academy is currently threatened by the politics of austerity and securitization that have emerged in the EU and the US as a preemptive response to the threat of global terrorism.

The second allusion in my title is to Lee Edelman's 2004 book, *No Future: Queer Theory and the Death Drive* (Edelman 2004). In remediating Edelman's title, I want both to borrow from and to expand his argument. I borrow from his project the concern to destabilize or "queer" normative notions of time, particularly of reproductive futurity. Both the nation state and the academy operate according to a temporal framework in which their institutional form is reproduced indefinitely into the future. Insofar as the contributors to this volume have been asked to imagine the institutional reproduction of an academy in which the study of culture will still have a place, the editors of this volume imagine the reproduction of an academic future that looks very much like the present. For Edelman a queer ethics negates the conservative political project of reproductive futurism. Queerness, Edelman argues, "comes to figure the bar to every realization of futurity, the resistance, internal to the social, to every social structure or form" (Edelman

2004, 4). I want, then, in this chapter, to try to "queer" the future of the study of culture. Although I follow Edelman in thinking about how we might queer the future of the study of culture in the twenty-first century, I want to move beyond the concerns with the human and the literary that Edelman's book foregrounds, and that the notion of culture customarily entails.

My thinking about the future of the study of culture in the twenty-first century was first formalized in September 2010 when I delivered a lecture to mark the commencement of my directorship of the Center for 21st Century Studies at the University of Wisconsin-Milwaukee, a position I held until August 2015. In that 2010 lecture I described the state of twenty-first-century studies at the start of the century's second decade to articulate what I saw as some of the key issues that should occupy scholars as they invented this emergent interdisciplinary field. So what did I mean by twenty-first century? My initial response was to define twenty-first century in the following three ways. These definitions focus mainly on the humanities, arts, and culture. But as I will suggest later in this chapter, the future of twenty-first-century studies cannot end with these disciplines. Nonetheless, it is here that twenty-first-century studies must begin.

First, twenty-first-century studies can be defined as the interdisciplinary studies of the humanities as they are currently being practiced in the twenty-first century, i.e., contemporary, cutting-edge study of the arts, humanities, and culture in all fields, and at all historical periods. In other words, twenty-first-century studies names the most up-to-date, contemporary manifestation of work in the academy. To help explain what this means I would borrow the words of Walter Benjamin who described the vocation of an educated journal as being "to proclaim the spirit of its age. Relevance to the present is more important even than unity or clarity" (Benjamin 1996, 292). In proclaiming the spirit of its age, twenty-first-century studies should not be about establishing a coherent program or intellectual project but rather about an active engagement with and representation of contemporary thought and criticism. This commitment to the contemporary does not mean that twenty-first-century studies is only concerned with the present moment. Indeed, as Benjamin would be the first to argue, it is often the past that can most effectively illuminate the present moment in its specificity.[1]

---

1 An interesting recent example of how the study of the humanities in any historical period can engage with issues of contemporary concern can be found in the decision of the Medieval Studies Association NOT to move its 2011 meeting from Arizona in protest of its current immigration laws but to use the conference to highlight the way in which the concerns at stake in Arizona's recent legislative attempt to criminalize immigrants could also be seen at work in different ways in the medieval period. As detailed in the online *Inside Higher Education*: "So rather than move the meeting, the academy plans 'to ensure that the program of the meeting reflects and relates to

In addition to being defined as the current state of interdisciplinary research in the arts, humanities, and social sciences, twenty-first-century studies can also be defined as the study of the key issues of the twenty-first century, which means in the first instance studies of the present and very recent past, of issues of pressing concern for the twenty-first century. But the study of the twenty-first century also means, now, that the study of the twentieth century has become historical in its incarnation as the century prior to ours as well as in its role as the final century of the second millennium. *In this sense, then, the twentieth century is the new nineteenth century.* Already in the first decades of the current century scholars have set out many of the key themes for the decades to come, almost all of which have emerged from the concerns of the late twentieth century but take particular forms when placed in relation to our twenty-first century future. Thus, issues like climate change, terrorism, finance, mobility, migration, security, sexual equality, water, food, health care, networked media, biotechnology, geo-engineering, and other concerns provide an ample field of research for twenty-first-century studies. Add to this the growing interest in future studies and it is easy to see that twenty-first-century studies is clearly a growth field.

Third, twenty-first-century studies can be defined as a synthesis of the first two definitions. That is, as I have found to be the case in my own work, particularly on digital mediation, by studying what is distinctive about the issues of the present or near future, we are also able to rethink or reconceptualize our study of the past, which in turn allows us to understand our present situation in a different light. This scholarly feedback system operates in almost every field of the humanities and social sciences, as, for example, contemporary concerns with questions of gender and sexuality prompted scholars to investigate earlier historical formations of queer, straight, and other sexualities, which in turn provided new insights on our own gendered and sexual formations. The same thing can be seen in the way in which, say, an enterprise like critical race studies has both provided new perspectives on and been strengthened by historical study of racial science. Similarly, interest in new forms of digital media has helped both to accentuate study of earlier media formations like photography, print, or linear perspective and to remind us that what was most new about our new media was the way in which they remediated prior media formations. This is one of the most interesting consequences of studying the concerns of the twenty-first century, the way in which they help to transform our received understandings of the history

similar issues at stake in medieval society, including such topics as race, ethnicity, immigration, tolerance, treatment of minority groups, protest against governmental policies judged unjust, and standards of judicial and legislative morality.'" [http://www.insidehighered.com/layout/set/print/news/2010/08/04/medieval].

of the humanities, arts, and social sciences, and how these new historical understandings help transform our current moment.

Indeed, one thing that should be evident from the definition of twenty-first-century studies I offered in 2010 is that for this emerging interdisciplinary field to engage the key issues of the day, it cannot focus on the humanities, arts, and social sciences alone. If twenty-first-century studies is to engage the crucial issues of the day, it is going to have to engage all of those other academic disciplines that have traditionally been seen as distinct from the concerns of the humanities. Because the dividing lines between science, technology, art, society, and culture are becoming increasingly blurred, it is essential for the humanities to collaborate with and reach out to scientific, technological, political, and economic disciplines. And conversely, if scientific and technological disciplines are going to answer the complex transdisciplinary questions that mark the first decades of the twenty-first century, they are going to have to engage and enroll the humanities, arts, and humanistic social sciences in doing so. Although in the twentieth century the study of culture has been distinguished from the natural, physical, and social sciences, from computation, engineering, or business, much as the human has been opposed to the nonhuman, in the twenty-first century these clear-cut divisions can no longer obtain. Borders are becoming increasingly difficult to draw, as questions of genetic engineering, for example, turn into questions of ethics, or questions of film, music, and literature become questions of sampling technologies and intellectual property, or questions of war and counterterrorism become questions of computer programming or the design of video games. The study of culture in the twenty-first century must expand the idea of culture if it is to remain relevant and vital not only to the academy but to the world at large.

2 No Future?

So, having just articulated a framework for the future of the study of culture in the twenty-first century, why do I want to argue that there is “no future” for the study of culture? Haven’t I just articulated a vision of what the future of the study of culture should look like? Well, yes and no. For if I am right about how we should pursue the study of culture in the twenty-first century, then culture as it has been understood for over three hundred years will no longer exist as an autonomous or self-sufficient realm. The future of the study of culture will in that sense be incoherent, and will be usurped by the study of media technologies, or the Anthropocene, or surveillance technologies, and so forth.

In singling out 'climate' and 'terrorism' I mean to signal not just the importance of interdisciplinarity but the increasingly vital threats that both ecological and geopolitical concerns pose to the continued future not just of the study of culture, but of culture itself. Indeed the idea that there is no future for human culture is today most often associated with questions of climate change, particularly with the concept of the sixth mass extinction or the Anthropocene, which imagines a time after humans are gone and when the impact of humans and their culture on the planet is evident only in nonhuman traces and effects. But arguments about the end of civilization have also been made in response to the proliferation of radical acts of terrorism in the post-9/11 world, and the preemptive responses by nation-states and transnational organizations to the threats posed by such acts. In my five-year tenure as director of C21, we pursued many of these ideas in trying to define and develop the field of twenty-first-century studies. Increasingly our treatment of the future of twenty-first-century studies moved from global to local issues, as we began to address the threats to the future of higher education in Wisconsin from our own state government. Ironically, doing so often entailed in part not the move away from the study of humanities and culture in the name of interdisciplinarity, but the attempt to protect the value of the study of culture for its own sake, without resorting to neoliberal arguments about utility and economic value. Doing so did not mean a refutation of the arguments for interdisciplinarity but a refusal to subordinate those arguments, as they often are in work that goes under the name of digital humanities, to instrumentalist claims for the value of the study of culture. In light of real threats to the reproductive futurism not only of the academic study of culture, but to culture and society as they have come to exist in institutional modernity, we need to do more than make the study of culture interdisciplinary.

Perhaps this is what Ulrich Beck had in mind when he claimed in *World at Risk* that "cultural criticism" was inadequate as a means of approaching the problems of climate, terrorism, and other twenty-first century threats: "Even the most radical cultural critiques look like caricatures compared to the catastrophic potentials of full-blown modernity. Indeed, one must even go an essential step further; in comparison to the horizon opened up by the negation of the basic principles of modernity, most cultural criticism looks outdated and 'idyllic,' i.e., blind to its own presuppositions or even downright affirmative" (Beck 2009, 229). In dismissing cultural criticism as a means to address the imminent catastrophes of global risk society, Beck takes it to task for failing to realize that the end of nationalist modernity, which the principles of modernity itself have brought about, is not the end of the world but the end of a particular historical formation. He accuses thinkers like Foucault, Adorno, or Weber for failing to "realize that where [cultural criticism] sees a world coming to an end the world order is in fact being transformed, that the rules and structures of power and domination are being renegotiated in the global age" (Beck 2009, 219). This renegotiation is prompted by the global risk brought about by such imminent threats as climate change and terrorism: "For there is no greater threat to the Western way and quality of life than the combination of climate change, environmental destruction, dwindling energy and water resources and the wars they could spark" (Beck 2009, 65). Beck sees cultural criticism as unable to get past its immanent critique of culture, "which presupposes and reinforces the basic principles of modernity as a measure of value without questioning them" (Beck 2009, 229).

While I share Beck's concern with the need to anticipate or premediate future risks, and his analysis of the way in which media proliferate potential global catastrophes, I am not entirely convinced that his invocation of the "'cosmopolitan moment' of world risk society" is necessarily the best alternative to the problems of climate or terrorism that threaten our future. Take, for example, the cosmopolitan anticipation of the always-imminent threat of global terrorism, Islamic or otherwise, which has generated the development of the largest transnational infrastructure of surveillance and securitization in world history. Such transnational cooperation has led to the widespread collection of as much communication and transaction data as technically possible, spearheaded by the NSA and Great Britain's GCHQ, but also supported by the BND in Germany and the DGSI in France, with the aim of preempting potential threats or catastrophes before they happen. But because such preemption is often unsuccessful, we have also witnessed the concomitant development of militarized securitization forces ready to respond immediately to a presumed terrorist attack. My concern here is that the everyday terror of such surveillance and securitization runs the risk of having a far greater negative impact on freedom and safety than the threats they are designed to counteract or preempt. Rather than dismiss cultural criticism for failing to keep up with these new global social formations, there is a role for the study of culture in resisting the abuses and overreaching of security and surveillance.

To put it differently, I would agree with Beck that there is no future for the study of culture if such study does not take as one of its foremost tasks the resistance to the forces of neoliberalism and austerity that threaten the very existence of the academy as we have come to know it. But this resistance cannot simply be a defense of reproductive futurism, of the institutional reproduction of the academic study of culture as it has existed for the past century and more. There is still, I would argue, pace Beck, Latour, and others, a role for critique in the study of culture, but that role must include taking on not only the discourse of humans but also the mediations of nonhumans, technical and otherwise (Latour 2004). While we are well on our way to doing just that, to breaking down the traditional barriers between humanistic and non-humanistic disciplines within the walls of the academy, we must at the same time defend the walls of the academy against the barbarians at the gate, the incursions of globalization and securitization brought about to protect us from the threats posed by climate and terrorism. While these threats are very real, and I would not in any way want to diminish them, I want also to challenge the way that these threats have been used as an occasion to strengthen and reinforce the control of global capitalism and militarized securitization over individuals and societies as well as academic institutions. More than twenty years of climate change summits have done little or nothing to reform or dismantle the economic structures of industrial and post-industrial global capitalism that have brought us the environmental and geopolitical catastrophes that we are now facing.

Indeed, we can see in the emergent paradigm of 'resilience' the acceptance of catastrophic change and the determination to create contingency plans to increase the chances that our economic and political institutions will survive in the face of environmental disaster. Barack Obama's historic "Climate Action Plan" of June 2013 was celebrated for being the first official acknowledgment by the US government that global warming has been scientifically proven to be real (Executive Office of the President 2013). But what I found most telling about this plan was not its commitment to cut carbon pollution in America but its recognition that no matter what changes might be able to be made, America (and by extension the world) must act now to prepare itself for the impacts of dramatic climate change by increasing the resilience of infrastructure, buildings, and communities, as well as economic and natural resources. Nowhere does the question of reducing production or commodification or development ever come into the equation, but only the question of developing new technologies that might reduce environmental damage in order to allow capitalism to continue unchecked. In other words, the significance of Obama's "Climate Action Plan" was less its commitment to reverse or even to end global warming than its commitment to prepare the United States to deal with catastrophic environmental change by increasing its 'resilience.'

Similarly, the paradigm of resilience is also at the heart of counter-terrorism planning. Despite the declared goal of using total data surveillance to preempt terrorist attacks a la the pre-crime unit in Steven Spielberg's *Minority Report*, the collection and mining of data and the unprecedented proliferation of hyper-militarized homeland security police forces is designed mainly to insure that communities and institutions will be resilient in the wake of inevitable acts of terrorism. For like Obama's 'Climate Action Plan,' his and other nations' plans to defeat terrorism begin with the acknowledgment that terrorism, like climate change, is inevitable, and therefore focus mainly on preparations for surviving these inevitable attacks. In the case of terrorism what this means is an increase in the militarization and securitization of society, evident in the US in the dramatically increased use of military equipment and tactics by local police forces, spurred on by the *National Defense Authorization Act* of 1997, which was signed into law during the Clinton administration. While the Obama administration moved to limit such military-style weapons, particularly in the aftermath of the riots in Ferguson, Missouri, in 2014, there has been little change in the degree of militarized police control in the United States, ready to be brought to bear at a moment's notice, as we saw in responses to shootings in places like the gay nightclub in Orlando or, to bring this issue closer to home, on college campuses.

I want to close by briefly addressing the overwhelming show of militarized police force that occurred in response to a murder-suicide at UCLA, on the first of June, 2016, as a way of underscoring my concern that securitization is as much or more to be feared than the acts of criminal violence that are lumped under the rubric of 'terrorism.' For those not familiar with this event, at around 10:00 AM police received the first of several 911 emergency calls reporting three shots fired and possible victims.

Fig. 1: Police force assembled on UCLA campus [public domain].

According to the *Los Angeles Times*, the shooting "prompted a massive response from state and federal law enforcement.

Fig. 2: Militarized police vehicle on UCLA campus [public domain].

Police officers in riot gear ran across campus, guns and battering rams drawn, while students exited buildings with their hands above their heads."

Fig. 3: Police officers in riot gear on UCLA campus [public domain].

Fig. 4: Students with hands up on UCLA campus [public domain].

In addition to shutting down the campus for two hours, police searched every student carrying a backpack, which as we all know means almost every student on campus.

Fig. 5: Students being searched by police on UCLA campus [public domain].

In retrospect this response proved to be massively disproportionate to the threat, which ended when the murderer, a former graduate student, killed himself after killing his advisor. Of course, one might argue that because there was no way to know this at the time, it was better for the police to be safe than to be sorry. And this argument makes a certain kind of practical sense. But what I find most troubling about this event is not the murder itself, as tragic as it was, nor the threat of future murders with greater numbers of victims. No, what is most troubling to me is the fact that such a massively militarized law enforcement infrastructure pre-exists and that the authority of the police to take complete control over any public space, including a college campus, is undisputed, particularly when, as was the case in this incident, every student on campus is transformed into a potential suspect.

Fig. 6: Students with hands up on their knees being searched by police on UCLA campus [public domain].

The metaphorical force of this event was quite powerful. Even in a supposedly "free" society like the United States, we live, work, and study on campus at the behest of the forces of securitization, and our permission to do so can be revoked at a moment's notice, and with very little evidence or explanation. In light of the mass school shootings that occur periodically in the US, calls to arm schoolteachers or to install armed guards in public schools from kindergarten through high school have only intensified. Indeed, since the Columbine school shootings in 1999 there are more than 10,000 additional police officers stationed in US public schools. Clearly these officers have not stopped the shootings – as evidenced, for example, in the February 2018 school shootings in Florida, which killed 17 students and teachers. But what they have done is to criminalize students, who are arrested at a pace of 70,000 per year for offenses that in the past would likely have resulted in disciplinary actions like detention, suspension, or expulsion. And like all police arrests in the US, these arrests have occurred disproportionately to black students ("Militarizing Schools, Criminalizing Students" 2018). Rather than safe spaces for the free exchange of ideas and the education of American youth, schools, colleges, and universities in the United States have become militarized zones to discipline and punish students. To return to the language of the 2016 GCSC/GGK symposium, and to the thesis of this paper, there appears to be no future for the study of culture that is free from the authority of militarized securitization and global capitalism, which means that unless we are vigilant in exposing and resisting these forces there may be no futures for the study of culture at all.

References

Beck, Ulrich. *World at Risk*. Cambridge: Polity, 2009.

Benjamin, Walter. "Announcement of the Journal Angelus Novus." W.B. *Selected Writings* 1. Eds. Howard Eiland and Michael W. Jennings. Cambridge, MA/London: Harvard University Press, 1996.

Edelman, Lee. *No Future: Queer Theory and the Death Drive*. Durham, NC: Duke University Press, 2004.

Executive Office of the President. *The President's Climate Action Plan*. 2013.

Latour, Bruno. "Why Has Critique Run out of Steam? From Matters of Fact to Matters of Concern." *Critical Inquiry* 30.2 (2004): 225–248.

"Militarizing Schools, Criminalizing Students." *The Real News Network* (26 February 2018). <https://therealnews.com/stories/militarizing-schools-criminalizing-students> [accessed: 10 October 2019].

Sex Pistols. "God Save the Queen." *Never Mind the Bollocks, Here's the Sex Pistols*. London: Virgin Records, 1977.

Hubertus Büschel

Beyond the Colonial Shadow? Delinking, Border Thinking, and Theoretical Futures of Cultural History

> It's as if the colonial event belonged to another age and another place, and as if it had absolutely nothing to teach us about how to understand our own modernity, about citizenship, about democracy, even about the development of our humanities. (Mbembe 2006, 4)

1 A Past Future: The Birth of New Cultural History and the Boom of Anthropological Theories

In the year 1989, Lynn Hunt proclaimed nothing less than the beginning of a *new* cultural history (Hunt 1989, 10) and pleaded for consequential anthropological theoretical receptions. Already eight years prior, Natalie Zemon Davis wrote, "anthropology can widen the possibilities, can help us take off our blinders, and give us a new place from which to view the past and discover the strange and surprising in the familiar landscape of historical texts" (Davis 1981, 275). In Germany in 1984, Hans Medick published his legendary – and subsequently updated – article *Missionaries in the Rowboat?* stating that anthropological knowledge and theories could help enlighten the "complex mutual interdependence between circumstances of life and the concrete practice" of historical actors – their "experiences and modes of behavior" (Medick 1995, 43).

All three authors, and many other cultural historians, argued against socio-historical approaches that they considered deficient, particularly due to their theoretical framework. During the late 1980s and early 1990s, new cultural history, as well as historical anthropology (a specific German version of cultural historical approaches), were coined by the constant pleas for an injection of anthropological theories and methods into historical theoretical and methodological approaches. Anthropology might help, the argument went, to see the past as a "strange foreign territory" and the everyday life of historical actors akin to those of "'primitive' or 'archaic' societies" (Davis 1981, 272). With this approach,

Note: I thank my research assistant A. Tancrède Pagès for his important remarks, corrections, and the editing.

https://doi.org/10.1515/9783110669398-008

new innovative research could be pushed forward in order to produce analyses less determined by socio-historical master narratives and present prejudices but rather by local experiences and negotations as well as multiperspectivity or counternarratives. Quite similar was the second major influence in the formation of new cultural history, French post-structuralism, notably Michel Foucault's work on discourse (see Burke 2008, 56).

In this article, I will critically discuss this "paradigmatic shift" (Kuhn 1970, 6) – this anthropological and poststructural turn – that was so fundamental in the formation of new cultural history. When Achille Mbembe said, in a 2006 interview, that colonialism seems to be far away and seems to belong "to another age and another place" (Mbembe 2006, 4), he hit the neuralgic point of central theories within new cultural history. While anthropologists had already considered the need for methodological and theoretical revisions in their discipline, in part due to the colonial heritage of their epistemic landscape (Comaroff 2010; Kohl 2010), cultural historians have paradoxically argued for a quite ahistorical reception of the very same anthropological theories derived from colonial research practices that have been criticized by the aforementioned anthropologists.

To illustrate this line of argument, I will analyze in the following article three central anthropological theories that have been thoroughly integrated into cultural history and cultural studies. By concentrating on these three examples, I wish to emphasize the colonial roots of anthropological theories and the ways in which they allow for, and cloak, analytical shortcuts and heuristic, theoretical, and methodological assumptions. The examples to be examined are as follow: ritual theory by Victor Turner, Pierre Bourdieu's concept of 'habitus,' and last, but not least, Michel Foucault's discourse analysis. I will also reflect upon the individual experiences these scholars had with colonialism – this might give us an idea about the "thought style" (Fleck 1979, 38–50) that pushed forward their thoughts and was subsequently reflected in their theories. At the end of the article, I will discuss scholarly efforts to overcome the colonial limits of cultural historical theories. Altogether, I would like to offer some space to discuss the futures of cultural history, allowing for the reconsideration and reformation of cultural historical theories. In the last decades, scholars have formulated tools for going beyond colonial shadows and eurocentric perspectives including 'delinking' and 'border thinking.' These tools might be helpful in efforts revolving around the decolonization and revision of cultural historical theories. I will shortly introduce and discuss these tools and ventilate their possible benefits for the theoretical future of cultural histories.

2 The Danger of Rituals: Victor Turner and the Homogenous Societies

Since the 1980s and 1990s, cultural historians have worked extensively with anthropological ritual theories (Bell 1997), particularly those of Victor Turner and his approach to liminality (Turner 1964, 1969; Deflem 1991). Turner, in developing Arnold van Gennep's (1960 [1909]) threefold structure of rituals of passage further, argues that rituals are usually based on a "liminal phase" with several "liminal spaces" followed one after another. These spaces are categorized in observed transitions performed over the course of a ritual. Turner was primarily interested in the transitions in which individuals were – as he wrote – "betwixt and between" (1964), exceeding and crossing boundaries of the liminalities of rituals that would create a sense of community among all involved (Turner 1967, 1969).

In reference to Turner, the sociologist and anthropologist David Kertzer proclaimed that rituals have a key role in the formation and cohesion of social groups. They make it possible to create social consensus even if there are substantive contradictions within or conflicts among the social groups. Through their constant repetition, rituals ensure the maintenance of social bonds. On the other hand, they are also to be constantly adapted to changing political conditions in order to maintain their socially stabilizing effect (Kertzer 1988, 189–196).

These assumptions made anthropological ritual theory, with all its symbolic interpretations, tremendously attractive for cultural historians; rituals seemed to represent key phases and elements of societal developments. Meanwhile, liminality appeared to give answers to questions of historical change and to the question of how societies are organized and restructured. Historical change can be seen as being condensed in the 'in between' phases of rituals.

Already in the 1970s, Natalie Zemon Davis worked with ritual theories for her analyses of violence and its religious roots in seventeenth-century France (1973). Subsequently, Roger Chartier (1984, 1989, 1990) investigated structures of experiences with the help of Turner's ritual theories and drew conclusions from the manifestations of rituals on the intentions of the organizers of these rituals and the perceptions, feelings, and thoughts of their participants. In Germany, Barbara Stollberg-Rilinger promoted ritual theories for the cultural historical analyses of early modern forms of governance (2013). Johannes Paulmann revisited the classical field of monarchical meetings in the long nineteenth century. In line with Turner and Kertzer's observations on rituals, Paulmann came to the conclusion that symbolic presentations of power would have had a direct impact on political relations (2000, 402–416). Both authors contributed to new and innovative interpretations within the field of political history and showed how rituals can establish, negotiate, and structure power relations. Altogether, the works mentioned above demonstrated how ritual theory can offer novel research findings. Most notably, it could cast analytical light on overlooked or ignored public rituals, ceremonies, and performances that were important elements of societal and political expression and negotiation. Although receptions of anthropological theories of ritual have undoubtedly led to new, innovative interpretations of more classical political-historical phenomena, the theoretical premises of ritual theory are problematic in that it becomes increasingly difficult to disentangle them from colonial ways of producing and accruing knowledge.

Similar to most of his contemporaries, Turner was deeply involved in the colonial establishment; he worked in Northern Rhodesia (modern day Zambia) as a research officer for the Rhodes-Livingstone Institute (RLI). The RLI, founded in 1938 to conduct demographic studies on the local population, was the first anthropological research facility in Sub-Saharan Africa. It was mainly financed by the Colonial Office in London (Schumaker 1996, 238–240). Therefore, it was clearly embedded within the colonial system of governance, contrary to constant official claims that the RLI would stand for liberal and anti-racist research promoting African independence (Brown 1973). Turner, like all researchers of the institute, was tasked with providing colonial authorities with useful information about the customs, needs, and thoughts of the natives to facilitate colonial rule by avoiding conflicts and frictions with and within colonized populations. It is within this context that we have to understand Turner's studies on the Ndembu people (Turner 1967). In his approach on 'social drama,' he took the experiences of colonialism and colonial civilizing missions into account to develop his theories about rituals as a form of symbolic conflict management and crisis resolution in the Ndembu villagers' society living between 'traditional' and 'modern' life – he was also interested in the experiences of cultural suppression and losses caused by the colonial encounters. Turner's research, in combination with the needs to serve the colonial authorities with information for their future policies, had to deal with the assumptions of a quite homogenous Ndembu society following the notion of the colonial concept of a static and, in its practices, 'mechanistic' (Comaroff 2010, 531) 'tribe' (Ekeh 1990). Turner's conception of the liminality of rituals and his focus on the 'tribe' can be explained with his training as a colonial expert; individuals were not of interest here. Instead, the focus lied on collective moods and events as these were the categories through which colonial experts could plan and calculate their policies. The figure of the 'tribe' as a constructed and imagined basic social unit was central. It conditioned the ways in which colonial administrators negotiated, planned, and eventually deployed their modes of governance entailing systematic cultural oppression through so-called 'civilizing missions.' In addition, Turner's theoretical roots and role models were profoundly colonial. Above all he was influenced by the founding father of British anthropology, Edward B. Tylor, who, among others, had developed the so-called "animistic theory" that emanated from the "soulfulness of magical-sacral presentations in primitive societies" (Tylor 1871, 424). By 1871, Tylor wrote that it was through magical rites and symbols in a "primitive" society that a collective belief could be evoked alongside a "sacralization and ensoulment" of all individuals in a society (Tylor 1871, 424).

As Suzanne Desan, among others, has pointed out, this notion seeped into the reception of anthropological theories of ritual in the new cultural history and formed itself primarily for "the methodological assumption that an analysis of the patterns of crowd activism will reveal its meaning and offers clues about community structures" (Desan 1989, 56). This approach was deeply criticized by scholars, such as the medieval historian Philippe Buc, who coined the slogan "dangers of rituals" and pointed to the heuristic problems of inferring the forms of rituals on their intentions, perceptions, and receptions. Buc claimed that individual resistance, reservations, or simply social performances without social consequences have to be taken into consideration (Buc 2000, 183–186, 2001, 8–11). While Buc and others objected to the methodological and heuristic shortcuts of anthropological theories of ritual, one can also assume that it would be possible to draw conclusions from the appearance of rituals that their intentions and effects were to be determined by colonial thoughts in two respects: First, the theories are reminiscent of typical colonial conceptualizations of native societies as homogenous, static, and passive; second, we can see the deeply colonial and even racist conception of ceremonies that everything has a cohesive meaning of the highest value within so-called 'primitive societies' or community. A similar, but slightly different, conception of community can also be found in Pierre Bourdieu's cultural historical theories, principally in his concept of "habitus" (Bourdieu 1977).

3 In the Imperial Gaze: Pierre Bourdieu's Colonial Concept of "Habitus"

In the aforementioned 1989 key publication *New Cultural History*, Aletta Biersack praises the usefulness of Bourdieu's theory of practice for cultural historians when they are interested in investigating the practical side of everyday life in the past (Biersack 1989, 90). Roger Chartier was influenced by Bourdieu's theory of practice as well as by ritual theory (Burke 2008, 59). Simon Schama, in his famous book *The Embarrassment of Riches* (1987), analyzed several "Dutch obsessions" like the ongoing preference of cleanliness through Bourdieu's concepts of "habitus" and "distinction" (see Gorski 2013).

Like Turner, Bourdieu had his own colonial experiences. In the year 1955 he arrived in Algeria during the war of independence. Once there, he absolved his military service and was first employed as a typist at the Air Force and then as a clerk at the news and documentation service of the General Government in Algiers. Bourdieu was fascinated by the country. He decided to stay and took over the position as a lecturer at the University of Algiers. In the following months, he became interested in anthropology and started to conduct fieldwork on the colonized Algerian society.

Bourdieu undertook two intensive, and sometimes dangerous, field studies: the first on the living conditions in cities during the war; the second on villages in rural Kabyle in northern Algeria hedged in between the Atlas mountain range on the one side and the Mediterranean See on the other. People from Berber societies and peasants who were imprisoned in French resettlement camps were the predominant demographic living in this region (Free 1996; Herzfeld 1987, 7–8; Lane 2000, 13–16). Around three million people – nearly half of the Muslim rural population – were relocated by force not only for 'security reasons' but also to push forward development programs of 'modernization.' These camps were structured like village communities in order to push forward a variety of colonial development programs including "modern agriculture," "rational economies," and "everyday life hygiene" (Wilder 2003, 2005; Büschel 2014, 169–171).

Bourdieu was aware of the colonial violence the establishment and operation of these camps represented as well as the ways in which the French civilization mission had caused deep political, social, and cultural ruptures within the Kabyle societies. He called the colonial development and education programmes "social vivisection" and "social surgery" (Bourdieu 1962, 131–133). There is no doubt that Bourdieu shared critical thoughts about colonialism as a racist system of suppression and domination, enforced through violence, manifest in both physical and psychological force, that destroys and restructures social relations (Go 2013, 49–74).

In spite of possessing deep sympathy for the colonized Algerians, illustrated by his profound critique of colonialism (Wacquant 2004), Bourdieu never adhered to the Third Worldism promoted by Jean-Paul Sartre, Frantz Fanon, and others (Kalter 2016; Le Sueur 2001; Ahluwalia 2010). In fact, he opposed Third World activists who seemed to him far too involved in promoting violence. The depth of Bourdieu's involvement in the mental horizons and style of thoughts of French colonizers becomes clear when analyzing his theoretical conceptions (Seibel 2004; Yacine 2003, 2004, 2008; Go 2013, 51). During his time in Algeria, and alongside his fieldwork in the Kabyle villages, Bourdieu developed his most central concepts for new cultural history in “symbolic capital” and “habitus” (Bourdieu 1977, 1990; Hammoudi 2009; Goodman 2009; Silverstein 2009). He did this while describing the Kabyle society in terms such as “collective consciousness” – very much in line with the epistemic culture of colonial expertise (Bourdieu 1977).

Critics argue that Bourdieu cultivated in his early writings a “romantic redemption of ethnic culture” (Burawoy 2011, 8) as well as a “structural nostalgia” about the local cultures, seen as passive and defensive instead of active and creative (Silverstein 2009). Edward Said and others have accused Bourdieu of developing Eurocentric theories and of being an example of a European scholar who constantly portrays non-Western societies as “different, static and homogenous” (Said 1989; Go 2013, 50). Furthermore, the concept of ‘habitus’ carries assumptions quite similar to ritual theory. Foremost, it makes it possible for one to draw conclusions about the social conditions of groups from a phenomenological analysis of incorporated collective intentions and thoughts (Callhoun 2006, 1403). Bourdieu has thus been criticized for sharing, or even reproducing, the imperial gaze of colonial knowledge systems (Go 2013, 50).

Indeed, the late and post-colonial humanitarian attitude of the colonial expert or development worker went hand in hand with deep colonialism implications as can be seen in Bourdieu’s theoretical assumptions. It should be taken seriously that concepts like ‘habitus’ are entangled with ethnocentric colonial collectivism and systematic silencing of the colonized ‘Others.’

4 Eurocentric Discourse and Knowledge: Michel Foucault and Colonialism

Similar to ritual theory and Bourdieu’s concept of ‘habitus,’ Michel Foucault’s theoretical assumptions became fundamental for new cultural history. Three aspects have been particularly influential. First, in his notions about discourse, Foucault seemed to serve the linguistic turn, which also deeply affected cultural history, whilst simultaneously asserting that discourse always means practice and that words are highly political and embedded in dispositives and governmentality. Therefore he also served the needs for theoretical framing of the study of practices in the past. Second, Foucault opposed ideas of teleological developments or progress in history. Instead, he emphasized discontinuities, ruptures, and shifts, which conforms to new cultural history’s opposition against socio-historical master narratives. Finally, Foucault’s attempt for ‘epistemes’ and ‘regimes’ of truth has been met with interest by many cultural historians investigating the cultural and historical development of ideologies and knowledge (Burke 2008, 56–57).

The important influence of Foucault on both cultural history and postcolonial studies can be traced in Said's study *Orientalism* (1978) that examines the West's patronizing and imperialistic representations of 'The East' – the people and their cultures who inhabit areas in Asia, North Africa, and the Middle East. Foucault's concept of 'discourse' was central for Said (1978, 48–49). Many scholars have criticized Said for employing a determining univocal and unidirectional notion of discourse. This was seen as a result of Foucault's reception within historical research, especially "The Order of Discourse" (1981), in which Foucault stressed the restrictive and homogenizing qualities of discourses as tools of power and violence (Foucault 1981; Young 1995, 60; Nichols 2010). The most prominent critique of Foucault's theories and their reception in cultural history was raised by Gayatri Spivak who wrote that Foucault's analysis would carry the "danger" of re-inscribing the West as a subject of analysis (Spivak 1985, 18, 1988, 291; Kaplan 1995, 90).

Similar to Turner and Bourdieu, Foucault had his own (post-)colonial experience. He arrived in Tunis in 1966 to take over a teaching position at the University of Tunis. Tunisia became a constitutional monarchy and gained independence from France on March 20, 1956 but was still determined by colonial elements in its governance, like an autocratic police state lacking any democratic element. Foucault lived in the coastal village Sidi Bou Said, known for its romantic atmosphere, occupied since 1900 by numerous European artists such as August Macke and Paul Klee. Shortly after his arrival in Tunis, there were violent clashes between students and authorities (Lazreg 2017, 176). Foucault's biographers claimed that the immense violence of the police entering the university campus, injuring students, and arresting them had a great impact on Foucault's thoughts who finally supported the protests in multiple ways including hiding a printing machine in his garden to clandestinely produce critical leaflets on behalf of students (Lazreg 2017, 177).

Taking these experiences into account, it is remarkable that Foucault's theoretical assumptions and his thoughts about discourse, dispositive, and governmentality almost completely lacked reasoning about race and colonialism (Young 1995, 5). Timothy Mitchell noted that the narrative of history in Foucault's writing is coined by a deep eurocentrism and is the story of Europe (2000, 3–7). Time and space are imagined homogenously; encounters, influences, cultural transfers, or entanglements from the non-West are irrelevant (Fernández and Esteves 2017, 141). The 'normalization' of suppression, discipline, power, punishment, and violence that Foucault describes in his writing is the narrative of "everywhere" (Foucault 1977, 304) and is thus a universal 'truth' that disregards, and ahistoricizes, its embedment within the Western classical Westphalian model of statehood and governance.

Considering this, it seems quite paradoxical that Foucault was the most cited theorist in studies about colonialism in the last decades (Stoler 1995, 1). One could say colonial studies with Foucault might carry the danger of Western asymmetries and silencing locals, thus necessitating innovative critical re-assessments and reflections of how his theories are implemented in the field of study.

5 Theoretical Futures? Decentering, Delinking, and Border Thinking

Taking into account the colonial limits of the aforementioned theories in new cultural history, it becomes clear that the theoretical futures in the study of culture need revisions and reformulations. In addition to the colonial shadows of these theories, one could argue alongside Dipesh Chakrabarty that new cultural history and the pleas for anthropology and post-structuralism carry typical elements of Western historiography's historicism of universalizing attempts to come closer to something that one may call the "truth about the past" (Chakrabarty 2000, 6–11; Young 1990, 142).

Since the early 2000s, scholars from all over the world have claimed a fundamental epistemic turn in the study of culture and therefore also in new cultural history (Grosfoguel 2007); they have called the result of this set of theoretical operations 'decoloniality.' 'Decoloniality' does not mean post-colonialism or adding post-colonial perspectives to research agendas and perspectives. Indeed, it does not seem to be fruitful to just add post-colonial theories to the canon of cultural historical concepts. 'Decoloniality' is a more fundamental approach with a whole set of operations to move "beyond the post-colonial" (Mignolo 2007, 452).

One of the operations gaining theoretical 'decoloniality' is 'decentering,' brought up by Natalie Zemon Davis as a response to Chakrabarty: 'Decentering' refers to new heuristic perspectives with particular emphasis on cultural crossings and counter-histories (Davis 2011, 190–191). Therefore 'decentering' includes the stance and the subject matter of the historian. The 'decentering' historian does not tell the story of the past from the "vantage point of a single part of the world or of powerful elites, but rather widens his or her scope, socially and geographically, and introduces plural voices into the account" (Davis 2011, 190).

A recent form of 'decentering' can be found in the initiative of Ulinka Rublack that has led to a collection of articles by historians. Rublack asked several experts to 'decenter' their own professional views and standpoints and to write about areas, approaches, and periods beyond their expertise in order to gain new and fresh analyses beyond the established self-confidence (2011). Davis' claims and

Rublack's initiative were important steps of 'decentering' on the way to 'decoloniality.' But both approaches still remained within the context of Western historiographical epistemology when it comes to methods, theories, and narratives. What we need is a more fundamental epistemological 'delinking' and 'decentering' from Western and sometimes even colonial dominated forms of thought.

An early example of this radical fundamental cultural historical 'decentering' is *First-Time: The Historical Vision of an African American People*, authored by the anthropologist Richard Price (1983). Oral testimonies, songs, stories, and artefacts of the Saramacca Marrons from the Suriname rainforest are composed together; the result is a masterpiece of historical bricolage, full of non-hierarchically presented records and thoughts about a formerly marginalized and silent society and their past experiences in violence, suppression, and slavery. 'Decentering' here means the absence of a master narrative and any voices of external Western observers, which makes it possible to perform a history of the peoples their own.

Walter Mignolo goes far beyond the non-hierarchical multiplication of voices in his attempt to de-silence the silenced. He speaks vehemently for a radical epistemic 'delinking' in the theories of cultural studies – the 'delinking' from heuristic, theoretical analytical, and methodological approaches from Western thought and culture (Mignolo 2007, 2009). This might entail the abandonment of Western modes of reasoning and epistemological practices and their replacement by 'indigenous' or 'local' thoughts. 'Delinking' implies the destruction of colonial networks of knowledge and power in order to give those who have historically been silenced a voice. This can happen with heuristic operations but also with theoretical 'delinking.' This introduces crucial theoretical questions: can we 'delink' ritual theory from its colonial framing sketched above and what is left then? Can we go beyond the colonial attitude of studying liminality with the holistic presumption of a crowd of individuals' feeling and thinking all the same in a ritual process (Desan 1989, 56)? Does bypassing the "dangers of rituals" with the heuristic operation of inferring the forms of rituals on their intentions, perceptions, and receptions mean that ritual theory have to come to an end (Buc 2000, 183–186, 2001, 8–11)? I do not think that this has to automatically and consequently be the case. 'Delinking' ritual theories from its colonial methodological and theoretical presumptions can mean 'decentering' the heuristic perspective and its analysis away from the form of rituals to its local conditions and commentators (Davis 2011, 190). Furthermore, in cultural history rituals are oftentimes seen as a form of societal cohesion and governance in a constant field of balance and imbalance between obedience and resistance (Desan 1989, 57). Already quite some time ago, Jean and John Comaroff coined the phrase of a "dialectically neomodern" analysis, which means a constant attention on transformations, processes, hegemonic practices, and cultural boundaries (Comaroff and Comaroff 1992, xi–24). Indeed, the colonial shadow of ritual theory also carries a potential; tracing it and taking it into account can help us to understand and criticize our own hegemonic practices and open our analyses for further critical reflections about the non-homogenic, non-linear, and non-conform. Therefore, theoretical 'decentering' and 'delinking' can have at least three dimensions. The first revolves around switching theoretical perspectives beyond hegemonic Western epistemic systems; the 'Other,' the supposed peripheral becomes central or at least a field of reasoning, discussion, and exchange. Second, there is a methodological-heuristical dimension that attempts to excavate and listen to marginalized and silenced sources of historical epistemology. Finally, there is the chance for constant critical revisions of hegemonic epistemic master narratives on the path for the multiplication of narrated analyses (Adichie 2009).

After 'delinking' might follow 'border thinking' (Anzaldúa 1987; Mignolo and Tlostanova 2006, 206), which focuses on all those theoretical and conceptual approaches that are so far excluded from the master narratives of cultural studies – these include different concepts of time, space, narration, plausibility, and epistemology. All these theories beyond the Western – or even beyond the colonial matrix of epistemic power – should help establish an "epistemology of the exteriority; that is, of the outside created from the inside" (Mignolo and Tlostanova 2006, 206). 'Border thinking' can also favor the creation of new analytical narratives beyond "epistemic ranking" and Western knowledge based hierarchies (Mignolo and Tlostanova 2006, 214). In practice, 'border thinking' could mean not just multiple narratives – like we can see it in 'decentering' – but also multiple theories with a new sensitivity of alternative knowledge traditions beyond Western hegemony as well as alternative theoretical inspirations – or at least the non-hierarchical multiplication of theories. Therefore 'border thinking' can produce a "redefinition/subsumption of citizenship, democracy, human rights, humanity, economic relations beyond the narrow definitions imposed by European modernity," how Ramón Grosfoguel points it out (Grosfoguel 2006, 178). [However] "border thinking is not an anti-modern fundamentalism. It is a decolonial transmodern response of the subaltern to Eurocentric modernity" (Grosfoguel 2006, 178–179). What can this mean in the practice of cultural historical analysis?

Can 'border thinking' help us to liberate concepts like Bourdieu's 'habitus' from its imperial gaze as it was described above? Bourdieu has been criticized for implementing a colonial view in concepts like 'habitus' (Go 2013, 50) by drawing conclusions of a human being's behavior and applying generalized projections onto inherent collective thoughts, feelings, or customs. Should we therefore give up 'habitus' and forget about an insightful tool for generations of sociologists, scholars of cultural studies, and historians whose goal is to trace the connection between human environment and behavior? Mignolo writes that 'border thinking' is "de-subalternizing knowledge" itself and can help to pluralize epistemic frameworks as well as make the "rigidity of epistemic and territorial frontiers" in assumptions visible (Mignolo 2000, 12–67). It might be useful to reread concepts like 'habitus' beyond the rigidity of Western presumptions and to understand that the phenomenology of the 'habitus' as well as the customs, feelings, and thoughts are localized and in constant fluidity between individuals and their live worlds, kinship, and societal relations. What could be given up by 'border thinking' and concepts like 'habitus' is the danger of holistic/universal conclusions; what could be preserved is a postcolonial turned tool for historical and societal analyses and the ambivalences, fluidity, locality, and last but not least wonderful diversity of human behavior in the past. Similarly, 'border thinking' offers great potential as a critical tool in reconfigurations of the Foucauldian concepts of dispositive and governmentality and its eurocentric determination (Young 1995, 5; Mitchell 2000, 3,7). 'De-subalternizing' mechanisms of power, influence, and governance might help to trace 'other' non-Western – but also "modern" (Grosfoguel 2006, 178–179) – governmental tools and show the pluralities of concepts in constant negotiations of societal domination, participation, and resistance far beyond the state and its Westphalian model.

In sum, I hope to have sketched that 'decentering,' 'delinking,' and 'border thinking' might be possible tools for a new theoretical future of cultural history beyond Western and colonial presumptions as "eternal and untouchable truths" (Rose 1997, 308). These approaches might help us to de-colonize important anthropological, social, and philosophical theories that were central for generations of cultural historians. Furthermore, they might help us to inspire re-readings and rethinking practical cultural historical research and prepare the theoretical design of cultural history more for the future of global exchange and local epistemic cultures.

References

Adichie, Chimamanda Ngozi. *The Danger of a Single Story.* https://www.ted.com/talks/chimamanda_adichie_the_danger_of_a_single_story?language=en. Ted Talk, 2009 [accessed: 22 March 2019].

Ahluwalia, Pal. *Out of Africa: Post-structuralism's Colonial Roots.* New York: Routledge, 2010.

Anzaldúa, Gloria. *Borderlands/La Frontera: The New Mestiza.* San Francisco: Aunt Lute Books, 1987.

Bell, Catherine. *Ritual: Perspectives and Dimensions.* Oxford: Oxford University Press, 1997.

Biersack, Aletta. "Local Knowledge, Local History: Geertz and Beyond." *The New Cultural History*. Ed. Lynn Hunt. Berkeley/Los Angeles/London: University of California Press, 1989. 72–97.
Bourdieu, Pierre. *The Algerians*. Boston: Beacon Press, 1962.
Bourdieu, Pierre. *Outline of a Theory of Practice*. Cambridge: Cambridge University Press, 1977.
Bourdieu, Pierre. *The Logic of Practice*. Palo Alto: Stanford University Press, 1990.
Brown, Richard. "Anthropology and Colonial Rule: Godfrey Wilson and the Rhodes Livingstone Institute." *Anthropology and the Colonial Encounter*. Ed. Talal Asad. New York: Humanities Press, 1973.
Buc, Philippe. "Ritual and Interpretations: The Early Medieval Case." *Early Medieval Europe* 9.2 (2000): 183–210.
Buc, Philippe. *The Dangers of Ritual: Between Early Medieval Texts and Social Scientific Theory*. Princeton: Princeton University Press, 2001.
Burawoy, Michael. *Sociology as a Combat Sport: Bourdieu Meets Bourdieu*. http://burawoy.berkeley.edu/Bourdieu/2.Combat.pdf. 2011 [accessed: 22 March 2019].
Burke, Peter. *What is Cultural History?* 2nd ed. Cambridge: Polity Press, 2008.
Büschel, Hubertus. *Hilfe zur Selbsthilfe. Deutsche Entwicklungsarbeit in Afrika 1960–1975*. Frankfurt a.M.: Campus, 2014.
Callhoun, Craig. "Pierre Bourdieu and the Social Transformation: Lessons from Algeria." *Development and Change* 37.6 (2006): 1403–1415.
Chartier, Roger. "Culture as Appropriation: Popular Cultural Use in Early Modern France." *Understanding Popular Culture: Europe from the Middle Ages to the 19th Century*. Ed. Steven Kaplan. Berlin/New York/Amsterdam: Mouton Publishers, 1984. 229–254.
Chartier, Roger. "Le monde comme representation." *Annales* 44.6 (1989): 1505–1520.
Chartier, Roger. *Les origines culturelles de la révolution française*. Paris: Le Seuil, 1990.
Chakrabarty, Dipesh. *Provincializing Europe: Postcolonial Thought and Historical Difference*. Princeton: Princeton University Press, 2000.
Comaroff, John. "The End of Anthropology, Again: On the Future of an In/Discipline." *American Anthropologist* 112.4 (2010): 524–538.
Comaroff, Jean, and John Comaroff. *Ethnography and the Historical Imagination*. Boulder, CO: Westview Press, 1992.
Davis, Natalie Z. "The Rites of Violence: Religious Riot in Sixteenth-Century France." *Past and Present* 59.1 (1973): 51–91.
Davis, Natalie Z. "Anthropology and History in the 1980s." *Journal of Interdisciplinary History* 12.2 (1981): 267–275.
Davis, Natalie Z. "Decentering History: Local Stories and Cultural Crossings in a Global World." *History and Theory* 50.2 (2011): 188–202.
Deflem, Mathieu. "Ritual, Anti-Structure, and Religion: A Discussion of Victor Turner's Processual Symbolic Analysis." *Journal for the Scientific Study of Religion* 30.1 (1991): 1–25.
Desan, Suzanne. "Crowds, Community, and Ritual in the Work of E.P. Thompson and Natalie Davis." *The New Cultural History*. Ed. Lynn Hunt. Berkeley/Los Angeles/London: University of California Press, 1989. 47–71.
Ekeh, Peter P. "Social Anthropology and Two Contrasting Uses of Tribalism in Africa." *Comparative Studies in Society and History* 32.2 (1990): 660–700.
Fernández, Marta, and Paula Esteves. "Silencing Colonialism: Foucault and the International." *Foucault and the Modern International: Silences and Legacies for the Study of World*

Politics. Eds. Philippe Bonditti, Didier Bigo, and Frédéric Gros. New York: Palgrave Macmillan, 2017. 137–153.
Fleck, Ludwik. *Genesis and Development of a Scientific Fact*. Chicago: University of Chicago Press, 1979.
Foucault, Michel. *Discipline and Punish: The Birth of the Prison*. New York: Vintage Books, 1977.
Foucault, Michel. "The Order of Discourse." *Untying the Text: A Post-Structuralist Reader*. Ed. Robert J. C. Young. London: Routledge, 1981. 51–76.
Free, Anthony. "The Anthropology of Pierre Bourdieu: A Reconsideration." *Critique of Anthropology* 16.4 (1996): 395–416.
Gennep, Arnold van. *The Rites of Passage*. Chicago: University of Chicago Press, 1960 [1909].
Go, Julian. "Decolonizing Bourdieu: Colonial and Postcolonial Theory in Pierre Bourdieu's Early Work." In: *Sociological Theory* 31.1 (2013): 49–74.
Goodman, Jane E. "The Proverbial Bourdieu: Habitus and the Politics of Representation in the Ethnography of Kabylia." *Bourdieu in Algeria: Colonial Politics, Ethnographic Practices, Theoretical Developments*. Eds. Jane E. Goodman and Paul Silverstein. Lincoln, NE: University of Nebraska Press, 2009. 94–132.
Gorski, Philip S., ed. *Bourdieu and Historical Analysis*. Durham, NC/London: Duke University Press, 2013.
Grosfoguel, Ramón. "World-Systems Analysis in the Context of Transmodernity, Border Thinking, and Global Coloniality." *Review* 29.2 (2006): 167–187.
Grosfoguel, Ramón. "The Epistemic Decolonial Turn: Beyond Political-Economy Paradigms." *Cultural Studies* 21.2/3 (2007): 211–223.
Hammoudi, Abdellah. "Phenomenology and Ethnography: On Kabyle Habitus in the Work of Bourdieu." *Bourdieu in Algeria: Colonial Politics, Ethnographic Practices, Theoretical Developments*. Eds. Jane E. Goodman and Paul Silverstein. Lincoln, NE: University of Nebraska Press, 2009. 199–254.
Herzfeld, Michael. *Anthropology Through the Looking Glass*. Cambridge: Cambridge University Press, 1987.
Hunt, Lynn. "Introduction: History, Culture and Text." *The New Cultural History*. Ed. Lynn Hunt. Berkeley/Los Angeles/London: University of California Press, 1989. 1–22.
Kalter, Christoph. *The Discovery of the Third World*. Cambridge: Cambridge University Press, 2016.
Kaplan, Martha. "Panopticon in Poona: An Essay on Foucault and Colonialism." *Cultural Anthropology* 10.1 (1995): 85–98.
Kertzer, David I. *Ritual, Politics, and Power*. New Haven, CT/London: Yale University Press, 1988.
Kohl, Karl-Heinz. "The End of Anthropology – An Endless Debate." *Paideuma: Mitteilungen zur Kulturkunde* 56 (2010): 87–98.
Kuhn, Thomas S. *The Structure of Scientific Revolutions*. Chicago: University of Chicago Press, 1970.
Lane, Jeremy F. *Pierre Bourdieu: A Critical Introduction*. London: Pluto Press, 2000.
Lazreg, Marnia. *Foucault's Orient: The Conundrum of Cultural Difference from Tunisia to Japan*. New York: Berghahn Books, 2017.
Le Sueur, James D. *Uncivil War: Intellectuals and Identity Politics during the Decolonization of Algeria*. Philadelphia: University of Pennsylvania Press, 2001.
Mbembe, Achille. "What is Postcolonial Thinking? An Interview with Achille Mbembe." *Edition Esprit* 12 (2006): 117–133.

Medick, Hans. "'Missionare im Ruderboot'? Ethnologische Erkenntnisweisen als Herausforderung an die Sozialgeschichte." *Geschichte und Gesellschaft* 10.3 (1984): 295–319.

Medick, Hans. "'Missionaries in the Row Boat'? Ethnological Ways of Knowing as a Challenge to Social History." *The History of Everyday Life: Reconstructing Historical Experiences and Ways of Life.* Ed. Alf Lüdtke. Princeton: Princeton University Press, 1995. 41–71.

Mignolo, Walter. *Local Histories/Global Designs: Coloniality, Subaltern Knowledges, and Border Thinking.* Princeton: Princeton University Press, 2000.

Mignolo, Walter. "Delinking." *Cultural Studies* 21.2 (2007): 449–514.

Mignolo, Walter. "Epistemic Disobedience, Independent Thought and Decolonial Freedom." *Theory, Culture & Society* 26.7/8 (2009): 159–181.

Mignolo, Walter, and Madina V. Tlostanova. "Theorizing from the Borders: Shifting to Geo- and Body-Politics of Knowledge." *European Journal of Social Theory* 9.2 (2006): 205–221.

Mitchell, Timothy. "The Stage of Modernity." *Questions of Modernity.* Ed. Timothy Mitchell. Minneapolis: University of Minnesota Press, 2000. 1–34.

Nichols, Robert. "Postcolonial Studies and the Discourse of Foucault: Survey of a Field of Problematization." *Foucault Studies* 9 (2010): 111–144.

Paulmann, Johannes. *Pomp und Politik. Monarchenbegegnungen in Europa zwischen Ancien Régime und Erstem Weltkrieg.* Paderborn: Schöningh, 2000.

Price, Richard. *First-Time: The Historical Vision of an African American People.* Chicago: University of Chicago Press, 1983.

Rose, Gilian. "Situating Knowledges: Positionality, Reflexivities and Other Tactics." *Progress in Human Geography* 21.3 (1997): 305–320.

Rublack, Ulinka. *A Concise Companion to History.* Oxford: Oxford University Press, 2011.

Said, Edward. *Orientalism: Western Representations of the Orient.* London: Routledge, 1978.

Said, Edward. "Representing the Colonized: Anthropology's Interlocutors." *Critical Inquiry* 15.2 (1989): 205–225.

Schama, Simon. *The Embarrassment of Riches: An Interpretation of Dutch Culture in the Golden Age.* New York: Alfred A. Knopf, 1987.

Schumaker, Lynette. "A Tent with a View: Colonial Officers, Anthropologists, and the Making of the Field in Northern Rhodesia, 1937–1960." *Osiris* 11.2 (1996): 237–258.

Seibel, Claude. "Les liens entre Pierre Bourdieu et les statisticiens à partir de son expérience Algérienne." *La liberté par la connaissance: Pierre Bourdieu (1930–2002).* Eds. Jacques Bouveresse and Daniel Roche. Paris: Odile Jacob, 2004. 105–119.

Silverstein, Paul. "Of Rooting and Uprooting: Kabyle Habitus, Domesticity and Structural Nostalgia." *Bourdieu in Algeria: Colonial Politics, Ethnographic Practices, Theoretical Developments.* Eds. Jane E. Goodman and Paul Silverstein. Lincoln, NE: University of Nebraska Press, 2009. 164–198.

Spivak, Gayatri Chakravorty. "Subaltern Studies: Deconstructing Historiography." *Selected Subaltern Studies.* Eds. Ranajit Guha and Gayatri Chakravorty Spivak. Oxford: Oxford University Press, 1985. 3–32.

Spivak, Gayatri Chakravorty. "Can the Subaltern Speak? Speculations on Widow Sacrifice." *Marxism and the Interpretation of Culture.* Eds. Cary Nelson and Lawrence Grossberg. London: Palgrave Macmillan, 1988. 271–313.

Stoler, Ann Laura. *Race and the Education of Desire: Foucault's History of Sexuality and the Colonial Order of Things.* Durham, NC/London: Duke University Press, 1995.

Stollberg-Rilinger, Barbara. *Rituale.* Frankfurt a.M.: Campus, 2013.

Turner, Victor W. "Betwixt and Between: The Liminal Period in Rites de Passage." *Symposium on New Approaches to the Study of Religion: Proceedings of the 1964 Annual Spring Meeting of the American Ethnological Association*. Ed. June Helm. Seattle, WA: University of Wisconsin Press, 1964. 4–20.
Turner, Victor W. *The Forest of Symbols: Aspects of Ndembu Ritual*. Ithaca: Cornell University Press, 1967.
Turner, Victor W. *The Ritual Process: Structure and Anti-Structure*. Chicago: University of Chicago Press, 1969.
Tylor, Edward B. *Primitive Culture: Researches into the Development of Mythology, Philosophy, Religion, Art and Costume*. Vol. 1. London: John Murray, 1871.
Wacquant, Loïc. "Following Bourdieu into the Field." *Ethnography* 5 (2004): 387–414.
Wilder, Gary. "Colonial Ethnology and Political Rationality in French West Africa." *History and Anthropology* 14.3 (2003): 219–252.
Wilder, Gary. *The French Imperial National-State: Negritude and Colonial Humanism Between the Two World Wars*. Chicago: University of Chicago Press, 2005.
Yacine, Tassadit. "L'Algérie, matrice d'une oeuvre." *Travailler avec Bourdieu*. Ed. Pierre Encrevé. Paris: Flammarion, 2003. 333–345.
Yacine, Tassadit. "Pierre Bourdieu in Algeria at War: Notes on the Birth of an Engaged Ethnosociology." *Ethnography* 5 (2004): 487–509.
Yacine, Tassadit. "Aux origine d'une ethnosociologie singulière." *Pierre Bourdieu, Esquisses algériennes*. Ed. Tassadit Yacine. Paris: Éditions du Seuil, 2008. 23–53.
Young, Robert. *White Mythologies: Writing History and the West*. London: Routledge, 1990.
Young, Robert. "Foucault on Race and Colonialism." *New Formations* 25 (1995): 57–65.

III Theorizing Pasts, Presents, Futures

Andreas Reckwitz

The Society of Singularities

Regardless of where we look in contemporary society, what is socially and culturally expected on both the local and global levels is not the *general* but the *particular*. What is increasingly being advanced and demanded and what has become the focus of people's hopes and longings is not the standardized and regulated but the unique, the singular.

Travel destinations, for example, can no longer simply be uniform vacation locations suited for mass tourism. It is the uniqueness of a place – a special city with an authentic vibe, an exceptional landscape, an unusual local culture – that attracts tourists' attention. A similar development has taken hold of the entire late modern global economy. True for material goods and services alike is the fact that the mass production of uniform products so characteristic of the old industrial economy has been replaced in the cultural capitalism of the present by events and objects that are not similar or identical but that strive to be singular. The passions of subjects are focused on extraordinary live concerts and music festivals, on sporting and artistic events, as well as on lifestyle sports and the imaginary worlds of computer games (see generally Rifkin 2000; Howkins 2001).

And yet the displacement of the general by the particular goes far beyond this, extending, for example, into the field of education. It is no longer sufficient, as it was 20 years ago, for schools to teach the material mandated by the state. Every school wants and is compelled to be different, to cultivate its own profile, to enable students to shape their own education, to have its own *spirit*. Or take the field of architecture, where the International Style, with its now purportedly dull *serial* buildings, has been cast aside in favor of the predictable surprises of star architects and their singular museum constructions, concert houses, residential buildings, and flagship stores (see McNeill 2009). The singular has quite clearly extended its reign over the subjects who move about in these different settings as well. In late modernity the subject is not just responsible for themselves, as is typically suggested by the term 'individualization,' but strives above all to be unique. Digital social media – perhaps paradigmatically the Facebook profile with its carefully curated and updated postings from one's personal life, with photos and likes and links not to be found anywhere else – offer a central location for the presentation and formation of this singular self and its performance of authenticity (see Miller 2011).

Note: This article is touching on a topic that I explore in greater detail in Reckwitz 2017a.

https://doi.org/10.1515/9783110669398-009

But this displacement of forms of generality by those of particularity also extends to the social, collective, and political realms. Formal organizations, major political parties, ultimately even the modern form of the bureaucratic state are on the defensive, having lost some of their appeal. On the rise are those particularistic and temporary forms of sociality that are not universally identical but claim instead to be unique. This is true of a wide range of forms of sociality, including professional and political projects, each of which is singular as an emotional entity with selected participants and an expiration date. It is true of scenes, events, short-lived aesthetic networks, and gatherings. And it is, finally, true in a different sense of neo-collectives – the new religious, national, or regional *imagined communities* that promise to endow members with identity in a way that bureaucracies or institutional churches do not seem capable of (see Castells 1997).

I have begun with a kaleidoscope of empirical phenomena that all point in the same direction. In late modernity, societies are being reconfigured based not on a social logic of generality but on a social logic of particularity – a particularity that I will attempt to define by means of the term singularity. This phenomenon involves a very crucial transformation of what defines *modernity* and *modern society*. I would like to sketch out this fundamental argument and then explore it in greater detail. I consider it of central importance that as a result of this logic of singularities, the structural principles of classic modernity, a modernity of industrially organized societies, are being eclipsed by new structural principles. The basic precepts of classic modernity were generalization and standardization, which were associated with the process of formal rationalization (see Wagner 1994). The antithesis to modernist rationalization is culturalization, and the phenomena of *singularization* and *culturalization* are inextricably connected to one another. In the first part I will therefore examine the oppositional differences between a social logic of generality and a social logic of particularity.

In the second part I will look more closely at two institutional mainstays at the center of late modern society. One is the transformation of the capitalistic economy from industrial mass production to cultural production, that is, to an *economy of singularities* (Karpik 2010), with the associated restructuring of markets, labor, professions, and forms of consumption. The second is the digital revolution of media technologies, which in turn also fosters singularities in subjects, images, texts, and other cultural elements. This is a decisive insight that I would like to emphasize: while in classic modernity the economy and technology were the most important motors of the standardization of the world, that is, of a social logic of generality, the most advanced forms of this same modern economy and this same technology have become powerful generators of singularities and culturalization.

1 The Social Logic of Singularization

To justify my diagnosis, I must first clarify how a modernity of rationalization and generalization and a modernity of singularities and culturalization are distinct from one another. To do this I will first describe the structural principles that gave rise to modern industrial society. Though quite easy to oversee, the fundamental trait of classic modernity is that it systematically strives to achieve the total generalization, schematization, standardization, and universalization of all elements. At the core of classic modernity is what I would call a social logic of generality. This standardization and universalization of social structures and processes, of subjects and objects, is closely related to the fundamental process of modernity that Max Weber (1968) referred to as *formal rationalization*.

The formal rationalization of classic modernity attempts to systematically foster a social logic of generality. The social logic of generality means that all potential elements of the world are observed, evaluated, produced, and adapted as copies or instances of generally valid patterns. The social logic of generality follows in part the principles of theoretical generalization (as required by the modern sciences) and in part those of normative universalization (as required by modern law with its precepts of equality). Yet, above all, formal rationalization is an expedient to achieve a comprehensive optimization of all societal conditions and an institutionalization of rules, which are intended to generate predictability, efficiency, and innovation. The reign of the general is to be found on all levels: objects are produced and used in a standardized and uniform manner. Disciplined subjects find orientation in functional roles and performance standards that apply to everyone. Space is utilized in invariable constructive series so that industrial cities appear interchangeable. Time also becomes an object of rationalization in the sense that it is systematically controlled and the future is, so to speak, colonized. Rationalized orders are objectified orders in which emotions are controlled and emotional intensity is minimized.

Of course, the modernity of formal rationalization and the reign of generality and uniformity are not dead. Many of these structural principles have been retained in late modernity, that is, in the period after 1980. Yet the countertendency that I mentioned at the outset is also to be observed: the spread of a social logic of singularities that is connected to a process of culturalization. To clarify this, I would like to more precisely define the term 'singularity,'[1] which up until now I have been using in a somewhat ad hoc manner based on different examples.

1 Two major sources of inspiration for this concept are Kopytoff 1986 and Karpik 2010.

1. An entity is singular in a sociocultural context when it is not produced, experienced, and evaluated as a uniform copy of a general type but as something particular. As such it appears to be unique, incomparable, and non-interchangeable. Singularity makes reference to a certain *quality* and cannot be reduced to quantitative properties,[2] which places it outside the schemata of generality. For in the realm of generality, entities can also differ from one another to some extent, but these differences can be described by such terms as better/worse, more/less, that is, they can be compared. Singularities, on the other hand, do not just vary to a greater or lesser extent, they have a completely different quality, they are distinct – and for this reason do not seem interchangeable. A Bach cantata seems fundamentally different from a Janis Joplin song. A trip to Venice is completely different from one to Nepal. And for the creative agency, employee X with his special profile and talents isn't just slightly different – the way applicants with different exam notes might be – but offers a critical qualitative advantage for the company. Of course, as Kant (2000) pointed out, there is always and inevitably the general and the particular, whereby – at least according to Kant – the general emerges from concepts (*Begriffe*) and the particular from intuition (*Anschauung*). But what is sociologically interesting is the fact that dependent on the form of society, a complete social logics of singularity can emerge, in which singularities are observed, evaluated, fabricated, and adapted in a certain way.

2. It is of central importance that singularities emerge in the form of very different entities and elements relevant to the social world. For this reason, singularity differs from the concept of individuality, though the two are, of course, related. As a rule, individuals are human subjects, yet to attribute singularities to humans alone would be to greatly underestimate their importance. Singularities can be observed on one initial level that I would like to put special emphasis on: in the realm of things and objects. This is true of fabricated things, which in modernity often assume the form of products and goods, but also of images and texts, of works of art or religious relics, and of three-dimensional things like architecture (as well as natural entities). Singularities can, however, also be identified on the level of spatial and of temporal entities. Spatial singularities are in the field of spatial analysis generally known as *places* (in opposition to spaces) – non-interchangeable, non-comparable locations. Temporal singularities are moments or episodes: a single instant perceived as such or a unique, discontinuous episode with a distinct beginning and end. Humans can of course also appear as singularities and be introduced to the world as such, here we are

2 See in this regard also Callon et al. 2002.

in the realm for which classically the term individuality was reserved. Finally, on a fifth and particularly interesting level certain kinds of collectives can also become singularities. Traditionally this is the case for what Ferdinand Tönnies called communities, but it also holds true for nations, and in late modernity it applies to such new socialities as projects, collaborations, and scenes.

3. In a social context singularities are ascribed a cultural value. In this sense they differ from the social logic of generality: while in the framework of the latter the individual element is attributed a derivative use or function in the framework of the rational order, singularities seem to have a value in their own right. This is true of works of art or relics, as well as of locations, moments, events, communities, scenes, and individualities. Singularities are to a certain extent not so much a means to an end but an end in itself. They are cultural in the active or robust sense of the term. This cultural autotelism of singularities can have an aesthetic dimension, but it can also have a hermeneutic, symbolic or narrative dimension, or a creative or ludic one. Yet, all in all, singularity always involves a certain performance, it is enacted in front of an audience. The intrinsic value of singularities is, however, not simply present: it depends on social processes of value attribution, on *valorization*.[3] Such valorizations can be consensual and hegemonic, but they can also be – at least in the modern period – extremely controversial, dynamic, dependent on discourses about valorization. Thus, there is a process of singularization taking place within the processes of valorization.

4. Singularities are generally associated with strong affects. It is not the general but the particular that leaves no one cold. While affective reactions to the universalities of modernity – rules and roles, mass-produced goods and statistics, serial buildings and serial cities – are minimal, affects related to singularities are all the more pronounced. These can include fascination, arousal, enthusiasm, and quiet satisfaction – or, on the other hand, such negative affects as aggression and hate. Closely connected to the emotional power of singularity is the fact that an intrinsic value is not just assigned but also experienced (or not experienced, as the case may be) in the participants' practical processes.

5. Singularities are in this sense to be distinguished from idiosyncrasies. Idiosyncrasies are unique traits that come about unintentionally and are often disregarded. They disappear or are viewed with indifference. Singularities, on the other hand, are socially and culturally fabricated. They are made, fabricated, intentionally shaped, or encouraged. In modernity this is true of works of art and design objects, for cities shaped by *cultural regeneration* and, of course, for

3 See for an analysis of valorizations Muniesa 2012 and Thompson 1979.

subjects who *are* not just individual but who *work* more or less consciously on their own individuality, who produce performances and profiles. Singularities are to be understood here as processes of singularization. Whether referring to objects (artefacts), subjects, events, or collectives – all of them are singularized through practices of making (*Verfertigung*), practices of observing, practices of valorization, and practices of perceiving. To speak of singularities as a noun can only be a snapshot. Henceforth, singularities exist solely in the process of singularization – whose downside is the desingularization, the loss of the unique status.

One additional explanation is also necessary: What do I mean by *culturalization* and how is it related to these singularities? It may initially seem strange to speak of the process of culturalization as antithetical to the process of rationalization. What can culturalization even mean if culture is everywhere, that is, if every activity depends on broader contexts and systems of meaning? Here I would like to distinguish between a general, weaker use of the term culture, and a more robust, narrower understanding of the term culture. In a general sense, of course – and this is an insight achieved by the study of culture – the social is always culturally determined, is based on often implicit systems of classification. In this way rationalization processes always have cultural preconditions, for example measures of efficiency or equality. This is the cultural realm. Against this backdrop, I would like to apply a more narrow yet more robust understanding of culture that allows for sociotheoretical distinctions. In this robust sense cultural objects and cultural practices only refer to those select objects and practices to which not a use or function is ascribed but rather an intrinsic *value*. Raymond Williams (1958) has correctly stressed this aspect of value as a component of culture. The antithesis to culture is in this case rationality, especially *purposive rationality*. While in the logic of purposive rationality an action, object, text, or image is the means to a further end and thus has instrumental significance, a cultural practice or cultural object has an intrinsic value in its social context. This intrinsic value can be and often is aesthetic in nature, yet it can also be narrative, hermeneutic, creative, or ludic.

In principle, cultural practices and objects can be quite varied, extending far beyond those related to art or religious rituals: playing football or collectively watching a football match, political ceremonies, experiences in nature, designing and decorating an apartment, or even work, provided it is not wage labor as a means to an end but work with an inherent value – all these things are cultural practices and objects in the strong sense of the term. In contrast to rational and normative practices, cultural practices thus contain a distinct element of lived experience and a distinct element of affective identification. To echo Georges Bataille's (1991) somewhat hyperbolic anthropological position: In comparison to purposive rationality with its dictate of efficiency and thrift, that is, in comparison to the world of necessity, the world of culture always contains an element of overexertion, of excess, of more than what is rationally needed.

We thus can see to what extent singularization and culturalization are related. Singularities are cultural in this robust sense of the word, laying claim to an intrinsic cultural value: the event and ritual, the specific location of a city or landscape, the singular object (be it work of art or of design), the individuality of the subject, the project, the scene, or the post-traditional community – these are not primarily purposive-rational institutions, rather to them an intrinsic value is *ascribed*. My principle argument is that following certain historical precedents that emerged from the end of the eighteenth century onwards, in late modernity the social logic of cultural singularities has spread both quantitatively and qualitatively. The social logic of singularities implies that at the center of society processes are taking place in which objects, subjects, collectives, locations, and temporal episodes are seen, evaluated, produced, and adapted as singularities, i.e. are singularized. All the examples that I cited at the beginning of the article are instances not just of the societal force of singularities but also of a process of culturalization.

2 Structures of the Late Modern Society of Singularities

Yet what form does a society that is oriented around cultural singularities assume? I will now list six traits that will be explained in greater detail in the second part of the article:

1. While historically cultural practices and objects and their singularities are often defined and shaped by the state, church, or a dominant social group, the widespread culturalization and singularization of late modernity is defined by an economy, a global cultural economy, that is also closely connected to a specific technological structure: the digital world. The structural framework is what I would call the global *cultural creative complex* (see Reckwitz 2017).
2. Cultural elements are valued highly in late modernity for their singularity because they are associated with the modern idea of authenticity.
3. The cultural creative complex seeks to continuously fabricate *new* singularities, which means that it is based on a regime of innovation, a regime of the culturally new, a regime of creativity.
4. In essence, cultural elements are negotiated in a social constellation made up of creators and an audience. Cultural elements are thus enacted and presented as performance.

5. Singular cultural elements are presented to an audience in a constellation of competition for attention in a hypercompetitive market of visibilities.
6. New forms of purposive rationality are emerging that are adapted to the interchange of cultural singularities. These forms are no longer based on a logic of generality but on one of particularity. The result is what I would call general infrastructures for particularities.

In fact, the spread of the logic of the culturally particular in late modernity can only be explained as part of the far reaching structural transformation of the economy from the mass production of industrial goods with utilitarian value to a post-Fordist fabrication of singularities, that is, of singular goods and services that contain the promise of something authentic and non-interchangeable. It has been possible to observe this incremental transformation since the 1970s. Yet the spread of the logic of the culturally particular also depends on a second phenomenon: the media technological revolution of computing, algorithms, and the World Wide Web, which, since the 1990s, has enabled not only the introduction of new cultural elements to the world (photos and stories, works of graphic art, films, games) in a historically unprecedented manner, but also the creation of a mobile realm of permanent competition for attention, in which singularities are to be made visible for potentially everyone and everything. The cultural creative complex encompasses the development of cities into *creative locations* by means of cultural regeneration as well as the global computer, internet, film, and music industries. It encompasses the development of such personalized services as individual care and counselling and the pervasion of everyday life by digital search engines like Google, by computer games and by social media like Facebook, Twitter, and YouTube. It now also encompasses virtually all consumer goods ranging from aesthetic design and so-called *moral consumption* to religious markets and the spiritual practices industry. Last but not least, it encompasses the vast touristization of global landscapes.

However, the emergence of the social complex of cultural singularities cannot be reduced to those economic and media technological structural conditions alone. Rather, the Western culture of authenticity is an ultimately discursive background for the triumph of such a social logic of singularities. Initially established within the social niches of artistic subcultures at the end of eighteenth-century Romanticism, it gradually spread throughout society (see Taylor 1989; Reckwitz 2006): Against the rationalism of mainstream modernity in the culture of authenticity, the idea and conviction emerged that the subject – if freed from all constraints – strives for authenticity, self-realization, and self-expression. To be authentic, however, means to be special, singular. In a second step, this search for authenticity is projected onto the whole world, which now is perceived in the expectation of the singular: a singularization of nature, places, communities, objects (artefacts), beliefs, and other subjects. Furthermore and ab initio it is closely linked to an ideal of creativity, a permanent self-creation and creative shaping of the world as well as a culture of intense emotions. Against this post-romantic background, which acted as an irritant countermovement of an 'other modernity,' classical, organized modernity of formal rationalization and the reign of generality seemed to suffer from a chronic lack of affect, authenticity, creativity, and singularity.

Now, Ronald Inglehart (1977) already described a fundamental change in values in the 1970s – certainly influenced by the counter culture of the late 60s – which was critical of rationalism and appreciated post-materialistic values, such as self-realization, the singular, the authenticity of a way of life, and the creative. Its social dominance could only be established by the onset of cultural capitalism and digitalization since the 1980s. This economic and media technological modelling generated a novel and very specific form of singularity. Late modern economy and media technology is driven by the subjects' orientation towards singularity, but is pushing it in an altered direction. This new social logic of the singular, which is institutionalized widely by the cultural creative complex, contains the characteristics that I already mentioned briefly above and which can be summarized as a constellation of *competitive singularities*. One prerequisite is the creator-audience-constellation: The cultural creative complex actively and purposefully produces singularities for an audience. The creation and design of singularities is thereby linked to a creativity orientation: it's all about the singular, which acts with a demand for novelty (see Reckwitz 2017a). These fabricated cultural elements with a claim for particularity can be aesthetically interesting artefacts as well as stories growing around goods, offered by therapists or narrated by an institution about itself. It can involve whole atmospheres fabricated for an experience of driving and living, live performances of various kinds, political-ideological models of identification or a moral value of a certain diet; it may concern luxury pleasure, beauty or sentiments of security, education, or – not least – the participation in a game (*gamification*).

It must be stressed that rather than disappearing, forms of purposive rationality are undergoing a transformation within this late modernist dynamic. Of its own accord, purposive rationality has begun to adapt from a logic of generality to a logic of particularity, or better: they develop into general infrastructures for the production of singularities. Here, systems of purposive rationality are developing an interest in and capacity for – and this is historically new – the production, analyses, and comparative assessment of singularities. With the help of software and 3D printers, unique products can be manufactured. Human resource management of the singular talents and potential of employees in the cultural economy and algorithmic data tracking of consumer profiles are focused not just on general patterns but on unique properties.

Decisive here is the fact that this creative cultural production of singularities is aimed at an audience made up of potential consumers. We have become so accustomed to the ubiquity of audience functions that it is easy to oversee how historically extraordinary they are. Yet the cultural elements produced in this context always exist as performance – a performativity for and in the presence of an audience. The cultural elements in the cultural economy, like those found on the Internet, are aimed at an audience. But in both the cultural economy and the Internet, there is now the constellation of a permanent competition between singularities for the attention of audiences (see Franck 1998). This is a constellation of competitive or even hyper competitive singularities, which are circulated on a market of visibilities. It is quite striking how the post-Fordist cultural economy and the Internet have institutionalized the same constellation of competitive singularities. Socioeconomic studies on *cultural markets*, that is, on markets for products of cultural singularity ranging from films to design objects, have shed light on this special phenomenon (see Caves 2000). In cultural singularity markets there is always a certain amount of overproduction of cultural goods, of which ultimately only a few will attract an audience's attention, though this attention is correspondingly massive. At the same time a great amount of cultural products will attract very little attention and have no appreciable success. This is precisely what is so peculiar about singularity markets: what appear to be minimal differences between products are perceived as absolute, qualitative, emotionally distinct differences between non-interchangeable items.

The culturalization of economic markets tendentially transforms them into *nobody knows*-markets as well as so-called *winner takes it all*-markets with strong asymmetries of visibility, attention, and success (see Frank and Cook 2010). Industrial economy pursued a standardized production, i.e. a standardized work process of standardized goods in a standard matrix organization of controlled markets for customers within a schematized *mass* consumption. The cultural economy on the other hand pursues a production of cultural singularities – goods or services – within a work process, which has itself a singularistic structure in ways of non-exchangeable projects on a market with consumers who work for a singular way of life by means of consumption.

A similar competition between the singularities regarding their visibility also structures the World Wide Web. Interpreting the process of digitalization only as another step to information and knowledge, society falls short. The discourse on knowledge and information society remains rooted in the logic of the industrial society, where texts and images could primarily be understood as cognitive and affect-neutral parts of information. However, the digital medialization means to a lesser extent an accumulation of cognitive knowledge but takes shape as a portable cultural space of images, narrations, game situations – a cultural hypertext, which constantly accompanies every subject and wherein an overproduction of cultural singularities is taking place (see also Stalder 2016). These singularities are under an on-going battle for the scarce attention by subjects and this battle is usually not one between pieces of information, but rather between affective intensities of images, narrations, and games with their aesthetic or hermeneutic offers. Not least, it is the subjects that are affected by the competition of visibility between the singularities who present themselves on the web, be it on YouTube, in Blogs, on Facebook, on Twitter, or future social media. The social media appear in fact as late modern generators of singularization.

The social media make particularly apparent how late modern subjects no longer take shape aligned to an organized modernity, but rather as singular subjects with a strive for authenticity and what it entails: this singularization converges in one social format, which is typical for the society of singularities in general – the *profile*. Digital subjects present themselves primarily through such profiles. In their profiles they compete with each other for visibility. Within these profiles there is a practice of what I would call a *compository singularity*: because here the subject becomes singular, especially in the composition, the configuration, the combination of various elements: news of the life of the self, likes showing certain cultural preferences, links relating to ones' interests, the timeline of biographical events of the past and not least, of course, the photographs from ones' own life. Singularity thus is not owned, it is *curated*. The authenticity of the singular subject here always adopts the paradox form of *performative authenticity*: authenticity has to be presented in front of an audience and hope to be perceived.

The exact same mechanisms of profile development can be seen in the cultural economy. Here again the singular cultural good has to develop a profile to attract attention as sustainably as possible – a whole brand is working on such a profile. In the cultural economy every single employee has to create a singular profile – beyond the standardized job requirements of the industrial society – to be of interest to projects of the working world. Overall, the culturalization of late modernity that has institutionalized a structure of competitive singularities leads to both an intensification of emotions and a dehierarchization of the cultural. While the formal rationalization of organized modernity has cooled off and minimized emotions, in the culturalization of late modernity we see an intensification of emotions and affects related to singularities. This is true of both the goods and services of cultural capitalism and of the events, experiences, claims to authenticity, and moralistic sensitivities that it fosters. Likewise, this holds true for singular labor in the creative professions and to a great degree for the emotional charging of images and narratives that circulate in the media, especially in digital form. It is true of the subjects whose performative authenticity emotionally correlates to their success or failure, and it is, of course, true of the cultural collectives that emerge within this same framework. Yet, at the same time, as a result of this focus on singularity, culture has been dehierarchized. Cultural hierarchies, such as the familiar stratification of high culture and popular culture, are being eroded. Every singularity can claim to have a legitimate value: football game or opera production, yoga retreat or computer game. By placing emphasis on the qualitative differences between singularities, the culturalization of late modernity has led to a *de jure* equality of singularities. *De facto*, however, there is an ongoing dynamic of inequalities and asymmetries among the singularities on the market for valorization and attention.

Two factors are primarily responsible for these asymmetries among singularities. The first is the antithetical processes of valorization and devaluation that affect cultural elements. The second is the self-reinforcing effect of the inequalities in attention mentioned above. That a singularity is recognized and experienced as such is neither self-evident nor obvious. Instead, in a society of singularities, societal processes of valorization and devaluation are of great importance. A cultural item in the cultural economy can, for example, lose its singularity and its cultural value if it does not appear or ceases to appear *authentic*. Locations or brand names – Ibiza or Adidas, for instance – that lay claim to an intrinsic cultural value can be de-singularized, reduced to little more than the expression of cheap mainstream consumption. In a society of singularities nothing is more fatal than to appear fake, a product of mass appeal, a mere expression of generality. The flip side of this kind of devaluation process is re-singularization, by means of which something that was once perceived as conformist or mainstream suddenly appears singular and non-interchangeable. The canned Hollywood movies from the 1950s will then be discovered as complex works of art and the nerd suddenly achieves the status of a hipster. Especially regarding subjects, the attribution or non-attribution of recognized, attractive singularity contains a considerable potential of cultural discrimination as well as glorification. Whereas the subject became problematic when subverting standards of normality during the organized modernity, it now risks – in a much subtler, but partly even more fatal way – to lose its recognition as being singular.

Concurrent to these processes of valorization and devalorization of qualities and singularities is the extremely disparate or unequal attention paid to elements on cultural visibility markets. This inequality of attention is at the outset of the career of a product, subject, location, etc. highly coincidental. Striking are the self-reinforcing effects of visibility that follow: once something manages to become visible it is not likely to lose this visibility very quickly. Analogous to the Matthew effect, attention is given to that which is already known, which is also the logic according to which such statuses as classic, celebrity, famous, cult, or star are assigned – exactly this is what winner takes it all-markets are about. Sighart Neckel (2008) observed in the context of the transformation of social inequalities that in late modernity the criteria that determine a subject's status have shifted from achievement-based to success-based. In fact, this shift from achievement-based to success-based criteria can by and large be explained in terms of the late modern structural transformation from industrial societies and their logic of generality to post-industrial societies and their logic of cultural singularity. In rationalistically organized modernity, gradual differences in objective achievement, especially in the professional world, lead to gradual differences in status. Yet, an economy that rewards absolute differences in exceptional singularity, visibility, and the successful accumulation of attention (regardless by what means) tends in fact to legitimize far more drastic social inequalities. These asymmetries in inequality affect products, companies, locations, and subjects alike. While achievement was defined by the fulfillment of general standards of better/worse or more/less, success results precisely from the seemingly non-rational properties of the singular performance that prevail on the attention market: the particular brand name, the particular location, the particular individual.

In contrast to organized modernity, the society of singularities gives thus rise to a new range of societal problems. The society of cultural singularities does not in any way imply that the classic modern realm of necessity has been replaced by a post-modern realm of liberty, free of cultural expediency. Instead, the societal preference for the unique is associated with a devaluation of the general, which yields, in turn, new problems: not least of which are problems of equality.

References

Bataille, Georges. *The Accursed Share: Volumes II and III*. New York: Zone Books, 1991.

Callon, Michel, Cécile Méadel, and Vololona Rabeharisoa. "The Economy of Qualities." *Economy and Society* 31.2 (2002): 194–217.

Castells, Manuel. *The Power of Identity. The Information Age: Economy, Society, and Culture*, vol. 2. Malden, MA: Wiley-Blackwell, 1997.

Caves, Richard. *Creative Industries: Contracts Between Art and Commerce*. Cambridge, MA: Harvard University Press, 2000.

Flew, Terry. *The Creative Industries: Culture and Policy*. London: Sage, 2012.

Franck, Georg. *Ökonomie der Aufmerksamkeit. Ein Entwurf*. Munich: dtv, 1998.

Frank, Robert H., and Philip J. Cook. *The Winner-Take-All-Society: Why the Few at the Top Get so Much More Than the Rest of Us*. New York: Virgin Books, 2010.

Howkins, John. *The Creative Economy: How People Make Money from Ideas*. London: Penguin, 2001.
Inglehart, Ronald. *The Silence Revolution: Changing Values and Political Styles Among Western Publics*. Princeton: Princeton University Press, 1977.
Kant, Immanuel. *Critique of the Power of Judgment*. Cambridge/New York: Cambridge University Press, 2000 [1892].
Karpik, Lucien. *Valuing the Unique: The Economics of Singularities*. Princeton: Princeton University Press, 2010.
Kopytoff, Igor. "The Cultural Biography of Things: Commoditization as Process." *The Social Life of Things: Commodities in Cultural Perspective*. Ed. Arjun Appadurai. Cambridge: Cambridge University Press, 1986. 64–91.
McNeill, Donald. *The Global Architect: Firms, Fame and Urban Form*. New York: Routledge, 2009.
Miller, Vincent. *Understanding Digital Culture*. London: Sage, 2011.
Muniesa, Fabian. "A Flank Movement in the Understanding of Valuation." *Measure and Value*. Eds. Lisa Adkins and Celia Lury. Malden, MA: Wiley-Blackwell, 2012. 24–38.
Neckel, Sighard. *Flucht nach vorn. Die Erfolgskultur der Marktgesellschaft*. Frankfurt a.M.: Campus, 2008.
Reckwitz, Andreas. *Das hybride Subjekt. Eine Theorie der Subjektkulturen von der bürgerlichen Moderne zur Postmoderne*. Weilerswist: Velbrück, 2006.
Reckwitz, Andreas. *The Invention of Creativity: Modern Society and the Culture of the New*. Cambridge: Polity, 2017.
Reckwitz, Andreas. *Die Gesellschaft der Singularitäten. Zum Strukturwandel der Moderne*. Berlin: Suhrkamp, 2017a (English translation: *The Society of Singularities*. Cambridge: Polity, 2020).
Riesman, David. *The Lonely Crowd: A Study of the Changing American Character*. New Haven: Yale University Press, 2001 [1949/1961].
Rifkin, Jeremy. *The Age of Access: How the Shift from Ownership to Access is Transforming Capitalism*. New York: Tarcher, 2000.
Rosanvallon, Pierre. *The Society of Equals*. Cambridge, MA: Harvard University Press, 2013.
Stalder, Felix. *Kultur der Digitalität*. Berlin: Suhrkamp, 2016.
Taylor, Charles. *Sources of the Self: The Making of the Modern Identity*. Cambridge, MA: Harvard University Press, 1989.
Thompson, Michael. *Rubbish Theory: The Creation and Destruction of Value*. Oxford: Oxford University Press, 1979.
Wagner, Peter. *A Sociology of Modernity: Liberty and Discipline*. London: Routledge, 1994.
Weber, Max. *Economy and Society: An Outline of Interpretive Sociology*. New York: Bedminster Press, 1968 [1922].
Williams, Raymond. *Culture and Society, 1780–1950*. London: Chatto and Windus, 1958.

Frederik Tygstrup

After Literature: The Geographies, Technologies, and Epistemologies of Reading and Writing in the Early Twenty-first Century

The notion of culture, as it circulates in contemporary studies of culture, combines two distinct meanings. On the one hand, culture designates an anthropological field of life forms, social relations and living conditions, and the habits, modes of interaction, and infrastructures that support them. On the other hand, there exists a more restricted notion of culture, as artistic and symbolic production: 'high' culture in its different forms as they have been examined and conceptualized by scholars of images, literature, and performative and spatial forms of creation. Traditionally, there has been a quite strict academic division of labor along the lines of this distinction, leaving the first for historians and social scientists of different specializations, and the second as the subject of aesthetic disciplines, heralded by an idea of 'the aesthetic' that parallels those notions of politics and economy at work on the other side. Against this backdrop, a recurrent and perhaps even defining feature of the study of culture as it has developed over the last decades has been to contest this distinction and its corollary distribution, between a realm of social phenomena on the one hand, and one of artistic objects and experiences on the other, to be understood and researched independently by scholars of different skills. As an academic endeavor, the study of culture has aimed especially at bridging these two approaches to culture (or indeed at criticizing the reifying consequences of the divide), and at developing a conceptual framework to gauge the relationship between the corresponding two levels of what we call culture.

If we think of these levels as pertaining respectively to the way we live and to the ways in which we picture this life, the relationship between them can be conceptualized as one of representation. Artistic and other symbolic forms represent the way we live, our conditions and experiences, our modes of seeing and our structures of feeling: in short, they display a menagerie of ways of inhabiting the world. But they are also, at the same time, *representative* of this world and the way we inhabit it; they respond to it, examine it, and, in the final analysis, provide it with intelligible, symbolic expressions. "A society," Émile Durkheim noted in 1912, "is not constituted simply by the mass of individuals who comprise it, the ground they occupy, the things they use, or the movements they make, but above all the idea it has of itself" (Durkheim 1995, 425). Cultural representations

https://doi.org/10.1515/9783110669398-010

cater to society with images of itself and thus suggest "[how] individuals imagine the society of which they are members and the obscure yet intimate relations they have with it" (Durkheim 1995, 227). To Durkheim, there is no culture (in the broad sense) that does not nurture its cultural artefacts (in the narrow sense), and inversely, there are no such artefacts that do not in some way or other contribute to building and consolidating the self-fashioning of culture and society at large.

The study of cultural representations thus necessarily shuttles back and forth between the social and political aspects of culture and the aesthetic artefacts and events it produces, reassessing, as it were, the constant process of suturing through which a society culturally reproduces itself. It lays bare the images that a society produces of itself, the "obscure yet intimate relations" that bind its members together, as well as the social dynamics, conflicts, and crises they represent. The future of the study of culture, understood in these terms, is of course contingent upon the recognition of the societal relevance of such an undertaking. On a more analytical note, it is also contingent upon the framework conditions of the practices and media of representation it scrutinizes. As some of its academic predecessors in the aesthetic disciplines have occasionally been accused of doing, cultural analysis cannot restrict itself to study the forms, imageries, and historical lineages of artistic works and events. To understand the cultural work of representation, we need to pay equal attention to the institutional scaffolding of its practices, and to the conditions of possibility of its modus operandi: the framing, in short, of its representing and representative activity.

This essay is an attempt to gauge some of the specific conditions of possibility that presently undergird the practice that we call 'literature,' and particularly the contemporary challenges and transformations that arguably confront it. A starting point for this enquiry can be found in the Nobel Laureate lecture given by Svetlana Alexievich in 2015:

> So what is it that I do? I collect the everyday life of feelings, thoughts, and words. I collect the life of my time. I'm interested in the history of the soul. The everyday life of the soul, the things that the big picture of history usually omits, or disdains. I work with missing history. I am often told, even now, that what I write isn't literature, it's a document. What is literature today? Who can answer that question? (Alexievich 2015, n.p.)

As a first step towards an answer, one should note – however truistically – that the word 'literature' itself has two different meanings. In a broad sense, it refers simply to letters and written texts: This literal sense of the word literature still prevailed in the eighteenth century, and it is still the usage of reference in, say, scientific literature, or opera literature. But by now we are also accustomed to a more restricted sense of the word, referring to artistic texts: *schöne Literatur, belles lettres*, or fiction as opposed to nonfiction. Such texts have become objects of scholarly scrutiny in academic studies of literature where they are read not only as texts, but as artworks, presupposing that they possess a particular kind of aesthetic value and consistency. Literary scholars have scrutinized such works individually and comparatively with a particular interest in their form and structure, their meaning and function, and how their forms and themes have developed historically, and the scholarly community has tailored specific concepts and approaches to address these questions of literary analysis, literary interpretation, and literary history.

Our modern idea of literature, and the academic protocols for studying it, thus focuses on a particular and delimited set of texts, which are given a particular status, and which are produced, circulated, and consumed in particular ways. This specific literary realm is situated in a wider context of other texts. So what is literature? How is its realm constituted, and when it interacts with other realms of making and using texts, how is this metabolism regulated? The definition of literature is a distribution of the sensible, identifying a subset of texts and assigning those texts to a particular societal sphere where they are understood and used in a special way. This distribution has been in place from the mid-eighteenth century until today. Answering the question "What is literature?" along these lines does not put any considerable emphasis on the proper characteristics of the 'literary' texts, moving the focus instead to understanding the distribution of the sensible – or the systemic differentiation – through which the particular realm of 'literature' in the modern sense of the word has come about. The "what" of literature hence becomes an institutional issue, a question of *framing* that enables us to distinguish between literature in general from literature as artistic texts.

The institution of literature has emerged over three centuries as a system of social conventions and protocols for the production, circulation, and consumption of a select set of texts. Michel Foucault famously identified a literary "author-function" (Foucault 1994) radically distinct from the scientific author-function; and similarly, we have developed a notion of the literary work with a specific juridical status and concomitantly a particular attitude we are expected to observe when understanding and interpreting it. And around this is built a corresponding economy and system of handling procedures executed by publishing houses, schools, newspapers, libraries, and so on. This entire institutional frame, ranging from social expectations and habits to very tangible principles for categorization and canonization, executes and reproduces the existence of this social thing that we call literature. We don't recognize the 'whatness' of literature in well-defined textual qualities, but rather in that small tag that this system has attached to it declaring it to be literature. This mode of being is of the same nature as Marcel Duchamp's famous urinal becoming a work of 'art' from the moment it is installed in the gallery space. The idea of literature in the modern, restricted sense stems from what Jacques Rancière in a different context has baptized "the aesthetic regime" (Rancière 2000, 31) where different artefacts retain a homogenous appearance by circulating in a particular economy.

Answering Alexievich's question in this way, i.e. by not identifying some essence of the literary thing as that which assures its identity to itself, but analyzing it as a product of an institutional framing of the historical production, circulation, and consumption of texts, we are in turn also invited to ask how the modern framing of literature might change over time, contingent on political, technological, and cultural transformations at large. We also eventually must consider if Alexievich's question indicates that literature might be in the process of changing its historical guise. In the following pages, I will try to make this question a bit more tangible by discussing some recent changes within three seminal aspects of the institutional framing of literature – what, in the title of this piece, I have referred to as the geographies, technologies, and epistemologies of literature.

1 Geography: After the Nation State

That the geography of literature is changing has been clearly indicated by the new relevance and wide circulation of Johann Wolfgang von Goethe's concept of world literature, developed nearly two hundred years ago. The concept itself is still somehow contested, and although many definitions have been suggested – ranging from a new understanding of literature in general to an exclusive canon of allegedly world-literary works – none have really won a general consensus. Nonetheless, this does not preclude a widespread acclaim for the concept as seminally important for the present time. In this sense, 'world literature' has become something like a fetish among scholars of literature, a still ambiguous expression of the feeling that something important is happening to the relationship between literature and the nation state that will hold consequences for our profession. What is happening, then, is probably first of all that a historic and otherwise solid alliance between national languages and their literatures is being demounted. The bland observation that literature is written in a national language was an important instrument for the forging of ideologies based on the nation state that accelerated after the Napoleonic wars. "Imagined communities," using Benedict Anderson's famous title, provided a shared cultural identity and political adherence to the national territory that were imperative to the consolidation of the modern nation states (Anderson 1991). Here, literature eventually came to play a prominent role, partly as a medium for cultivating the particularities of the national tongues, and partly as a repository of national mythologies, not least through the canonization of a literary pre-history that could contribute to the process of what Eric Hobsbawm has aptly called "the invention of tradition" (Hobsbawm and Ranger 1992).

This amalgamation between territory, tradition, and language has been under progressive liquidation since the Second World War, under pressure from the increasing circulation of people, commodities, technologies, capital, and information usually conveniently wrapped up in the notion of globalization. As the nation state and its genealogies become contested as a relevant frame of reference for the still more cosmopolitan space of our everyday experiences, this inevitably affects the ways in which literature is produced, circulated, and consumed. Stuart Hall coined the concept of "epistemic spaces" to portray the connections between our localization and our ways of thinking and feeling, how a spatial position also serves as a frame of reference for our expectations and predispositions, our patterns of agency and ways of seeing the world (Hall 1996, 396). During the nineteenth century, the nation state was the prime epistemic space for literature, but throughout the twentieth century we have seen the advent of more differently organized spaces. Thus, postcolonial epistemic spaces emerge from the conflictual superposition between imperial and local orders, and new urban spaces emerge as global and relational spaces with completely different geometries than territorially bundled spaces. Such epistemic spaces function on a different scale than the ancient national territorial spaces and gradually outline a new context of experience. Concomitantly, we tend to focus more on strictly local or regional epistemic spaces than on national territorial formations, and on trans-local spaces – using Arjun Appadurai's helpful term (Appadurai 1996, 192) – where different local horizons merge without any need for the coordinates of the nation state, like Bangladesh and East London, or Anatolia and Kreuzberg.

Such post-national epistemic spaces gain still more importance as a backdrop for contemporary literary creation, ranging from Salman Rushdie to Kamila Shamsie, and from Junot Diaz to Gary Shteyngart, where different versions of globalized epistemic spaces are evidently more relevant than the spaces of national communities. These four writers all write their books in English, but it is a globalized kind of English, which in stunning and effective ways captures and articulates the codes of globalized communication. Or to put it differently: even when the nation state is no longer the primary epistemic space, literary texts are still written in languages that have a national index, but this language is also shot through with lexical and idiomatic hybridizations to a degree where literary writing itself already seems to be partly also a work of translation. In turn, this is probably also why precisely the idea of translation has occupied such a prominent position in the cultural and literary studies of the latest decades. Translation, traditionally considered as a craft of transposing a work from one national language to another, is no longer a mere matter of post-production, but becomes an intercultural literary force of production in its own right with a steep increase in significance. Literary creation has increasingly become a question of *tele-poesies*, as suggested by Gayatri Spivak in her farewell to literary studies from 2003, *Death of a Discipline*: Creation by way of transpositions in time and space exactly mirrors the complex and layered time-spaces of our globalized epistemic space (Spivak 2003, 29).

The vitality of the contemporary literary scene surely testifies to the fact that it does indeed thrive beyond the epistemic space of the nation state, and that it is an ideally suited companion to getting a sense of direction in a globalized culture. It is questionable, however, if literature will be able to retain the central role it had in national culture and education under these new circumstances. The consecration of literature to become a privileged medium for culture and education – the proper word in this context is the German *Bildung* – does actually stand out as something quite extraordinary, in the educational system, where the teaching of language and of literature have been inseparable, in the propagation of national canon formations, in the endowments for public libraries, and in the leading public media. In this sense, the institution of literature has been intimately entangled with the political project of the nation state as an important point of reference for a shared tradition and a shared language. And to the extent that literature and the nation state actually come to parting ways – as literary creation encroaches itself in differently organized epistemic spaces, and as the weight of cultured literacy withers away from the societal sectors of education and of culture – literature eventually risks being left behind, devoid of the underwriting it has been privileged with, as a ghost in want of its blanket: another marginalized art form afloat on an aggressive global marketplace for mass-produced text.

2 Technology: After Gutenberg

The technology of literature has long been an underrepresented topic in literary studies, most often relegated to a corner of literary sociology; it first really flourished with the young discipline of book history, whose materialist corrective to the traditional spiritual air of the discipline has been both provocative and benign. Moreover, this direction of research seems to have surfaced at exactly that point in time when its object, the material book, started to lose its perceived obviousness, and when literature started to become mediated through new small screen technologies. In not too many years, we have come quite a long way in the transition from book to screen, from codex to Kindle, from Gutenberg to Google.

It remains contested what this transition actually entails, and it is a favored subject of conversation between reading people to ponder what it actually means for the experience of literature. Friedrich Kittler once remarked that "the new literary recipe for success" in the nineteenth century was "to surreptitiously turn the voice or handwriting of a soul into Gutenbergiana" (Kittler 1999, 9). Making "Gutenbergiana" into a language of the soul, addressed to the dispersed national communities of readers, and consequently considering reading as a mode of being attentive to the soul's voice, is probably an important albeit somehow intangible characteristic of how literature became institutionalized. It has been associated with a particular emphatic attitude, as observed by Georges Poulet, where one finds oneself thinking the thoughts and seeing the sights of somebody else: "The extraordinary fact in the case of a book is the falling away of the barriers between you and it. You are inside it; it is inside you; there is no longer either outside or inside" (Poulet 1969, 54).

In October 2014, the British writer Will Self published a piece in *The Guardian* entitled "The Transformation of our Gutenberg Minds," arguing that the transfer from book to screen would erode a significant component of our attitude towards literature, namely what he called "deep reading" – the absorptive devotion to a made-up universe that requires seclusion and contemplation, something which network-connected screens negate almost by definition. Even if it is by no means obvious that he is right about this, and even if the argument has more than a taint of nostalgia to it, it does highlight something like a historical phenomenology of the reading human body by taking into account how an entire array of rituals, expectations, and experiences pertaining to reading keep resonating in our understanding of what literature is.

Digital text is also, on a different note, instrumental in changing the routines of academic literary studies. If 'deep reading' has been the so-to-say civilian attitude to the reading of literature, the hermeneutic scrutiny of literary texts in academia has been one of 'close reading' (not necessarily to be associated with the programmatic myopia that characterized the North American new critics, who originally coined the term). Contrary to this, Franco Moretti has forcefully launched the concept of "distant reading" (Moretti 2013), which is no longer based on the interpretation of textual finery, but on parsing and statistical metrics based on big textual data that now has become available thanks to the large data repositories of the written cultural heritage. This is of course first of all a methodological change of perspective, where we have been given control of a powerful apparatus that we might still not be completely aware of how to handle. But it surely contributes to a gradual reorientation in literary studies, away from analyzing literary works in order to analyze text, text in large quantities, which enables us to extricate precise and numeric acute answers to even the vaguest questions we might ask.

In addition to his observations on the changing phenomenology of reading, Will Self also points to the broader societal infrastructure built around the material book and the way it is inflicted by new technologies:

> The relationship between words and revenue has become a debatable one – we can wax all we like about the importance of the traditional gatekeepers and the perspicacity of editors and critics in separating out the literary wheat from the pulpy chaff, but the fact is that these professions depend on the physical book as a commodity. It is the sad bleat of the book world that we'll be sorry once they're gone – and with them all the bookshops, literary reviews, libraries and publishing houses that supported their endeavors – but it was their mistake to assume their acumen to be inelastic. I mean by this, that a certain kind of expertise was understood to have a value to its consumers that was both constant and capable of being monetized at a fixed rate. The web has grabbed hold of this inelasticity and stretched it until it has snapped back in the myopic faces of the literati. (Self 2014)

Self here delivers a diagnostic of how the literary ecology is being rationalized by way of a more efficient and lean business model bringing the commodities directly to the consumers without costly intermediaries, and where whatever is saved is probably being shared in equal parts between the entire reading community on the one side and Jeff Bezos on the other. But this is also a potential eradication of the entire, ramified, and complex societal infrastructure that came with the historic technology and afforded the reproduction of what we called literature. Digitization and the substitution of the book with a file displayed on a screen emaciates the institutional eco-system around the literary thing. Moreover, this new apparatus also intervenes in the production of texts. In an industrial perspective, digital social networks open up a new production line: *Fifty Shades of Grey*, for example, was originally a blog, which morphed into literature when it had rounded a critical number of hits; something similar seems to be the case for the increasing production of fan fiction. Content production now becomes veritably industrialized in a way that short-circuits the slow grinding mechanisms of the historic literary system, partly by tapping into the immense text production that takes place on social platforms, partly by systematically (i.e. algorithmically) surveying the patterns of our purchasing and reading of electronic books and using this information for marketing as well as for literary production proper.

On a short-term basis, it is no doubt the commercial exploitation of the changing media technology that draws the largest profits from this transformation (all while Will Self's myopic literati get snaps on their noses), but the digital production environments eventually also become an important spur for new and experimental modes of literary creation. Thus, Mark Danielewski's novels would be unthinkable without a thorough familiarity with the workings of digital media, just like the new, platform-specific formats of links, tweets, and real-time postings contribute to altering our sensibility to the literature's media ecology and to the creation of new non-linear forms of composition and modes of reading.

These different and admittedly quite heteroclite tendencies obviously interact in a number of different ways, but most notably, I will argue, they converge in denaturalizing the 'book' as a literary object and the 'work' as an aesthetic category. For one thing, we come to realize what book history has already attempted to teach us: namely that a text is actually a quite fuzzily delineated object that cannot easily be disentangled from what we have otherwise tried to contain as 'para-textual' features, i.e. the social and technological protocols and processes through which the text is materialized as a cultural object. This becomes increasingly evident when dealing with digital text, where the material and technological para-textual stuff we have become accustomed to over a long stretch of time is discarded and replaced by other features, now pertaining to hardware and software and providing scalability, searchability, linking, sharing, statistics, algorithmic parsing, and much more. Eventually, we might become less inclined to think about 'a' text and instead prone to recognize 'some' text that cannot ideally be separated from the material forms and temporal processes that undergird its actual appearance. To this effect, N. Katherine Hayles remarks:

> Perhaps it is time to think the unthinkable – to posit a notion of text that is not dematerialized and that does depend on the substrate in which it is instantiated. Rather than stretch the fiction of dematerialization thinner and thinner, why not explore the possibilities of texts that thrive on the entwining of physicality with informational structure? (Hayles 2003, 275)

Literature as we have understood it rested on a certain regime of the text whose material form was the Gutenbergian codex and whose corresponding aesthetic form was that of an institutionalized work of art. With the denaturation of the inherited material form, a more generalized sense of textuality is unleashed: in terms of writing, in guise of processing a highly malleable scroll on a screen; in terms of processing by different forms of capture and postproduction; in terms of distribution based on data files; and in terms of reading, no longer handling a book object, but reading a fragment of text on a screen which is virtually surrounded by endless expanses of more text, as a haphazard spotlight flickering over an endless surface of written stuff.

In the early nineteen-seventies, Roland Barthes prophesized the transition from work to text; perhaps we are now witnessing a surprising version of this transition, where commercial producers no longer trade in works, but capture and repackage text, where reading is not confined to a volume with two covers but to patches of luminous flicker on a screen. In this situation, we are obviously still somehow dealing with literature; it is produced in unprecedented quantities and is amazingly accessible, but still it somehow differs from the product we knew that had its material foundation in the ancient printing press. The technology of the book helped to frame that thing we called literature and the literary works we came to cherish; this thing is now coming unframed in the digital propagation of a new kind of generalized textuality that modifies the traditional modes of production, circulation, and consumption of literature and instigates new practices, new business models, and new sensibilities, eventually leaving us less with 'literature' and literary works than with much more generic forms of text that are captured, distributed, and used in slightly different ways.

3 Epistemology: After Fiction

Between 2004, when Michael Moore won the Golden Palm at the Cannes Film Festival, and 2015, when Svetlana Alexievich was awarded the Nobel Prize in Literature, the question of how to distinguish reliably between documentary and fiction has gradually gained increasing importance. We have learned, throughout the modern history of literature, that literature belongs to the realm of fiction, where utterances are not supposed to – or rather, are supposed not to – have a real referent. Fiction, here, is really to be understood rather as a legal category: a particular discursive practice where beings that are designated do not have a referent. This is why Flaubert was not indicted in the trial on the morality on Madame Bovary, and why we so abhorred the *fatwa* on Salman Rushdie. Representations are not designations: Within the modern regime of literature, we have applied the *caveat* of fictionality to allow imaginary tales about our world to circulate as a testing ground for conjecture and speculation. The trade-off of fiction, magisterially condensed into a formula by Catherine Gallagher, is that it combines *non-reference* with *probability* (Gallagher 2006, 344).

This contract of fictionally, however, seems to have come undone. On June 21, 2013, at the publication of the American translation of the second volume of Karl Ove Knausgaard's *My Struggle*, the reviewer for *The New York Times* wrote: "immediately striking is the ways in which fiction is born of fact,“ and then thoughtfully added, "and the question whether this is fiction at all.“ This question has occupied literary scholarship and the literary public to a quite remarkable degree over the last years, not only in the case of Knausgaard, but in the panoply of instances where contemporary writers in different ways transgress the ancient contract of fiction, from W.G. Sebald to Michel Houellebecq, from Marie Darrieussecq to Rainald Goetz. All the prominent theories of fiction in store have been invoked to accommodate this new situation, and new sub-generic classifications have laboriously been devised, but somewhat, it seems, in vain. In the different experiments undertaken by writers like these, we are no longer dealing with new, subtle negotiations of the contract of fiction, but rather with a practice to which the contract and the divide it implies is simply becoming increasingly irrelevant. With characteristic, unfailing perspicacity, James G. Ballard already in 1995 stated:

> I feel that the balance between fiction and reality has changed significantly in the past decades. Increasingly their roles are reversed. We live in a world ruled by fictions of every kind – mass-merchandizing, advertising, politics conducted as a branch of advertising, the pre-empting of any original response to experience by the television screen. We live inside an enormous novel. It is now less and less necessary for the writer to invent the fictional content of his novel. The fiction is already there. The writer's task is to invent the reality.
> (Ballard 1995, i)

Now, twenty years later, the tendencies spotted by Ballard have come to full-blown fruition, and we are presently dealing with a double-faced erosion of the threshold that used to separate fiction from the document. From one side, creative literary practices defy and provocatively transgress the borderline set down in the traditional contact of fiction, overtly juxtaposing and intertwining real references and invented things. And from the other side, we see media and political discourse flooded with 'alternative facts' and generally blending alleged accounts of states of facts with improbable figments of imagination, in the increasingly toxic conglomeration of storytelling and statistics, branding and bigotry.

The differentiation of the public sphere, as magisterially theorized by Jürgen Habermas, implied a thorough separation of discursive modes and their spheres of validity, combined with a gradual development of different rationalities for political deliberation, for the organization of production, for the market place, for art and literature, for scientific inquiry, and so forth (Habermas 1962). The institution of literature took place under the aegis of such a societal differentiation; the present situation, however, seems to be one of rolling back, of de-differentiation, inflicting the ability for – or the interest in – distinguishing between facts and fiction. Matters of fact are becoming rarefied, as Bruno Latour has noted, leaving us in an acute perplexity about how then to identify and negotiate the matters of concern we need to face up to (Latour 2005, 29). So one thing is that literature today seems still more preoccupied with toying with sometimes ludic, sometimes dead-serious references to mundane reality, and with blending discourses that we are accustomed to identifying as fiction and documentary, respectively, into new hybrid forms. But this also stands out as a reaction to the way in which our present reality is becoming saturated with what we would otherwise have expected to find only in the inoffensive realm of fiction. Referential stuff of all kinds now abounds inside this realm, and outside of it, reference is shot through with fiction.

This new distribution of discourses – and the hybridization of otherwise distinct rationalities it brings along – evidently has serious repercussions when it comes to how we (are to) understand what literature is. Literature no longer holds the privilege of being the one, specific discourse that could produce imaginary versions of the world and hold them up against the actual world outside. Within the new disorderly distribution of discourses in society, such properties now surface everywhere, whether in politics, advertisement, or journalism. In this sense, the new and sometimes confusing literary involvement with something we identify as documentary is not primarily an attempt on the part of literature to break out of the institutional enclosure it has found itself confined to, but more accurately a reaction to the undermining of the architecture of social discourses that upheld this enclosure. The divides that scaffold the epistemological *differentia specifica* of literature are coming down, not through pressures from within, but washed away from the outside. This can be regarded as a waning of the power that was once endowed to literature, as it now loses its privilege do be the discourse that can say something which is not true, but is still much more than a lie. But it can also be regarded as an unbinding of this very power, the speculative power of fabulation that has been bred within the confines of the literary institution for a couple of centuries, now eventually in a position to directly address – beyond analogy, beyond allegory – the narratives and imaginaries that make up our cultural space of experience.

4 Literature After Literature

When studying the cultural practices of representation and their role and function in the life of society at large – and indeed when gauging the future of this academic endeavor – we need to factor in also the framework conditions upon which the representations under scrutiny are contingent and the particular historical transformations they undergo. In the case of literature, and admittedly based on a haphazard set of observations that does not allow for too far-reaching conclusions, I have nonetheless tried to sketch out how changes are underway that might eventually affect three fundamental pillars supporting our modern notion of literature and perhaps alter the representational regime of literature in the twenty-first century. Thus, with the new geographies, technologies, and epistemologies of reading and writing I have touched upon, literature as we have known it for some three hundred years could be morphing into something slightly different: a literature based on a new trans-local idiomatic, beyond the nation state and the national language; a new proteiform textuality, beyond the book and the work; and a new art of fabulation, beyond the particular category of fiction as opposed to the referential.

The literary creativity of the present in no way seems to be impeded by these transformations of the framework within which it operates; on the contrary, it is stimulated by the new vistas that open up and the novel possibilities they entail. And literature remains, to be sure, a powerful medium of representation. Yet the cultural logic of representation it is about to develop has different coordinates than the one we knew from the classical modern regime of literature. The publics it caters to and to whom it offers the images it confects are less those who can be circumscribed by way of a geography of nations, but instead by dispersed communities engaged in intricate processes of translation and negotiation based on new trajectories and encounters. Moreover, literary representations are not only subjected to new and aggressive business models, they are also morphing beyond the book and the circuit of books as we know it. This textual practice in turn connects to other forms of live-ness and to the practices of the other arts, as it nests itself in ramified guises within a different media ecology, which in turn also facilitates new encounters and different platforms for producing stories and images of the way we live. And finally, the mimetic faculty of literature is bleeding into neighboring discursive practices of wording our experience within different forms of post-fictional fabulation, intervening in the discursive fabric of the present in more subtle ways, taking risks outside of the comfort zone of what is 'mere fiction' and renegotiating its legal and political interaction with a host of other representations among which it is increasingly enmeshed.

One challenge to the study of culture today is of course to chart such complex processes that presently affect the framework conditions for aesthetic representations. We should keep a broad outlook not only on what is happening to the cultural practices as we know them, but also on the social and infrastructural changes at work, in a continuous dialogue with an array of other sciences of society, technology, and culture. The trans-disciplinary horizon of the study of culture is going to expand further, and we will have to keep learning new disciplinary languages and research designs, and probably engage even more in collective research projects that are equipped to properly handle the polyvalence of the objects we are dealing with.

The social institution of literature does seem, as argued above, to undergo quite dramatic changes in respect to its inner geometry and its interfaces with the world around it. In this process, the relations it builds to its audiences are diversified, all while it aims to gain a different foothold in a mobile, ubiquitously mediated and de-differentiated discursive sphere. Literature, and the other arts, are becoming less distinctive as objects, more flexible in their articulation and their address, operating more transversally in an increasingly complex and amalgamated sphere of social practices. As such, literature is about to become less distinctly identifiable as an object of scholarly inquiry, eventually also challenging our intellectual habits of object construction and scientific protocol. The study of culture will find itself increasingly in the position of a partner in dialogue and interaction with art practices rather than an impartial observer and interpreter. In this dialogue, artists and scholars are going to navigate the same waters, and most likely in a common endeavor of an ultimately political nature: to make sure that there remains sufficient room for maneuver for a creative and critical inquiry into the way we live now.

References

Alexievich, Svetlana. *On the Battle Lost*. 2015. <https://www.nobelprize.org/nobel_prizes/literature/laureates/2015/alexievich-lecture_en.html. 2015> [accessed: 22 October 2019].

Anderson, Benedict. *Imagined Communities: Reflections on the Origin and Spread of Nationalism*. London: Verso, 1991.

Appadurai, Arjun. *Modernity at Large: Cultural Dimensions of Globalization*. Minneapolis: University of Minnesota Press, 1996.

Ballard, James G. *Crash*. London: Micador, 1995 [1974].

Durkheim, Émile. *The Elementary Forms of Religious Life*. Transl. Karen E. Fields. New York: The Free Press, 1995.

Foucault, Michel. "Qu'est-ce qu'un auteur?" *Dits et écrits*. Vol 1. Paris: Gallimard, 1994 [1969].

Gallagher, Catherine. "The Rise of Fictionality." *The Novel*. Ed. Franco Moretti. Princeton: Princeton University Press, 2006.

Habermas, Jürgen. *Strukturwandel der Öffentlichkeit*. Darmstadt: Luchterhand, 1962.

Hall, Stuart. *Critical Dialogues in Cultural Studies*. Eds. David Morley and Kuan-Hsing Chen. London: Routledge, 1996.

Hayles, N. Katherine. "Translating Media: Why We Should Rethink Textuality." *The Yale Journal of Criticism* 16.2 (2003): 263–290.

Hobsbawm, Eric, and Terence O. Ranger. *The Invention of Tradition*. Cambridge: Cambridge University Press, 1992.

Kittler, Friedrich A. *Gramophone, Film, Typewriter*. Transl. Geoffrey Winthrop-Young and Michael Wutz. Stanford: Stanford University Press, 1999.

Latour, Bruno. "From Realpolitik to Dingpolitik." *Making Things Public: Atmospheres of Democracy*. Eds. Bruno Latour and Peter Weibel. Karlsruhe: ZKM, 2005.

Moretti, Franco. *Distant Reading*. London: Verso, 2013.

Poulet, Georges. "Phenomenology of Reading." *New Literary History* 1.1 (1969): 53–68.

Rancière, Jacques. *Le Partage du sensible: Esthétique et politique*. Paris: La Fabrique, 2000.

Self, Will. "The Transformation of Our Gutenberg Minds." *The Guardian* (04 October 2014).

Spivak, Gayatri Chakravorty. *Death of a Discipline*. New York: Columbia University Press, 2003.

Andressa Schröder

The Integrative Potentials of Arts-based Research for the Study of Culture: A Reflection on *The Lagoon Cycle* by Helen Mayer Harrison and Newton Harrison

The future of the study of culture can generate reflections from several different perspectives involving issues related to contents, methodologies, and political stances, as well as academic and/or social relevance, and epistemological ethics and responsibility. For early career researchers like myself, this can be ambiguously motivating, as a set of opportunities to engage in new and challenging endeavors, but also can sometimes seem a tiring and frustrating practice, particularly when one encounters outdated academic structures that are too rusty to be moved and transformed.

In this sense, one crucial consideration in the future of the study of culture is related to the future (or the 'crisis') of the humanities. This topic has been gaining increasingly concerned attention, mainly from scholars concerned not only with the future of their field but also with the course and extent to which human knowledge is framed and used as a tool for power and manipulation (Nussbaum 2010; Bono et al. 2008; Delbanco 2012). This concern arises from a growing and perceptible decrease in investment in the humanities and the arts in recent decades, and the consequent reduction in disciplines offered in and projects related to these fields (Nussbaum 2010; Delbanco 2012). Other symptoms of this crisis refer to mechanization processes in the production of knowledge, and investment solely in technical programs or projects that might bring immediate economic results but fail to generate a critical self-reflexive practice among students, researchers, professors, and other professionals in academic and non-academic worlds (Nussbaum 2010; Kristeva and Davidson 2014). This issue is reflected in the field's constant effort to validate the study of culture as 'science,' and its emphasis on the adoption (or adaptation) of what is accepted as a 'standard scientific method.' At the expense of this effort, in many cases, comes the disregarding of methods that are considered more ambiguous or subjective because they fail to reach the criteria of scientific measurability and reproducibility.

In an effort to address this issue, this article discusses the integrative potential of arts-based research for the field of the study of culture, in its encouragement of self-reflexive thinking processes about the statuses of academia and of

https://doi.org/10.1515/9783110669398-011

research themselves. Accordingly, this is not an empirical study based on concrete empirical evidence, but an invitation to reflection, based on the example of an artistic project that opens the possibility of thinking about research from different perspectives.

The interest in arts-based research as a potential methodology for qualitative research originated within the field of art education in the 1970s. Since the 1990s, it has been growing in connection to different academic areas, primarily within the social sciences, educational sciences, and the study of culture, but also in other scientific areas like psychology, psychotherapy, and environmental science.[1] In many cases, arts-based approaches are subordinated to fit into expected scientific accounts of measurability, applicability, predictability, reproducibility, credibility, and so on, despite that these are not always compatible to what arts-based research stands for. Scholars Natalia Eernstman and Arjen E.J. Wals describe the role of arts-based research in sustainability studies: "The arts techniques involve improvisation, intuition, spontaneity, lateral thinking, imagination, co-operation, serendipity, trust, inclusion, openness, risk-taking, provocation, surprise, concentration, unorthodoxy, deconstruction, innovation, fortitude, and an ability and willingness to delve beneath the surface, beyond the present, above the practical and around the fixed" (Eernstman and Wals 2013, 1648). These qualities are not only important for arts-based research, however. On the contrary, they are relevant to the process of innovative thinking and self-reflexive practice in any field of research, even as they require some distancing from the limitations of predictability and mechanical reproducibility. The arts, in this sense, can play a fundamental role in enabling the possibilities for overcoming a scientific-mechanical-rationality, and allow for the emergence of more humane and unexpected characteristics in the process of doing research.

First, it is important to highlight that it is not the goal of this paper to dismiss the importance of scientific research or the standards of scientific methods. Its focus lies, instead, on trying to bring unpredictable characteristics of research to the surface, and rather than denying or suppressing them, indicating their value as legitimate steps of the research process that could and should be embraced in the study of culture.

Second, there are different forms in which arts-based research can be understood, interpreted and applied and the goal here is to briefly indicate this variety

1 References to arts-based research in relation to art education can be found in Eisner 1991; Barone and Eisner 1997; Irwin and deCosson 2004. It was also expanded to other areas of the social and cultural sciences as in the examples of Pink 2001; Pickering 2008; Knowles and Cole 2008, as well as to art therapy in McNiff 2013 or environmental sciences and education in Mantere 1998; Curtis et al. 2014; Eernstman and Wals 2013.

of approaches and to focus on one specific kind of arts-based research that is not often given credit within the scientific community, which is defined in this article as *research through art.*

1 Multiple Forms to Think About in Arts-based Research

As previously mentioned, in the last decades there has been a growing number of different approaches to arts-based research: trying to understand it, define it, and integrate it into qualitative research (Barone and Eisner 1997; Knowles and Cole 2008; McNiff 2013). From a recent literature review of arts-based approaches used in fields related to the study of culture and the humanities, there are at least three different basic forms in which the role of the arts can be interpreted in relation to research. The first and more commonly identified one can be defined as research *about* art, in which the 'products' of artistic practice are analyzed according to the methods of other disciplines: For example, works of art that are analyzed using art-historical methods, or some forms of anthropological interpretations of cultural manifestations. In this kind of practice, the researchers are usually not artists themselves and they are not involved in the process of production of such artistic/cultural expressions. Instead, they engage theoretically with different works of art, or cultural manifestations, and combine these personal encounters with previous knowledge and theories to produce different readings and interpretations of that work, expression, or practice.

A second role for the arts is in research *with* art, in which the artistic practice is taken as a method and is incorporated into the interpretive processes of other disciplines. This happens, for example, in approaches of visual anthropology or art therapy (Pink 2001; McNiff 2013), wherein artistic practice is turned into a process of data production or collection and is combined with other research methods to achieve the desired results of a specific project. In these cases, the researchers can also come to be the producers of the analyzed artistic expressions, or they work closely with the people producing such expressions. However, the focus usually lies on analyzing the 'final result' or applying the 'final result' as an analytical tool to attend to a specific research question. The processual aspects of the artistic production are not completely disregarded, but they do not hold as a significant role as the 'end product.' In many practices and accounts of art therapy, the line between research *with* art and research *through* art, is not clearly defined (McNiff 2013): Both aspects can be identified but slight differences in focus or expectations of the analysis are perceivable.

Finally, another way in which arts-based research can be understood is in relation to the very process of artistic thinking. This can be defined as research *through* art, and this is the understanding of arts-based research that is explored in this article. In research *through* art, the artistic thinking process is understood as a methodological procedure in itself and the different stages of the creative process can be identified as distinguished methods for the development of the research (Irwin and deCosson 2004; Sullivan 2005). If one thinks about the centuries-old artistic habit of writing about creative processes in personal journals or in letters shared with fellow artists, this is not a completely new idea. The practice has often simply been obfuscated in the academic world (holding some recognition perhaps only in art schools) and it is now getting more visibility in other fields of research as well. Other common practices among artists that indicate the importance of procedural development include: the sketchbooks that many artists develop during the creative phase of their work; portfolios that often highlight finalized works but that also indicate a certain 'line-of-thought' among these works; as well as the journal publications that invite the contribution of artist-produced texts that reflect upon creative processes. These practices demonstrate that creative thinking, although it may be chaotic and subjective, is also systematic (Sullivan 2005), and not only can it be incorporated in the process of research related to various fields, it is actually inherent to the process of developing research. However, it is usually underestimated and not taken into consideration as a method or a part of the methodology applied to research.

As already indicated, these three different understandings of arts-based research are not mutually exclusive, and varied forms of arts-based research can be identified simultaneously in different research practices. Furthermore, there are other authors that might define these (and other) interpretations of arts-based research with different concepts, indicating the multiplicity of forms in which the arts can be integrated into research.[2]

2 Sullivan (2005), for example, focuses on the practice of research in visual arts. He develops different sets of complex triangulations between agency, structure, and action, which work in connection with different levels of research practice and theory. Furthermore, he emphasizes the importance of the role of the artist as researcher and provides a distinction among arts-based research and other quantitative and qualitative methodologies, denoting the distinguished characteristics of arts-based research. He does not conceptualize the visual-arts-research practice in the same sense as it is done in this article, but the emphasis on the processual characteristics of the artistic research shows a similar interpretation for it.

2 Three Integrative Potentials of Arts-based Research

With these considerations about arts-based research in mind, there are some integrative potentials enhanced by research *through* art that are emphasized in this section of the paper, and that can make meaningful contributions to discussions about the future of the study of culture and the humanities.

An example of the 'artworld' that initiates such reflections and makes connections involving the practical and theoretical potential of arts-based research is *The Lagoon Cycle*, by Helen Mayer Harrison and Newton Harrison, more commonly known together as the Harrisons.[3] While not the most contemporary work of art to emphasize in this context, nor the most contemporary work produced by the Harrisons, it has several important qualities that are suitable for this paper. Also, since it is not such a recent work, it allows the possibility for readers to find other references and reviews of this work that might be enriching for the development of their own critical analyses of the integrative potentials highlighted in this article.[4]

The Harrisons are well-known environmental artists and activists, and their artistic research usually addresses ethical issues of the human relationship with the environment. *The Lagoon Cycle* is an examination of the processes and linkages between food production and watersheds along the perimeter of the Pacific Rim, developed by the artists in collaboration with other professionals between 1973 and 1985 (Harrison and Harrison 1993). The 'final product' is a portable mural, which is about 100 meters long and is divided into sixty sections that show representations of their research in drawings, paintings, photographs, collages, and poems. The artistic process, however, entailed a lot more than what is perceivable in this 'final product.' It involved many different encounters between the artists and the environments they investigated; performative actions; the creation of poetic metaphors based on processes of self-reflection about the conditions for life; and the production of imagined maps and poetic dialogues between two characters that represented the artists themselves, the Lagoon Maker (Newton) and the Witness (Helen) (Harrison and Harrison 1993). The artists describe the work as follows:

---

3 I would like to take this moment to respectfully acknowledge the passing of Helen Mayer Harrison on 24 March 2018 and express my condolences to her loved ones.

4 Notable references for the work of the Harrisons and *The Lagoon Cycle* are: Matilsky 1992; Ingram 2013; and the writings of the artists themselves, many of which can be found at their official website: <http://theharrisonstudio.net/> [accessed: 25 July 2017].

> This 360 foot long and eight foot tall mural is an extended semi-autobiographical dialogue, with stories and anecdotes, plays between two characters, a 'Lagoon Maker' and a 'witness', and serves to establish the philosophical basis for the ecological argument in many later works. Beginning in Sri Lanka with an edible crab and ending in the Pacific with the greenhouse effect, it seeks ever-larger frames for a consideration of survival. It looks at experimental science, the marketplace and megatechnology, finally posing the question, "What are the conditions necessary for Survival" and concluding that it is necessary to reorient consciousness around a different database. [sic][5]

Furthermore, through their international recognition, which allowed them to expose parts of this work, accompanied by performances and manifestoes, in different museums, the artists managed to bring these local environmental issues to the attention of an international community (Matilsky 1992). In 1984, *The Lagoon Cycle* was also published in the form of a handmade book titled *The Book of the Lagoons*. The publication presents the story of the seven lagoons examined in *The Lagoon Cycle* with poems, hand-colored photographs, collages, and drawn imaginary maps: *The First Lagoon: The Lagoon at Upouveli*; *The Second Lagoon: Sea Grant*; *The Third Lagoon: The House of Crabs*; *The Fourth Lagoon: On Mixing, Mapping and Territory*; *The Fifth Lagoon: From the Salton Sea to the Pacific*; *From the Salton Sea to the Gulf*; *The Sixth Lagoon*: *On Metaphor and Discourse*; and *The Seventh Lagoon: The Ring of Fire; The Ring of Waters*.

The first integrative potential that I would like to emphasize in relation to this work is that there is no 'unique' model of the 'ideal' artistic research. It is always re-modeled according to the specifics of each case, as in many projects in the social sciences and the study of culture. However, in arts-based research, neither non-reproducibility nor errors and flaws are seen as negative aspects of the research process. On the contrary, they are often taken as new points of departure and knowledge creation (Sullivan 2005), enabling the artist/researcher to reconsider particularities and perceive the object of research from an unexpected point of view.

This is one of the fundamental aspects of *The Lagoon Cycle*. In the process of developing their work, the artists realized that because their perspectives were constantly affected by their experience, they needed to constantly reconsider the conceptual frames of their work:

> The story concerns two characters who begin a search for a 'hardy creature who can live under museum conditions' and who are transformed by this search. The characters define themselves in *The First Lagoon* by the differences in their values and perceptions, with one

5 Information available at the Harrisons' website: <http://theharrisonstudio.net/the-lagoon-cycle-1974-1984-2> [accessed: 25 July 2017].

> naming himself Lagoon Maker and the other naming herself Witness. Both proceed to live up to their names although they finally surrender them as circumstances push the two characters into constructing ever-larger frames for their discourse.
>
> (Harrison and Harrison 1993, 371)

Furthermore, in order to be able to engage with each of the different explored environments in a meaningful manner and find the connections among these environments for the larger bioregional connection, it was necessary for the artists to develop a renewed singular examination and expression based on the specific characteristics of each particular region. Their brief description of the *Fifth* and *Sixth Lagoons* exemplifies this:

> *The Fifth Lagoon* deals with the Salton Sea, which was formed by flood flow released by human error from canals along the Colorado River. *The Sixth Lagoon* treats the entire Colorado River basin. Lagoon Maker and Witness reflect on the insights they have gained through observing aquatic systems. They expand the scale of their thinking from the Salton Sea to the Colorado River watershed, which has been changed by lifestyles that demand vast amounts of electricity and irrigation. (Harrison and Harrison 1993, 372)

One important reflection upon issues of uniqueness and universal models in research can be related to the concept of creativity. In the academic world, creativity is usually overestimated as a form of generating completely innovative research questions, methods, and/or results that should nonetheless also be reproducible and aim for universal applicability. However, creativity is an endless exercise of combining and re-combining the material of previous experiences, re-orienting structures of thoughts, and re-creating knowledge. In artistic research, this even includes 'stealing' forms of knowledge from other disciplines and recombining them in new formats without following the strict methodological structures of such disciplines, since the artistic research feeds from other disciplines but does not necessarily have to give them back a functional product.

In this sense, artistic research is a processual and unfinished form of research. Even if there is a 'final product,' there are also always infinite new ways in which it could be expressed, re-thought, and transformed. It does not move to enhance certainty or universality. On the contrary, it looks for ambiguities and different forms to uncover the questions that have been buried by answers.

Sensibility provides a second integrative aspect of the artistic thinking process, in its basis in the multisensorial experiences and imaginative capabilities of the artist/researcher and the perceiver. Since there is no urgent need to prove or disprove any thesis, it is also open to the imagination. Its persuasive techniques seduce its perceivers and instigate both the perceivers and the artists themselves to revisit the world from a different perspective. Reimagining the world and reality departing from their own sensorial experiences. This is nicely expressed in *The Lagoon Cycle* through the dialogues created between the two characters, the Lagoon Maker and the Witness. These dialogues involve different levels of imagination with the real experiences that the artists encountered in the explored environments, which are poetically combined and summarized in the symbolic metaphors of the cycle of the lagoons with the cycle of life:

> For us it was a moment
> We didn't know it had begun
> until we were already in the middle
> Then we looked forward
> And knew how it should end
> but we didn't know how to get there
> You could as well say that knowing the ending
> we worked backward to what we must have been to begin it
> as forward to what we must become to end it
>
> I said
> What would happen if I told
> the story just as it occurred
> You said
> How could you
> Every time we recreate the past
> it is different
> I said
> Then let us reinvent ourselves
> You said
> We are always doing that anyway
> I said
> Let's do it publicly
> You said
> From one point of view or another
> everything is visible and public
> I said
> Let us experiment with a moment
> You said
> A moment may have no existence whatsoever
> I said
> A moment may have no boundaries and
> may be expanded indefinitely
> You said
> A moment is like an atom and can be exploded
> I said
> Then let us choose a radioactive moment
> with a ten year half-life.
>
> (Harrison and Harrison 1984a)

The (self-)reflexive practice and poetic forms of expressing it can have transformative influences on individual and collective perceptions of the world and, consequently, can instigate collective actions. As indicated by Sullivan:

> Reflexive practice is a kind of research activity that uses different methods to work against existing theories and practices and offers the possibility of seeing phenomena in new ways. [...] a reflexive practitioner will question content and contexts as problematic situations are revealed within particular settings. Issues-driven inquiry of this kind not only identifies problems but also opens up areas whereby participants become responsive to potential change. (Sullivan 2005, 110)

In some of the cases of this work of the Harrisons, the process of poetic self-reflection did generate practical outcomes. Through their metaphorical, poetic, and imaginative expression of the situation of these watersheds, they generated political reactions to engage in finding solutions for some of the environmental problems of the areas that were part of the research. Many of the Harrisons' projects have indeed become long-term community projects (Ingram 2013). Nonetheless, that is not the primary goal of artistic research. The functionality of such research may arise in the process and generate great impacts (and that is of course a good thing), but the very process of artistic thinking is about finding new poetic forms of expressing something and not necessarily solving it.

The last potential that I want to emphasize for this article is that artistic research, in the sense of this example, is based on transdisciplinary collaborative work. That is not necessarily always the case, but it happens very often mainly in contemporary artistic practices. Thinking processes are developed in forms of collaborative works, which can instigate an integrative relation to the artistic 'products' and slightly dissolve the sense of authorship and ownership over a piece of artistic expression, or over the research project itself. The process of thinking depends on the act of expressing it and sharing it, in order for it to gain meaning, and be reinterpreted and transformed. Furthermore, the experience as a whole becomes a meaningful exchange of knowledge between different collaborators in non-hierarchical forms. There is no division between the value of the knowledge of the artist, the local community, or the other professionals involved in the project.

In the case of *The Lagoon Cycle*, the artists developed a first-hand study through interviews with ecologists, biologists, and community planners of the specific visited areas. After that, they created photographic narratives that identified the problem, questioned the systems of beliefs (that allowed the problem to develop) using specific aesthetic and poetic strategies like irony and sarcasm, and proposed initiatives to counter the damages departing from the process of self-reflection and imagination expressed in the poetic dialogues (Matilsky 1992). As stated by the artists themselves: "Our work begins when we perceive an anomaly in the environment that is the result of opposing beliefs or contradictory metaphors. Moments when reality no longer appears seamless and the cost of belief has become outrageous offer the opportunity to create new spaces – first in the mind and thereafter in everyday life."[6] The Harrisons actually define themselves as a collaborative team and try to turn their initiatives into community projects that can become independent of their presence:

> Their work process is singular. It begins with the question, 'How Big is Here?' Here may be a street corner, as in California Wash or a sub-continent, such as Peninsula Europe. They only do work that is the outcome of an invitation to engage a particular place or situation. Typically, they agree to go to such a place to see, think, speak, research and engage a broad spectrum of people and groups. They will only take on a work if there is a general agreement that their actual client is the environment itself. The agenda is created by the artists in discourse with the larger community. *Thus, the Harrisons see themselves simultaneously as guests and co-workers*. They stay only as long as the invitation continues, or until they deem that they have done all that is possible for them to do.[7] [Italics mine]

Furthermore, they emphasize the importance of bridging different kinds of knowledge among the artistic, the scientific, and the 'popular' spheres of human life:

> The formation of ideas about complexity appears to us as a complicated process primarily because these ideas often do not lend themselves readily to translation into other forms of communication. We suspect that complexity theorists also need more grounded modes of comprehension. Our opinion is that if complexity science groups wish to make more comprehensible and concrete the imagery of complex systems, two things are necessary. First, ennobling issues need to be taken up directly. By 'ennobling' we mean envisioned actions that most people would accept as prima facie good to do, whether or not they believed they could be done. Second, we think that new language is needed that makes clear the juxtaposition of culture and ecology in a way that can be easily understood in the context of everyday discourse. (Harrison and Harrison 2007)

Although this particular quote does not refer to *The Lagoon Cycle* but to a more recent work of the artists, *Peninsula Europe*, it nicely summarizes the problematic gap that can be very common between academic research (and not only the work of complexity theorists) and other dimensions of social behavior and human knowledge.

---

6 Information available at the Harrisons' official website: <http://theharrisonstudio.net/> [accessed: 25 July 2017].

7 Information available at the Harrisons' official website: <http://theharrisonstudio.net/> [accessed: 25 July 2017].

3 Concluding Remarks

This article aimed to instigate a reflection about the integrative potentials of arts-based research for the future of the study of culture, primarily concerning methodological procedures and epistemological production. For this matter, it began with a critical reflection about the future of the humanities, its entanglements with the study of culture, and the possibilities for arts-based research. It did not intend, however, to provide an extensive review of the current state of arts-based research, nor the 'crisis' of the humanities. There are many controversial ideas about the benefits for the arts (or the subversive potentials of artistic thinking processes) of being given academic credit and institutionalized, then suffering a process of 'academic commodification' that is counterproductive to the subversive potentials and aims of artistic practice (Holert 2011). Nonetheless, this article aimed to highlight the benefits that the scholars in the field of the study of culture might have if they were to embrace some of the distinguished characteristics of arts-based research (in the sense of *research through art* mentioned above).

The article then moved to the exploration of the multiple forms in which arts-based research can be interpreted, providing a brief description of three basic forms in which the arts can be integrated in research: research *about*, *with*, and *through* art. Exposing different characteristics of each of these forms of arts-based research emphasized that they are not mutually exclusive and that different forms of arts-based research can be identified and applied within a single project. The distinction among these forms of research exemplifies the different possibilities in thinking about arts-based research without determining any hierarchical distinction among them. The focus was then directed to research through art because it is a form of research that is not commonly recognized in fields that are not necessarily related to the arts.

Finally, three integrative potentials of arts-based research were explored through the example of the artistic project *The Lagoon Cycle* by the Harrisons. The reflections about the potentials of arts-based research that emerged from specific characteristics of the work of the Harrisons were expanded to examine their benefits for the practice of research in a more general sense and in relation to the study of culture.

I would like to conclude by stating again that this article did not intend to deny the importance of scientific methods in research (which is not even the goal in artistic research); rather, it meant to indicate the ways in which the study of culture could benefit from such processes. The product of research in the study of culture is expected to meet scientific standards of verifiability, but perhaps, at least in the process of doing research, it would be beneficial to embrace some of the ambiguities, openness, and imaginative, or even utopian strategies of the artistic thinking processes. Such strategies may even open the eyes of cultural researchers to new possibilities in their own research that they would be otherwise unable to see because they do not fit their chosen methodological models.

References

Barone, Tom, and Elliot W. Eisner. "Arts-Based Educational Research." *Complementary Methods for Research in Education*. Ed. Richard M. Jaeger. Washington, DC: American Educational Research Association, 1997. 73–116.

Bono, James J., Tim Dean, and Ewa P. Ziarek, eds. *A Time for the Humanities: Futurity and the Limits of Autonomy*. New York: Fordham University Press, 2008.

Curtis, David J., Nick Reid, and Ian Reeve. "Towards Ecological Sustainability: Observations on the Role of the Arts." *S.A.P.I.E.N.S* 7.1 (2014): 1–15.

Delbanco, Andrew. *College: What It Was, Is, and Should Be*. Princeton: Princeton University Press, 2012.

Eernstman, Natalia, and Arjen E.J. Wals. "Locative Meaning-making: An Arts-based Approach to Learning for Sustainable Development." *Sustainability* 5.5 (2013): 1645–1660.

Eisner, Elliot W. *The Enlightened Eye: Qualitative Inquiry and the Enhancement of Educational Practice*. New York: Macmillan, 1991.

Geannopulos, Lawrence. *The Future of the Humanities: A Conversation with Julia Kristeva and Arnold Davidson*. France Chicago Center. 2014. <https://fcc.uchicago.edu/directory/future-humanities> [accessed: 25 July 2017].

Harrison, Helen M., and Newton Harrison. "The First Lagoon: The Lagoon at Upouveli." *The Book of the Lagoons*. Self-published. 1984a. <http://theharrisonstudio.net/wp-content/uploads/2011/03/botl_part1.pdf> [accessed: 25 July 2017].

Harrison, Helen M., and Newton Harrison. "The Fifth Lagoon: From the Salton Sea to the Pacific, From the Salton Sea to the Gulf." *The Book of the Lagoons*. Self-published. 1984b. <http://theharrisonstudio.net/wp-content/uploads/2011/03/botl_part5.pdf> [accessed: 25 July 2017].

Harrison, Helen M., and Newton Harrison. "The Sixth Lagoon: Metaphor and Discourse." *The Book of the Lagoons*. Self-published. 1984c. <http://theharrisonstudio.net/wp-content/uploads/2011/03/botl_part6.pdf> [accessed: 25 July 2017].

Harrison, Helen M., and Newton Harrison. "The Seventh Lagoon: The Ring of Fire, The Ring of Water." *The Book of the Lagoons*. Self-published. 1984d. <http://theharrisonstudio.net/wp-content/uploads/2011/03/botl_part7.pdf> [accessed: 25 July 2017].

Harrison, Helen M., and Newton Harrison. "Shifting Positions Toward the Earth: Art and Environmental Awareness." *Leonardo* 26.5 (1993): 371–377.

Harrison, Helen M., and Newton Harrison. "Public Culture and Sustainable Practices: Peninsula Europe from an Ecodiversity Perspective, Posing Questions to Complexity Scientists." *Structure and Dynamics* 2.3 (2007): 1–38.

Holert, Tom. "Artistic Research: Anatomy of an Ascent." *Texte zur Kunst* 82 (2011): 38–63.

Ingram, Mrill. "Ecopolitics and Aesthetics: The Art of Helen Mayer Harrison and Newton Harrison." *Geographical Review* 103.2 (2013): 260–274.

Irwin, Rita L., and Alex deCosson, eds. *A/r/tography: Rendering Self Through Arts-Based Living Inquiry*. Vancouver: Pacific Educational Press, 2004.

Knowles, J. Gary, and Ardra L. Cole, eds. *Handbook of the ARTS in Qualitative Research*. London/Los Angeles/New Delhi/Singapore: Sage, 2008.
Mantere, Meri-Helga. "Art and the Environment: An Art-based Approach to Environmental Education." *Rapporter om utbildning*. 3. Ed. L. Rubinstein Reich. Malmö: Lärarhögskolan, 1998. 30–35.
Matilsky, Barbara C. *Fragile Ecologies: Contemporary Artist's Interpretations and Solutions*. New York: Rizzoli International, 1992.
McNiff, Shaun, ed. *Art as Research: Opportunities and Challenges*. Bristol/Chicago: Intellect, 2013.
Nussbaum, Martha C. *Not for Profit: Why Democracy Needs the Humanities*. Princeton: Princeton University Press, 2010.
Pickering, Michael, ed. *Research Methods for Cultural Studies*. Edinburgh: Edinburgh University Press, 2008.
Pink, Sarah. *Doing Visual Ethnography: Images, Media and Representation in Research*. London: Sage, 2001.
Sullivan, Graeme. *Art Practice as Research: Inquiry in the Visual Arts*. Thousand Oaks/London/New Delhi: Sage, 2005.

Uwe Wirth

After Hybridity: Grafting as a Model of Cultural Translation

"All translation," Walter Benjamin states in his essay *The Task of the Translator*, "is only a somewhat provisional way of coming to terms with the foreignness of languages" (Benjamin 2007 [1921], 75). The notion of cultural translation as it was suggested by postcolonial studies in the past decades stands, as it seems, under the heading of a similar idea, namely coming to terms with the foreignness of 'other cultures.' This is, to mention just one famous example, especially true for Mary Pratt's ethnographic notion of "contact zone," when she is stressing – quite similar to Benjamin – that contact zones refer to a situation in which the modalities of contact are not yet determined, "where disparate cultures meet, clash, and grapple with each other, often in highly asymmetrical relations of domination and subordination – like colonialism, slavery, or their aftermaths as they are lived out across the globe today" (Pratt 2003 [1992], 4).

Here, a second aspect comes into play, which is not about the fact of foreignness but rather about the need for negotiation: Whenever the contact of different cultures is influenced by the circumstance that one party wishes to sell something to the other – whether this 'something' is merchandise, technology or ideology – the situation of cultural contact implies a translational process that adapts and integrates the 'own' to the 'foreign' – different traditions, different life styles, different world views. Peter Burke addresses this point in his book *What is Cultural History* with regard to Christian missionary attempts. Missionaries often tried to present their message in such a way that it would seem to be in harmony with the local culture. In other words, "they believed Christianity to be translatable." At the same time indigenous individuals and groups in China, Japan, Mexico, and Africa, "who were attracted by particular items of western culture, from the mechanical clock to the art of perspective, have been described as 'translating' them in the sense of adapting them to their own cultures, taking them out of one context and inserting them into another" (Burke 2004, 121–122).

In this paper (see also Wirth 2015), I would like to understand this adapting and inserting with reference to the concept of grafting, and thereby differentiate my approach from others that conceive of cultural translation simply as a process of hybridization. In addition, I would like to contrast adaptation and insertion with two notions that have been highly problematic in the debate concerning transcultural relations: namely assimilation and integration. The term 'assimilation' implies not only that one *adapts* to foreign ways of life – or that one gets

https://doi.org/10.1515/9783110669398-012

forced by others to adapt their way of life; it also implies that the *self* and the *other* are to be understood as identifiable, homogeneous entities. This is hinted at in the definition of assimilation offered by Haim Hillel Ben-Sasson in the *Encyclopedia Judaica* – according to which assimilation is a socio-cultural process

> [...] in which the sense and consciousness of association with one national and cultural group changes to identification with another such group, so that the merged individual or group may partially or totally lose its original national identity. (Ben-Sasson 2007, 605)

Three questions arise from this definition, which are, first of all: What are the implications of "original national identity" when such an identity is determined by a fixed cultural frame-set consisting of one language and specific national traditions?

The second question pertains to the understanding of a "merged individual": to merge means 'to incorporate,' 'to conglomerate,' 'to fuse' – and thus refers to the concept of 'hybridity.' Robert Park pioneered the modern notion of hybridity with his influential essay "Human Migration and the Marginal Man," which began with the premise that: "Every nation, upon examination, turns out to have been a more or less successful melting-pot" (Park 1928, 883). This implies both: fusions at a bodily-sexual level and the mixture of traditions, which Park calls "cultural hybrid" (Park 1928, 891). As already mentioned in the beginning: In postcolonial studies hybridity has become a central concept, in order to negotiate the 'foreignness of the other.' Hereby the main thesis is that the relations between different cultures can be described as cultural contact between bodies, languages, and worldviews of highly different backgrounds, whose mixture generates something new, something 'third' (see Bachmann-Medick 2006, 200). The classical terms for describing this dynamics of fusion are, as García Canclini points out in his book *Culturas Hibridas*: creolization, synchretism, *mestizaje*. These fusion dynamics are personified by Malinche, the indigenous translator of the Spanish conqueror Cortés, who bore Cortés' child to become the primordial mother of the so-called 'mestizos.' To this day she is a highly controversial figure – and of course, the term 'mestizo' is very problematic, too (see García Canclini 2005, xxxii). What seems interesting about Malinche, however, is that she became an allegory of all the interferences of bodily and conceptual aspects that cultural contacts carry with them. One could even see her as a protagonist in a constellation that Homi Bhabha calls *colonial mimicry* (see Bhabha 1994, 75–76): a notion that refers to situations in which the colonized appropriate the mixing of their own culture with that of the colonizer as a subversive strategy. Here, of course, 'mixing' stands for the mere pretense of assimilation.

This brings a third question to the fore, which proceeds from these considerations; it has to do with the relationship between assimilation and translation: To what extent do translation processes presuppose a 'making similar,' an imitation, an assimilation of the foreign language into one's own language? A question that extends beyond purely linguistic problems of translation, and touches upon what Homi Bhabha is referring to as *cultural translation*, basing his considerations on Walter Benjamin's essay "The Task of the Translator." Bhabha understands cultural translation as a process in which we can no longer assume that there is something like an original self with a static identity.

> In that sense there is no 'in itself' and 'for itself' within cultures, because they are always subject to intrinsic forms of translation. This theory of culture is close to a theory of language, as part of a process of translations – using that word [...] not in a strict linguistic sense of translation [...], but as a motif or trope as Benjamin suggests for the activity of displacement within the linguistic sign. (Bhabha 1990, 210)

According to Bhabha's reading, translation becomes a "way of imitating [...] an original" (Bhabha 1990, 210), in which the predominance of the original dissolves through the possibility of being copied and transformed, and thereby reveals a notion of an original "that is never finished or complete in itself. The 'originary' is always open to translation" (210). This implies a concept of the 'original' that does not appear as a homogeneous, static unit, but rather as something unfinished, as something still in motion. Such a conception of a constructed, non-homogeneous, not-yet-complete original impacts the understanding of both assimilation and cultural translation: The original is no longer considered an unchangeable archetype around which processes of adjustment and translation orient themselves; instead, the original itself becomes an object that changes during the translation process – an original in progress. Or maybe even: an original in motion.

The idea that the original itself undergoes change during the translation process, that it leads a life of its own and is not oriented around the principle of equivalence or fidelity, falls in line with Benjamin's thoughts on the task of translation, namely that

> no translation would be possible if in its ultimate essence it strove for likeness to the original. For in its afterlife – which could not be called that if it were not a transformation and a renewal of something living – the original undergoes change. (Benjamin 2007 [1921], 73)

With this passage, Benjamin clearly rejects the idea of an assimilating attempt to achieve similarity between a translation and its original. Instead, the translation is described as a living process, which is capable of changing the original in the course of translation. Within this context, the question emerges how such a concept of translation can be applied to the different definitions of hybridity described above. Doing so leads to a striking realization: namely, the extent to which the historical semantics of translation theories has been based on biological and organological metaphors. Above all, we find the image of 'transplanting' languages, which we also see in Benjamin's work when he writes, "thus translation, ironically, transplants the original into a more definitive linguistic realm since it can no longer be displaced by a secondary rendering" (Benjamin 1991 [1921], 75). Apparently (and ironically) Benjamin joins here a tradition in the history of translation theory that sees transplantational displacements and relocation as a form of translation: a tradition that engages in cultural translation.

In the following I would like to attempt to illustrate the implications of this engagement as it pertains to the concepts of assimilation and integration. In doing so, I will also show why the concept of 'hybridity' alone is not sufficient to do justice to the complexity inherent in processes of cultural translation. Indeed, my thesis is that we need a second concept as well, namely that of 'grafting' (see Wirth 2011a, 2014, 2017).

1 'Hybridity Model' and 'Grafting Model' in the Sphere of Translation

At this point I should make plausible, why I think the concept of grafting is relevant for addressing problems of translation, before I will go into some details about the actual notion of grafting.

In his treatise *On the Different Methods of Translation* Friedrich Schleiermacher compares the task of translation to an exotic, agricultural intervention when he writes that the "diverse transplantation of foreign plants have made our soil richer and more fertile" (Schleiermacher 1973 [1813], 69, my translation). The decisive question is, of course, what actually happens in this translational transplantation. Schleiermacher described the method of translation as a "composure of language that is not grown freely, but is rather bent over towards a foreign similarity" (Schleiermacher 1973 [1813], 55, my translation). What is that supposed to mean? Apparently Schleiermacher refers to a form of translation that attempts, as accurately as possible, to "match the phrases of the original writing," thus giving the reader the feeling that "they have something foreign in front of them" (Schleiermacher 1973 [1813], 54, my translation). Schleiermacher's theory thus represents an example of what has been called an 'alienating translation.' Anne Bohnenkamp – drawing from Mikhail Bakhtin – has suggested that this kind of translation should be classified as a "phenotypically hybrid translation" that seeks

> to mix elements of the original text and the original language in such a way that the mixture remains recognizably a mixture – that is, demonstrates the heterogeneity of parts. The antithesis *naturalizing* [*einbürgernd*] – *alienating* [*verfremdend*] would thereby be replaced by a model that makes not only the differences, but also the similarities of different translation processes obvious. (Bohnenkamp 2004, 20, my translation)

The point of this reformulation affects the linguistic surface of the translation, because it is obvious that from a genotypic perspective, translations are always 'hybrid.' Here, the reference to Bakhtin comes into play, who analyses the interdependence between "hybrid culture and hybrid literary forms" (Bakhtin 2002, 63) in relation to operations of "translation," "reworking," and "re-conceptualizing" (Bakhtin 2002, 377–378). According to Bakhtin, languages change historically primarily by hybridization, "by means of a mixing of various 'languages' co-existing within the boundaries of a single dialect, a single national language, a single branch" (Bakhtin 2002, 358–359). In contrast to the non-intentional form of an, as Bakhtin calls it, "historical, organic, obscure language hybrid" (Bakhtin 2002, 360), the artistic hybrid does not function as a mere melting pot, but rather as an arena, where different points of view on the world collide: These two points of view are not mixed, but "set against each other dialogically" (Bakhtin 2002, 360). For Bakhtin, this becomes particularly obvious in parodistic novels, where we find intentional hybrids as "hybrid compounded of two orders," for instance two contradictory styles ("low" and "high"). In these cases, Bakhtin states, "two 'languages' (both intra-lingual) come together and to a certain extend are crossed with each other" (Bakhtin 2002, 75). In my view this is not only true for parody but also for translation: Translation is a process, where different styles of thinking, different ways of speaking, different points of view on the world collide.

This ambivalent aspect of linguistic and cultural hybrids is also an issue in Schleiermacher's essay, when he writes: "who would not rather have children that represent the fatherly lineage purely, rather than as *Blendling*?" (Schleiermacher 1973 [1813], 54, my translation). With the term *"Blendling"* Schleiermacher explicitly introduces the concept of hybridity into his theory of translation – but also a curious discourse of purity (see Latour 1993, 59–60). With reference to the Greek expression *hibrida*, *Grimm's Dictionary* defines a *Blendling* as a, "bastard and hermaphrodite in whom the pure, natural type is blurred and mixed, from humans, animals, and plants" (Grimm 1854, my translation). Interestingly, Schleiermacher shifts to another metaphorical register as soon as the "strange similarities" between two languages are overlaid by the cultural differences between worldviews, and the question of contamination is overshadowed by the question of how to make foreign styles of thought compatible. In *Hermeneutics and Criticism*, Schleiermacher states that the "Christian spirit" in the *New Testament* "emerges from a mixture of languages in which Hebrew is the root within which the new is originally conceived, while the Greek is grafted on to it [*das Griechische aber aufgepfropft*]" (Schleiermacher 1977 [1838], 90, my translation).

Apparently, the agricultural technique of grafting – as a form of 'transplantation' – is used as a metaphor in which 'transfer' into a new syntactic context becomes a conceptual and linguistic 'translation' into a different cultural context.

At the same time, grafting becomes literally a model for the adaptation processes of foreign elements – be it language, be it thoughts, be it bodies – in which the 'self' and the 'foreign' become connected, but do not mix. In grafted compounds foreign elements get naturalized but not completely assimilated, since they remain visible as foreign elements. At the same time, we can also read Schleiermacher's two-sided description of translation as an indication that recourse to a model of hybridity alone is not sufficient. Indeed, the situation surrounding translation seems to be characterized by the interferences and interactions between a model of grafting and a model of hybridity.

2 Agricultural Implications of 'Grafting Model' and 'Hybrid Model'

At this point it is time to explain the notion of grafting in contrast to the notion of hybridity: Together with selective breeding, hybridizing and grafting can be conceived as *fundamental* techniques of culture (see Siegert 2011). While selective breeding is the purposeful strengthening and weakening of specific genetic traits within a biological species, hybridization is the mixing of different species. The crucial point of hybridization is the genetic mixing of heterogeneous elements. This is how the mule is generated by horse and donkey, as well as how new breeds of fruit trees spring from successful crossings. In contrast to hybridization, grafting does *not* result in a genetic mixture. It is not a blending, but a binding of two different parts into one organic unit: a combination in a literal sense, whereby the bound parts remain genetically different, and that also means each compound "maintains its own genetic identity" (Mudge 2009, 440), even when they are grown together. While hybridization follows the logic of sexual reproduction (and hence, the logic of sexual contact), namely: *a third is made from two*, grafting boils down to the idea: *make two into one.*

A look at a special issue of the *Time-Life Encyclopaedia of Gardening* on *Pruning and Grafting*, makes clear what the formula means:

> In essence, grafting involves the wounding of two growths and the arranging of them so that they heal together. One of the two growths is called the stock, understock or rootstock. It is the host plant, rooted in the soil and providing nourishment for the other growth, the dependent top section, which is called the scion. (Allen 1978, 60)

The graft is described as a "friendly parasite" (Serres 2008, 65) that grows with its host. This requires an accurate cut – with the help of special tools – that allows the injured cambium of the rootstock to come into direct contact with the injured cambium of the grafted scion. Both parts are subsequently united and adhered with tree wax. Thus, grafting proves to not only be a botanical bricolage that unites foreign bodies in processes of cut and paste; it also proves to be a concept that pertains to a cultural technique of intervention in which the natural circulation of plants juices is not cut off but rather 'rechanneled.' At the same time grafting – like cloning – is a technique of copying, which aims to maintain the purity of a species: hundreds of scions of the same sort (perhaps even twigs from the same tree) are grafted onto suitable roots. The result is a fragile entity of heterogeneous, artificially joined, non-assimilated parts that are still organically integrated with one another. Still, such grafting requires a minimal level of 'compatibility,' which is referred to in biological terms as "vegetative affinity" (see Hertwig 1923, 505).

Particularly in the eighteenth century, grafting also gained a biopolitical relevance (see Foucault 2004, 70): It became an economic figure of amplification, stressing the possibilities of maximization of natural powers, namely the qualitative and quantitative maximization of output. This is what we read in Zedler's *Encyclopedia*:

> Tree grafting is also called *impffen*, *pelzen*, and *zweigen*, and in gardening refers to the work through which a wild and infertile tree-trunk is combined with another that is set upon it, and which is improved by the broken branch of a fertile tree or the so-called graft scion. This is a glorious invention through which wild trees are tamed, the infertile are made fertile and prolific [...], indeed even the color and taste of these trees is transformed and changed.
> (Zedler 1961 [1753], 762, my translation)

Although most of these claims are – bio*logically* speaking – false, they express a bio*political* ideology: by combining different parts, by creating a new element, the forces of nature are simultaneously deployed, enforced and controlled. Thus, around 1800, the making fertile, transformation, change and improvement of nature become codes for an attitude in which the cultural technique of grafting appears as a kind of governing technique. This is made especially clear in the encyclopedia of Diderot and D'Alembert. Under the lemma "Greffe," grafting is called the "triomphe de l'art sur la nature" (D'Alembert 1757, 921), because this process enables one to force nature into producing new kinds of plants. This means that nature is modelled on culture understood as practice.[1] Grafting is

---

1 This also applies to hybridization, as long as this term is used to refer to crossbreeding initiated by human intervention. The best example to illustrate hybridization in this sense is the mule. Although a mule can result from an 'evolutionary coincidence,' its artificial crossbreeding is determined by an economic motive: A mule is more robust than a horse but less stubborn than a donkey. This controlled procreation implies a biopolitical concept in which hybridization

Fig. 1: Robert Sharrock, various methods of grafting, in: *The History of the Propagation & Improvement of Vegetables by the Concurrence of Art and Nature*, 1659, p. 59.

becomes what Foucault calls a *dispositif* [device] (Foucault 2004, 70). What is at stake here are natural resources that can be controlled and even improved by grafting or hybridization with the aim of economic exploitation. This is also the way to overcome the nature/culture split: The total, or even totalitarian economic framing of everything we refer to as nature puts nature at our disposal and into a 'stand-by mode' – what Martin Heidegger referred to as "*Bereitstellung*" [state of preparedness] (Heidegger 1967, 16). This stand-by mode is the very basis of what is called biopower (Foucault 1984, 257; Bertilsson 2003, 120).

herewith ascribed the potential to transform primordial plant species into something new, and to transform by means of bringing one species into contact with a foreign graft scion: refinement and cultivation through 'contact' with other plant cultures. At the same time – and until today – one can observe another tendency, namely the metaphorical employment of the agricultural technique of grafting to describe the symbolic cultural technique of 'writing' (see Böhme 1999). Since the eighteenth century one finds 'grafting' used within the framework of poetological discourses as a metaphor for practices of quoting, copying, and commenting.[2]

Indeed, in French, the term *greffer* signifies not only grafting in the sense of botany and surgical transplantation; the graft is also the concept for an agency of transcription. The "greffier" is, as we learn in the *Encyclopédie*, a notary scrivener who copies pieces of writing, registers, and archives (D'Alembert 1757, 924).

The German Romantic author Jean Paul even wrote an entire novel about grafting as a plagiarizing, collage-like process of text production: *Leben Fibels* (1811). The novel depicts the life of a certain Gotthelf Fibel who presents himself as a gifted literary writer even though he actually works as a writer in the literal sense: His passion is copying the Alphabet. Through a series of coincidences, he surprisingly succeeds in becoming famous as the author of an ABC primer (in German commonly called *ABC-Fibel*). As a consequence of his fame, a 40-volume biography (written by some of his employees) is published. Within the turmoil of the Napoleonic Wars, however, the pages of this biography disperse, and only fragments of the original remain – becoming part of every-day life and serving for various purposes such as wrapping paper or spice bags.

The premise of the story is that the editor-narrator Jean Paul attempts to gather and glue together these "fliegende Blätter" as "flying pages" or "loose leaves." The result is a fictional cut and paste in which the titles of the individual chapters announce the location where the fragments are found. In the "20th or Pelz chapter," we read:

> This entire chapter was found in the grass of an Impf- or Pelz-garden and seems to have been used to bind Pelz wounds, which the reader could interpret as subtly allegorical if he so desired. (Jean Paul [1811], 464)

Since "pelzen" is (as we saw in the Zedler-Article) an old-fashioned German synonym for 'grafting,' the model of grafting evoked here becomes an allegory for a poetics that draws the lines between original and copied writing. And there

2 See, for instance, Shaftesbury, who describes in his *Characteristicks of Men, Manners, Opinions, Times* authors who do nothing other than to write commentary in the following way: "They have no original Character [...]; but wait for something that may be call'd a Work, in order to graft upon it [...] at second hand" (Shaftesbury 1711, 269).

is more: An expert in print by the name of *Master Pelz* shows Fibel how one can imprint one's own name on the title pages of anonymous works. Thus, the *Pelzing* as an act of engrafted insertion becomes a gesture of appropriation as well. With this idea Jean Paul gives somehow advance notice to a concept that Jacques Derrida will call *greffe citationelle.* In his influential essay "Signature Event Context", Derrida introduces grafting as a metaphor for the "essential iterability" of all sign. Due to its iterability,

> a written syntagma can always be detached from the chain in which it is inserted or given without causing it to lose all possibility of functioning, if not all possibility of 'communicating,' precisely. One can perhaps come to recognize other possibilities by inscribing it or grafting it onto other chains. (Derrida 1988, 9)

Here, grafting represents the possibility of a 'force of rupture' with external (historical, spatial, social) contexts, but also internal, linguistic-syntagmatic contexts. And in this sense, as Jonathan Culler puts it, "the graft is the very figure of intervention" (Culler 2007, 141).

Whilst John Austin's speech-act theory is based on the assumption that the process of citation results in a loss of "illocutionary force" in what is said and that citation is a "parasitic" form of language use (Austin 1975, 21), Derrida's conception turns grafting as a citational graft [*greffe citationnelle*] into a figure that fosters the circulation of communicative power through an act of removing and re-inscribing signifying bodies, that is: through an act of displacement. At the same time one has to admit: re-inscribing or re-inserting signs into other contexts is an ambivalent gesture. Especially with the act of re-integration, the differences between graft and rootstock are marked: The new entity emerges from heterogeneous, artificially combined, unassimilated parts that are forced together.

3 Implications for a Notion of Cultural Translation

Here, two aspects of the grafting model are of significant relevance for cultural studies: on the one hand the role of grafting as a figure for describing ambivalent cultural integration and translation processes, in which the foreignness and difference of the translated remain visible; on the other hand the role of grafting as a figure of describing political constellations marked by nonsymmetrical power-relations. It is in this latter sense that Max Weber speaks of "grafted social orders" (Weber 1988 [1918], 516) and describes the relationship between European and Chinese culture as a merely extrinsic grafting of foreign mentalities (see Weber 1986 [1920], 440). Hence, grafting turns out to be a dominant gesture of the Western civilizing mission, since it contains "in germ the idea of transplanting the

European spirit" onto the traditions of other cultures (Acheraïou 2008, 33). Robert Young followed a similar line of thought in his book *Colonial Desire: Hybridity in Theory, Culture and Race*, in which he traces the political, and most centrally the terminological politics, of using organic metaphors to describe social forms of organization. The concepts of hybridity and grafting stand at the center of interest, because both terms denote the phenomenon "of forcing incompatible entities to grow together (or not)" (Young 1995, 4).

Maybe it is time to pursue a research project that analyses all the metaphorological (*sensu* Blumenberg) implications that models of hybridity and models of grafting bear (working-title could be: *Graftology*). The aim should be to develop a notion of cultural translation that is taking into account all the interactions and interferences between the 'hybridity-model' and 'grafting-model' – such as for instance the dynamics of forcing incompatible entities together, or the process of negotiating the modalities of how these entities come into contact. The argument could be this: As we saw in the initially quoted passage by Bhabha, the term 'translation' became a trope for the "activity of displacement within the linguistic sign" (Bhabha 1990, 210). The term 'displacement' obviously signifies two things in this context: First, similar to the way it is employed by Derrida, it refers to the internal sign structure of statements and is then projected onto the external dynamics of sign usage. Second, 'displacement' stands for a political dynamic of 'rupture' in which people are torn from their 'original' home contexts and forced to migrate to new contexts, 'foreign' to them.

In my view, such an interpretation of the concept of cultural translation displays the same dynamic as that which Derrida calls *greffe citationelle*: The displacing 'rupture with context' and the grafted insertion into another context finds its re-entry here in form of a *greffe culturelle*.

At the same time some of the precarious political implications that the grafting model carries with it become visible: It carries traces of the state of being torn out and placed as a mark of cultural difference onto another, foreign context; and it becomes herewith a model not only for justifying the possibility of integrating signs and people, but also for mobilizing forces that repel migrants as 'foreigners.' Maybe one could say that grafting is a parasitic model of what Pratt calls "contact zone," where disparate cultures "meet," often in "highly asymmetrical relations of domination and subordination – like colonialism" (Pratt 2003 [1992], 4). Indeed, the model of grafting also implies a specific form of cultural dominance: in the context of colonial constellations, it gives expression to a superior position of power. It relegates the colonized to the role of a wild substratum that can be cultivated by quasi-horticultural interventions. The hybridity model, on the contrary, encounters this play of power with a semantics of subversion, through which an interference between the logic of grafting and the counter-logic of hybridity emerges.

In consequence, a relationship of strained interference between a logic of hybridity and a logic of grafting arises. This interference is indicative of a highly complex situation similar to the situation Bruno Latour describes in his book *We Never Have Been Modern*. According to Latour, the project of modernity is initiated by an ambiguous dynamic; the word 'modern' designates, as he points out, two entirely different sets of practices:

> The first set of practices, by 'translation,' creates mixtures between entirely new types of beings, hybrids of nature and culture. The second, by 'purification,' creates two entirely distinct ontological zones: as for instance that of human beings on the one hand; that of nonhumans on the other. (Latour 1993, 10–11)

While the "modern critical stance" (Latour 1993, 11) always tried to keep these two sets of practices separate, the pre- and postmodern styles of thinking have confused the practices of translation and purification. I would like to argue that this confusion can also be understood as a certain kind of interference between 'hybridity-model' and 'grafting-model.' Since in grafting the combined genetic elements are not altered, it is an operation to produce a 'pure copy' – the concept of combining different bodies is connected to an emphasis on the genetic difference of the elements. Hence, grafting stands for a strategy of purification. In hybridization, on the other hand, the genetic differences are overridden by 'crossing' and 'translation.' But the most interesting aspect is, I would like to suggest, the style of confusion between 'hybridity-model' and 'grafting-model.'

In Homer's *Iliad* we find a famous description of a hybrid called *chimera*: Its front part is that of a lion, the back is that of a dragon and in the middle it is a goat. In a chimera, apparently, the style of confusion between logic of grafting and logic of hybridization is significant: The connected parts are not represented as mixed together, but as placed together: a phantasmagorical cut and paste. To put it in another way: Maybe fairytale creatures, monsters like chimeras or centaurs, are figurations of hybridity represented in the representational mode of grafting. Today, we have become aware again of such phantasmagorical and monstrous creatures (see Paré 1841, 23), which raise the question of the borders and limits of the human body in the context of biopolitical ideologies. For instance, we have representations of a chimera-like combination of human and machine – think of the movie *Robocop* (1987) – and maybe our high-tech culture is a culture in which the hybridization of machines and organisms is becoming a normality. Taking into consideration the various possibilities of using prostheses or of transplantation medicine (see Hamilton et al. 2012), we find ourselves, as Donna Haraway stated in her "Cyborg Manifesto," "to be cyborgs, hybrids, mosaics, chimeras" (Haraway 1991, 177).

In fact, the various modes and styles of confusing grafting-model with hybridity-model are quite puzzling, since it is by no means clear why and when they occur – maybe they indicate situations similar to what Hans Blumenberg called "Logical Embarrassment" [*Logische Verlegenheit*] (Blumenberg, 2005 [1960], 10). It is very often literature that becomes the playground where the consequences of these logical embarrassing confusions between grafting and hybridity are represented and negotiated. Just to mention one example, well known in postcolonial studies: In Salman Rushdie's *The Satanic Verses* (1988), the relation between graft and hybrid is re-negotiated in a very obvious way. It is not only that the "genetic possibility of centaurs was being seriously discussed" (Rushdie 1988, 467); on another occasion we even find the confusion of graft and hybrid as a keyword. One of the protagonists, Otto Cone, a Polish Jew who immigrated to England after he survived Nazi-Concentration Camps, tries to assimilate to his new home-land: not only by changing the family name from Cohen to Cone, but also by starting to adapt to one of the most popular hobbies in England: gardening. Rushdie chooses to let a tree become the symbol of Otto's wish to assimilate in an "incompatible world" (Rushdie 1988, 471):

> After Otto's death Alicja [...] planted vegetables in what Otto had insisted should be an English floral garden (neat flowerbeds around the central, symbolic tree, a 'chimeran graft' of laburnum and broom). (Rushdie 1988, 476)

Apparently, for Rushdie the "chimeran graft" also becomes a metaphor for the paradoxes and logical embarrassments of cultural translation. The crossing of laburnum and broom implies a special kind of cultural contact taking place 'in-between' the logics of graft and hybrid. At the same time, it transposes and translates two central horticultural – and biological – questions of the nineteenth century into twentieth century discourse about assimilation, integration, and intercultural contact.

When Charles Darwin was dealing with the problem of explaining the physiological processes involved in heredity and reproduction in his book *The Variation of Animals and Plants under Domestication* (1886), he chose the *Laburnum Adamii* as an example of the rather puzzling phenomenon of transition: "Mr. Adam inserted in the usual manner a shield of the bark of *C. purpureus* into a stock of *C. laburnum;* and the bud lay dormant, as often happens, for a year" (Darwin, 1886, 390). In other words, Adam practiced a certain technique of grafting, called oculation.

> If we admit as true Mr. Adam's account, we must admit the extraordinary fact that two distinct species can unite by their cellular tissue, and subsequently produce a plant bearing leaves and sterile flowers intermediate in character between the scion and stock, and

> producing buds liable to reversion; in short, resembling in every important respect a hybrid formed in the ordinary way by seminal reproduction. Such plants, if really thus formed, might be called graft-hybrids. (Darwin 1886, 390)

For Darwin, graft-hybrids point to an extraordinary fact. Between grafts and hybrids there are intermediary forms: figures not only of transition, but of what I suggest can be called translation, to which both the formulas *two into one* and *two into three* apply. In Rushdie's text, the graft-chimera (a term that in biology is still used synonymously with graft-hybrid) becomes a model for the connection of multiple parts of different origin. It not only stands for the "body eclectic" in an "incompatible world" (Rushdie 1988, 647), but also represents the condition of an all-encompassing 'in-betweenness.' In both cases – Darwin and Rushdie – the graft-hybrid functions as the ambiguous figuration of a classificatory undecidability. It incorporates and embodies the condition of 'in-between,' being subject exclusively neither to a logic of grafting nor to a logic of hybridization. This condition questions the plausibility of the dichotomy of grafting and hybridization and, at the same time, marks an ambiguous situation that configures not only the split between nature and culture but also the gap between one culture and another.

4 Cultural Translation

Against the backdrop of what was said so far, I would like to come back to Schleiermacher and pose the question what it means when he describes the act of translation as a "transplantation of foreign plants" that "have made our soil richer and more fertile" (Schleiermacher 1973 [1813], 69). In my view it is noteworthy that Schleiermacher did not introduce this motif of reciprocal influence between foreign plants and native ground in the sense of a concept of *terroire* in which the ground influences the plant through the roots. Instead, his conception is reversed: The ground is influenced by the plant. This is an opposing model to the concept of originality made prominent by the European genius-aesthetic as proposed by Edward Young. In his 1759 work *Conjectures on Original Composition*, Young makes the assumption that the original is rooted in the fertile ground of a natural genius: "an Original may be said to be of a vegetable nature; it rises spontaneously from the vital root of Genius" (Young 1759, 12). In contrast, the imitator appears as a, "transplanter of Laurels, which sometimes die on removal, always languish in a foreign soil" (Young 1759, 10). In other words: The process of copying and imitation is described here as transplantation to new ground, which leads the transplanted plant to become weaker.

There is a third possibility to reformulate the problematic relationship between assimilation and transplantation, namely with recourse to a concept of originality that uses botanical metaphors without paying homage to the idea of a homogeneous primitive nature as found in Young's writings. Such a concept can be found in Herder's *Fragments on Recent German Literature* (1768), where he first writes, following Young, that, "every book is a bed of flowers and growths; every language an immeasurable garden of plants and trees" (Herder 1985 [1868], 552, my translation). Shortly after this, however, Herder provides a totally different kind of linguistically critical summary when he – in contrast to Schleiermacher – strikes rather xenophobic tones while remaining in the garden paradigm:

> Which revolutions did the German language undergo, partially within its own nature and partially through the admixture of foreign languages and ways of thinking in order to change its mind while its body remained the same? How full is the language taught in foreign colonies [*fremde Kolonien*], which have taken on German dress, German civil rights, and German habits? How many foreign branches have been grafted onto the tree-trunk of our literature – how they are on the trunk where it is not degenerated, but rather changed and often refined? (Herder 1985 [1868], 567, my translation)

The question is, of course, what Herder meant when he used the term 'foreign colonies.' I would like to suggest that this expression is referring to the so-called *Antiqua-Fraktur-dispute*. In my view it is an allusion to the typographical convention of printing all foreign words in *Antiqua*, while the German text was printed in *Fraktur*. According to this convention, the difference between *Antiqua* and *Fraktur* becomes a cipher for ostentatious non-assimilation: Foreign words remain foreign within a field of native *Fraktur* script (see Wehde 2000, 216). At the same time, foreign words also represent an externalized 'foreign similarity' that forms the basis of Schleiermacher's concept of an alienating translation.

But if foreign words are 'colonies' that are marked as 'foreign words' by a different typographical form (*Antiqua*), then don't the German words written in *Fraktur* take on the systematic position of an indigenous people? And when Herder claims that foreign colonies have taken on "German dress, German citizenship, and German customs," then the formulation "German citizenship" obviously plays with the possibility of a naturalizing form of translation. The naturalization of a foreign word, the translational adoption of a 'foreign similarity,' appears here as a form of colonial mimicry within the framework of written culture: on the one hand, an assimilation of the typographical dress-code; on the other hand, a linguistic assimilation of foreign words with regards to 'German grammar.' In this respect, Herder's metaphorical reference to foreign colonies points to the sensitive political problem of linguistic incorporation that remarkably parallels the so-called 'integration debate' seen in Germany today.

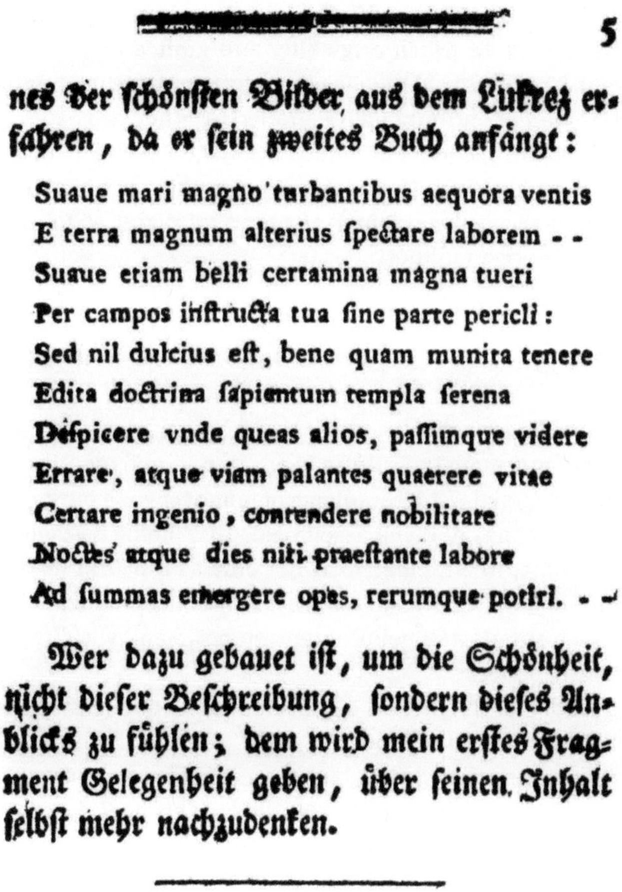
5

nes der ſchönſten Bilder aus dem Lukrez erfahren, da er ſein zweites Buch anfängt:

Suaue mari magno turbantibus aequora ventis
E terra magnum alterius ſpectare laborem - -
Suaue etiam belli certamina magna tueri
Per campos inſtructa tua ſine parte pericli:
Sed nil dulcius eſt, bene quam munita tenere
Edita doctrina ſapientum templa ſerena
Deſpicere vnde queas alios, paſſimque videre
Errare, atque viam palantes quaerere vitae
Certare ingenio, contendere nobilitate
Noctes atque dies niti praeſtante labore
Ad ſummas emergere opes, rerumque potiri. - -

Wer dazu gebauet iſt, um die Schönheit, nicht dieſer Beſchreibung, ſondern dieſes Anblicks zu fühlen; dem wird mein erſtes Fragment Gelegenheit geben, über ſeinen Inhalt ſelbſt mehr nachzudenken.

Fig. 2: Page from Herder's *Fragmenten über die neueste deutsche Literatur* (1768) (Third Collection).

What I find most remarkable about Herder's writing, though, is that despite the somewhat lachrymose tone with which he first characterizes the German language as an entity that has been grafted together, he also recognizes that the "original, peculiar" nature of the national language is a result of exactly this historical grafting: a language that, as he puts it, "is as it is, after its branches have been trimmed and transplanted, with all of its grafted foreign twigs, but still standing as a self-grown tree-trunk, injured but not dismembered by bare hands" (Herder 1985 [1768], 571, my translation). To the extent that around 1800 (but also for the remainder of the nineteenth century) language as 'national language' constituted the definitive point of reference for what was described above as "original national identity" (Ben-Sasson 2007, 605), Herder describes in this passage a conception of national language in which originality and grafted-togetherness interfere with one another. The national language that has been grown on native soil proves to be a pieced together entity, namely a grafted together language tree that is remarkably unstable and incomplete – and remains, "never finished or complete in itself" (Bhabha 1990, 210). National language, thus, appears to be an 'original in motion' that becomes 'an original' as soon as it is grafted together with linguistic branches from other cultures.

5 Conclusion

I would like to conclude by claiming that grafting should be considered to be an indispensible component of any model of culture in which national identity and originality are no longer considered to be paradigms of homogeneous purity, but rather as always unfinished modes of being assembled. This 'being assembled' is, as I tried to make plausible, to be understood not only as a hybrid mixture but also as a grafted combination. This also implies a new conceptualization of 'cultural translation,' which stands at a point of tension between hybridity and grafting. This has two consequences:

Firstly, grafting is a model for the process of cultural translation in the sense of the transplantation of branches of language.

Secondly, the grafted tree becomes a model for an original in motion.

This view gets support from a rather unexpected side: In his book *Word and Object*, the analytical philosopher Willard Van Orman Quine raised the question of how processes of understanding can be conceived of as processes of translation. In trying to answer this question, he draws from an eminently botanical register in the beginning of his investigation, when he writes:

> Different persons growing up in the same language are like different bushes trimmed and trained to take the shape of identical elephants. The anatomical details of twigs and branches will fulfill the elephantine form differently from bush to bush, but the overall outward results are alike. (Quine 1960, 7)

What is quite surprising in this context is the strange interference between a natural rootedness of language on the one hand, and its cultural formation via gardening interventions on the other hand. It seems that Quine introduces a ruthless French gardener who trims the flora until it can be used for the presentation of fauna (a colonial fauna, by the way).

There is a second passage in which a botanical register is employed; a passage in which the problem of cultural translation is explicitly addressed – more specifically, it addresses a situation of cultural contact in which one does not understand the language of the other at all. In order to assign meaning to words of a foreign, unknown language, we need, as Quine calls it, a "radical interpretation" that begins with the positing of analytical hypotheses about possible meanings:

> The method of analytical hypotheses is a way of catapulting oneself into the jungle language by the momentum of the home language. It is a way of grafting exotic shoots on to the old familiar bush [...] until only the exotic meets the eye. (Quine 1960, 69)

I wonder, how this metaphor relates to the translation theories of Schleiermacher, Herder, and Benjamin (but, of course, also to the theories of Bakhtin and Derrida). Is the "old familiar bush" possibly understandable as a functional analogy for the "self-grown tree-trunk" of Herder upon which one has grafted all sorts of "foreign twigs"?

If this were the case, what would the poem of Yoko Tawada, which appeared in her book *Abenteuer der deutschen Grammatik*, [*Adventures of German Grammar*] mean?

Die 逃走 des 月s

我 歌 auf der 厠
da 来 der 月 herange転t
裸
auf einem 自転車
彼 hatte den道 mitten 通 den 暗喩公園 ge選
um 我 zu 会

戸外 die 道 entlang
散歩e 歯磨end eine 美女
auf der 長椅子 im 公園
飲 ein 男 in 妊婦服 林檎汁
Am 末 eines 世紀s ist die 健康 eben 適

Im 天 穿 ein 穴
Die 月的不安 Der 月的苦悩 sind 去
全「的」飛翔 活発
um das 穴 herum

Die 皺 des 深淵s 平
Auf der 光滑en 表面 der 苦悩
登場 die 詩人 auf 氷靴 an
月 我的 neben我

Fig. 3: "Die Flucht des Mondes" by Yoko Tawada from: *Abenteuer der deutschen Grammatik*, p.41.

A poem which, as she writes in a comment, is the transcription of the translation of her poem "Flight of the Moon" – written according to the same method that is used when combining Japanese and Chinese ideograms and phonetic transcriptions.

> In order to write Japanese, one must write the roots of the meanings with Chinese ideograms and everything else (hands and feet of words) with a phonetic script. The poem shows that one can also write German with this mixing method. (Tawada 2000, 41)

Here, the foreign colonies are no longer foreign words typeset in *Antiqua* that stand out against the *Fraktur* of 'native words.' The foreign colonies are now Chinese ideograms that are combined with German phonetic transcriptions according to the Japanese method of mobilizing foreign typographical characters. This is a conceptual and bodily form of the transplantation of characters that aims to bring them into an adventurous grammatical situation of a 'grafted-script-culture-contact,' and at the same time gives birth to a model of cultural translation adapting and inserting the elements of a foreign language as a 'graft.'

References

Acheraïou, Amar. *Rethinking Postcolonialism: Colonialist Discourse in Modern Literatures and the Legacy of Classical Writers*. New York: Palgrave Macmillan, 2008.

Allen, Oliver. *Pruning and Grafting: The Time-Life Encyclopedia of Gardening*. New York: Holt, 1978.

Austin, John L. *How to Do Things with Words*. Oxford: Oxford University Press, 1975 [1962].

Bachmann-Medick, Doris. *Cultural Turns. Neuorientierungen in den Kulturwissenschaften*. Reinbek: Rowohlt, 2006.

Bakhtin, Mikhail. *The Dialogic Imagination: Four Essays*. Ed. Michael Holquist. Austin: University of Texas Press, 2002.

Benjamin, Walter. "Die Aufgabe des Übersetzers." *Gesammelte Schriften IV.1. (Kleine Prosa, Baudelaire Übertragungen)*. Ed. Tillman Rexroth. Frankfurt a.M.: Suhrkamp, 1991 [1921]. 9–21.

Benjamin, Walter. "The Task of the Translator." *Illuminations*. Ed. Hannah Arendt. Trans. Harry Zohn. New York: Schocken, 2007 [1921]. 69–82.

Ben-Sasson, Haim, et al. "Assimilation." *Encyclopedia Judaica*. Ed. Fred Skolnik and Michael Berenbaum. Farmington Hill: Macmillan Reference USA, 2007. 605–613.

Bertilsson, Thora. "The Social as Trans-Genic: On Bio-Power and Its Implications for the Social." *Acta Sociologica: The Knowledge Society* 46.2 (2003): 118–131.

Bhabha, Homi. "The Third Space: Interview with Homi Bhabha." *Identity: Community, Culture, Difference*. Ed. Jonathan Rutherford. London: Lawrence & Wishart, 1990. 207–221.

Bhabha, Homi. *The Location of Culture*. London/New York: Routledge, 1994.

Blumenberg, Hans. *Paradigmen zu einer Metaphorologie*. 3rd ed. Frankfurt a.M.: Suhrkamp, 2005 [1960].

Böhme, Hartmut. "Vom Cultus zur Kultur(wissenschaft). Zur historischen Semantik des Kulturbegriffs." *Literaturwissenschaft – Kulturwissenschaft: Positionen, Themen, Perspektiven.* Eds. Renate Glaser and Matthias Luserke. Opladen: Westdeutscher Verlag, 1999. 48–68.

Bohnenkamp, Anne. "Hybrid statt verfremdend? Überlegungen zu einem Topos der Übersetzungstheorie." *Linguistische Aspekte der Übersetzungswissenschaft.* Eds. Peter Colliander, Doris Hansen, and Ingeborg Zint-Dyhr. Tübingen: Julius Groos, 2004. 9–26.

Burke, Peter. *What is Cultural History?* Cambridge: Polity, 2004.

Culler, Jonathan. *On Deconstruction: Theory and Criticism After Structuralism.* Ithaca: Cornell University Press, 2007 [1982].

D'Alembert, Jean Le Rond, and Denis Diderot, eds. Lemma "Greffe." *Encyclopédie*. Vol. 7. Paris: Briasson, 1757. 921–924.

Darwin, Charles. *The Variation of Animals and Plants Under Domestication*. London: John Murray, 1868.

Derrida, Jacques. "Signature Event Context." *Limited Inc.* Evanston, IL: Northwestern University Press, 1988 [1972]. 1–23.

Foucault, Michel. *The Foucault Reader*. Ed. Paul Rabinow. New York: Pantheon Books, 1984.

Foucault, Michel. *Naissance de la Biopolitique: Cours au Collège de France. 1978–1979.* Paris: Seuil, 2004.

García Canclini, Néstor. *Hybrid Cultures: Strategies for Entering and Leaving Modernity.* Minneapolis/London: University of Minnesota Press, 2005.

Giuriato, Davide. "'Blendlinge.' Zur Theorie der Übersetzung bei Friedrich Schleiermacher." *Impfen, Pfropfen, Transplantieren*. Ed. Uwe Wirth. Berlin: Kadmos, 2011. 121–134.

Grimm, Jacob, and Wilhelm Grimm. *Deutsches Wörterbuch*. Leipzig: S. Hirzel, 1854–1971.

Hamilton, David, Clyde Barker, and Thomas Starzl, eds. *A History of Organ Transplantation: Ancient Legends to Modern Practice.* Pittsburgh: University of Pittsburgh Press, 2012.

Haraway, Donna. "A Cyborg Manifesto: Science, Technology, and Socialist-Feminism in the Late Twentieth Century." *Simians, Cyborgs, and Women: The Reinvention of Nature*. Ed. Donna Haraway. New York: Routledge, 1991 [1985]. 149–181.

Heidegger, Martin. "Die Frage nach der Technik." *Vorträge und Aufsätze.* Ed. Martin Heidegger. Tübingen: Neske, 1967 [1953]. 5–36.

Herder, Johann Gottfried. "Über die neuere deutsche Literatur. Fragmente. Erste Sammlung. Zweite völlig umgearbeitete Ausgabe." *Werke in zehn Bänden.* Vol. 1 (Frühe Schriften 1764–1772). Ed. Ulrich Gaier. Frankfurt a.M.: Deutscher Klassiker Verlag, 1985 [1768]. 541–650.

Hertwig, Oskar. *Allgemeine Biologie*. Jena: Gustav Fischer, 1923 [1906].

Latour, Bruno: "Visualisation and Cognition: Drawing Things Together." *Knowledge and Society: Studies in the Sociology of Culture Past and Present.* Ed. Henrika Kuklick. Oxford: JAI Press, 1986. 1–40.

Latour, Bruno. *We Have Never Been Modern*. Ed. and transl. Cathrine Porter. Cambridge, MA: Harvard University Press, 1993.

Mudge, Ken, Jules Janick, Steven Scofield, and Eliezer E. Goldschmidt. "A History of Grafting." *Horticultural Reviews* 35 (2009): 437–493.

Paré, Ambroise. *Œuvres Complètes d'Ambroise Paré.* III. Paris: Baillière, 1841 [1573].

Park, Robert E. "Human Migration and the Marginal Man." *American Journal of Sociology* 33.6 (1928): 881–893.

Paul, Jean. *Leben Fibels. Werke in zwölf Bänden*. Vol. 11. Ed. Norbert Miller, Munich: Hanser, 1975.

Pratt, Mary Louise. *Imperial Eyes: Travel Writing and Transculturation*. 2nd ed. London: Routledge, 2003.
Quine, Willard Van Orman. *Word and Object*. Cambridge, MA: MIT Press, 1960.
Rushdie, Salman. *The Satanic Verses*. New York: Penguin, 1988.
Schleiermacher, Friedrich. "Über die verschiedenen Methoden des Übersetzens." *Das Problem des Übersetzens*. Ed. Hans Joachim Störig. Darmstadt: Wissenschaftliche Buchgesellschaft, 1973 [1813]. 38–69.
Schleiermacher, Friedrich. *Hermeneutik und Kritik*. Frankfurt a.M.: Suhrkamp, 1977 [1833].
Serres, Michel. *Der Parasit*. Frankfurt a.M.: Suhrkamp, 2008.
Shaftesbury, Anthony. *Characteristicks of Men, Manners, Opinions, Times*. Treatise VI. London: Printed by John Darby, 1711.
Sharrock, Robert. *The History of the Propagation & Improvement of Vegetables by the Concurrence of Art and Nature*. Oxford: Lichfield 1659.
Siegert, Bernhard. "Kulturtechnik." *Einführung in die Kulturwissenschaft*. Eds. Harun Maye and Leander Scholz. Munich: Fink, 2011. 95–118.
Tawada, Yoko. *Abenteuer der deutschen Grammatik*. Tübingen: Konkursbuchverlag, 2000.
Weber, Max. "Die Wirtschaftsethik der Weltreligionen." *Gesammelte Aufsätze zur Religionssoziologie*. Ed. Max Weber. Tübingen: Mohr, 1986 [1920]. 238–442.
Weber, Max. "Der Sozialismus." *Gesammelte Aufsätze zur Soziologie und Sozialpolitik*. Ed. Max Weber. Tübingen: Mohr, 1988 [1918]. 492–517.
Wehde, Susanne. *Typographische Kultur. Eine zeichentheoretische Studie zur Typographie und ihrer Entwicklung*. Tübingen: Niemeyer, 2000.
Wirth, Uwe. "Gepfropfte Theorie. Eine 'greffologische' Kritik von Hybriditätskonzepten als Beschreibung von intermedialen und interkulturellen Beziehungen." *TheorieTheorie. Wider die Methodenmüdigkeit in den Geisteswissenschaften*. Eds. Mario Grizelj and Oliver Jahraus. Paderborn: Fink, 2011a. 151–166.
Wirth, Uwe. "Kultur als Pfropfung. Pfropfung als Kulturmodell. Prolegomena zu einer *Allgemeinen Greffologie* (2.0)." *Impfen, Pfropfen, Transplantieren*. Ed. Uwe Wirth. Berlin: Kadmos, 2011b. 9–27.
Wirth, Uwe. "Between Hybrid and Graft." *From Literature to Cultural Literacy*. Eds. Naomi Segal and Daniela Koleva. London: Palgrave Macmillan, 2014. 232–249.
Wirth, Uwe. "Pfropfung als Kulturkontakt." *Kulturkontakte. Szenen und Modelle in deutsch-japanischen Kontexten*. Eds. Yuichi Kimura and Thomas Pekar. Bielefeld: transcript 2015. 29–48.
Wirth, Uwe. "Poetisches Paperwork. Pfropfung und Collage im Spannungsfeld von *Cut and Paste*." *Paperworks. Literarische und kulturelle Praktiken mit Schere, Leim, Papier*. Eds. Magnus Wieland and Irmgard Wirtz. Göttingen: Wallstein, 2017. 7–29.
Young, Edward. *Conjectures on Original Composition*. London: A. Millar; R. & J. Dodsley, 1759.
Young, Robert J. C. *Colonial Desire: Hybridity in Theory, Culture and Race*. London/New York: Routledge, 1995.
Zedler, Johann Heinrich, ed. *Großes vollständiges Universal-Lexikon aller Wissenschaften und Künste, welche bißhero durch menschlichen Verstand und Witz erfunden und verbessert worden*. Vol. 3. Graz: Akad. Dr.- u. Verl.-Anst., 1961 [1753].

Dirk van Laak

Liquid Spaces in Modern Historiography

It is astonishing how much the ranges and spaces of history have changed and extended in the past generation of historical research. In what follows I will present, and attempt to explain, some key categories of recent historical writing in 'the West.' In hindsight, they document a tendency toward spatial concepts and disciplinary boundaries becoming more and more *liquid*. These are, in order of their appearance: the history of everyday life, the comparative history of nations, international history, history of international organizations, history of globalization, colonial history, transnational history, entangled history, global history, universal history, area studies, glocalization, and finally big history. My discussion will be conducted from a Central European viewpoint, and, I admit, this may narrow or confine my scope. To conclude, I will add some observations about the intersections of general and cultural histories, and will dare to look upon what appears to be relevant in the near future with regard to methodology and to contents.

1 Histories Going Global

When, in the early 1980s, I began to study the science of history, the specter of the Cold War still lingered. It was marked by an immobile, almost static conception of time, and an apparent distinctness of spatial borderlines and ideological frontiers (Leffler and Westad 2010). Germany was divided into separate political systems, and East- and West-German historians were primarily concerned about their respective national backgrounds, researching their own pasts and their liberal or socialist traditions. Historical interest tended to focus more on smaller sections of the German society and minor units of national spaces: The order of the day was the history of everyday life, granting common people their share in suffering, and in the shaping of history.

In this historical examination of one's local region, city, neighborhood, or workplace, many amateur and professional researchers discovered – not to everybody's surprise – that women had also participated in history, as did people beyond the ruling or propertied classes, members of minorities, and outsiders. In these years, it seemed that anything concrete and exemplary for a distinct historical place, person, or case study roused public interest. Little was changed concerning national boundaries in the 1980s, but historical spaces were opened up in a more and more refined way, zooming into smaller levels of investigation.

https://doi.org/10.1515/9783110669398-013

While a few methodological suggestions from beyond national borders were taken up and incorporated into German research strategies during the 1980s – for instance, the *history of everyday life* or *from below* came from Sweden or the United States, France, Italy, or Great Britain (van Laak 2003) – German historical research of that era still remained largely within national confines. Very few German historians actually followed the history of international relations and even fewer were interested in world history, which, rather, was assessed a special interest of eccentrics. In the later 1980s, a long-lasting dispute among German historians was staged as to the legitimacy of comparing National Socialism to Stalinism. For different reasons, many renowned representatives of the historians' guild rejected the acceptability and logic of such comparisons (see Augstein 1987).

This, however, changed in the years following the dramatic upheaval of 1989–90. When the wall between East and West Germany came down, I was at home studying for my final exams at the university. Accordingly, I paid very little attention to the epic events taking place in Berlin and elsewhere: Even historians are not always aware of history as it actually happens. My academic thesis was devoted to certain aspects of German fascism and to the persecution of the German Jews, and it was exemplified by the cities in which I had grown up. In my ensuing dissertation, which was again dedicated to an aspect of German history but also went beyond a German setting, I turned my attention to a broader scope, focusing on a law professor who was famous throughout Europe, the United States, and even Japan (van Laak 1993). My first professional position at a university led me to Jena, a city that prior to 1989 had existed under the socialist system. In Jena, I gathered many experiences and gained insights on what the dissolving of borderlines entails: for instance, how this leads to cultural amalgamations and hybrids, but also to new kinds of social segregation. Seen from today's perspective, what happened in Jena was something like a transnational integration within a national framework.

For a certain period during the early 1990s, people tended to believe that the fatal history of the twentieth century – with all its brutal wars and ideological confrontation – would come to a definite end, and expected that an era would arise in which nations and states would cooperate peacefully, and would not challenge existing borders. The concept of the nation-state and its associated nationalism was rarely questioned even by historians, although they already had recognized that since the eighteenth century, national traditions had often been invented from close to scratch (Gellner 1983). More and more historians did, however, embark upon comparing with each other the histories of nations as they had established since the nineteenth century (Kaelble 1999).

For the "age of extremes" – that is, the short twentieth century, as Eric Hobsbawm labeled it – it actually did make sense to compare national histories

(Hobsbawm 1995). In 1989–90, a distinct era appeared to reach its conclusion, and was in anticipation of becoming historicized. During my working years in Jena, I witnessed many historical myths of the former German Democratic Republic being challenged and liquefied (see, for instance, Niethammer 1995). Those who had to adapt to the Western political and economic system constantly compared the histories of East and West Germany (Kleßmann, Misselwitz, and Wichert 1999). When they asked what had kept the histories of both societies together despite more than 40 years of separation, the answer turned out to be much more than many analysts initially believed (Wengst and Wentker 2008). Additionally, during the 1990s, the European (Economic) Community – which since the 1950s had evolved and developed rather slowly – suddenly expanded vigorously. Following the *Treaty of Maastricht* in 1992, the European Union was established, and was quickly implemented by many (East-)European countries (Loth 2014). However this process is evaluated in detail, Europeans were quite amazed to learn about and actually experience falling borders, more freedom and mobility, and formerly separated societies moving closer together. In the formation of the European Union, Europeans came to understand that they shared political and economic interests with countries that previously had appeared quite distant and inaccessible. Together with immigration from very different countries – and influences from postcolonial thought (Said 1978, 1993) – this generated disputes on how to deal with and encompass cultural differences. It would be misleading, though, to state that Germans, Spaniards, or Italians of today are exceedingly interested in neighboring countries or their histories: this is just the case for selected people. But no European citizen can dismiss the fact that more Germans, Spaniards, and Italians travel to their respective countries than ever before, that work migration is rising, and that there now exists a soaring interdependence in economic and administrative systems.

Nevertheless, in European countries the prevailing concepts of history still follow national orientations and confines. This path dependency is due to the fact that modern historiography came along *with* and was marked *by* the creation of nation states in the eighteenth and nineteenth centuries (Conrad and Conrad 2002; Duchhardt 2006; Berger 2007; Carvalho and Gemenne 2009). The writing of history was even seen as a kind of auxiliary science of building up a nation state, and establishing grand national narratives effectively contributed to the creation of “imagined communities” (Anderson 1983) and “invented traditions” (Hobsbawm and Ranger 1992). Largely for this reason, many politicians and historians of today still pretend nation states to be something like the principal unit of history, based on a clear-cut territory, an integrated language and culture, and, last but not least, an ethnically coherent population (Smith 2001; Wehler 2011).

However, with the possible exception of isolated locations like Iceland, such consistency in a nation has never existed in history. Rather, it was the other way around: Attempts to unite all of these factors into unique and independent national bodies must be ranked among the most violent aspects of recent history. Such efforts have led to forceful relocations of borders, to the displacement of peoples, to absurd actions like 'ethnic cleansing,' and subsequently to much suffering. Recent research has quite bluntly revealed just how the idea of the nation state, and also the purported 'right of self-determination' proclaimed after World War I, would entail a lot of very violent outcomes (Manela 2007).

It is almost a natural consequence that the 1990s sparked renewed interest in *international history* in its classic sense, a history of international relations. The clear frontiers of the global Cold War had dissolved and were replaced by a situation that longed for the analysis and explanation that history could provide (Loth and Osterhammel 2000). Why, for instance, did a country like Yugoslavia split, break down so violently, and sink into war, in a process that Europe – luckily – had not seen for more than 50 years? The international situation changed fundamentally, not just for Europeans or the dissolving Eastern and Western Blocs, but for the countries in the Global South as well. Before decolonization, Eastern and Western Blocs had courted them for a generation; now, they were almost left alone to subsequently pursue a status of being non-aligned (Prashad 2008; Westad 2007; Sluga 2013). International history predominantly revolves around conflicts and cooperation among nation states, but during the 1990s it was faced with a historically incomparable situation. So, in the end, it could not really contribute much to our understanding, and it experienced just a temporary boom.

For the modern phenomenon of *internationalism* this was different. Internationalism was mainly rooted in the nineteenth century and was sparked anew by the social movements of the 1970s and 1980s. What is *not* meant here are political movements like communism or socialism. Internationalism as it is understood here was marked by the realization in the nineteenth century that increasing technology, science, travel, and exchange of goods had a tendency to integrate mankind, and that there should be people or agents to organize and moderate this process in a peacekeeping and humanitarian direction (Geyer and Paulmann 2001). Institutions like the International Red Cross, the International Postal Union, news agencies like Reuters or Agence France Press, or initiatives to implement a common international language like Esperanto or Volapük were particularly persuasive. Compared to the often-ostensible actions of statesmen and politicians, their activities were indeed obscured and disregarded by historians for too long (Gorman 2012; Housden 2012; Löhr and Wenzlhuemer 2013). After the Cold War's demise it became more and more evident that the security and stability of the international order was fundamentally based on and depended upon shared international interests, and these agencies of international cooperation have exactly promoted that for many decades (Schot and Lagendijk 2008; Badenoch and Fickers 2010). Consequently, the history of internationalism, of human rights, of the international fights against slavery, against epidemics, or for the protection of the environment presently ranks among the most vivid and expanding branches of historical research. Part of it is the *history of international organizations*, which are neither engaged nor authorized by governments. Instead, they act as independent individuals or agencies beyond and across national borders, e.g. technical or scientific associations or international conferences, which took place to exchange and coordinate knowledge internationally (Iryie 2002; Herren 2009; MacKenzie 2010).

This renewed interest in internationalism was accompanied by a *history of globalization*. In dictionaries, the term 'globalization' emerged not earlier than during the 1960s to denote processes that before had been characterized as 'world politics,' 'world interior policy,' 'internationalization,' 'one world,' or equivalents.[1] The history of globalization addresses actions and endeavors to progressively entangle nations, regions, or continents that had previously lived in a more separated state. This is first of all a history of travelers, discoverers, and explorers, mainly in parts of the world that until then were foreign and unknown; it is a history of collecting things and knowledge about places and people that Europeans often assessed as 'exotic' (Pratt 1992; Friedrichsmeyer, Lennox, and Zantop 1998). On another level, the history of globalization is a story of the intensified exchange of people, goods, ideas, and information across existing or imagined borders. It is a history of economic, political, and cultural transfers, a history of creative adaptation, adoption, and interaction, and also a history of backlashes within those groups of people that have initiated all this (Stuchtey 2004; Raj 2007). The history of globalization identifies periods in which these processes of entanglement and interaction tremendously accelerated – for example, the late nineteenth century – and periods in which such an exchange stagnated or even decreased, like in the Interwar years or during Cold War years with respect to the 'Iron Curtain.'

The history of globalization raises questions concerning the causes, the ranges, and the effects of networks and interconnectedness. It is world history in a specific sense; it takes its point of departure from the matter of fact that the globe of today is interconnected, and it asks how, historically, this came into reality (Mattelard 2000). Following a definition by Jürgen Osterhammel and Niels

1 The term 'global village' was coined by Marshall McLuhan 1964.

Petersson, it is concerned with the "expansion, consolidation and speeding-up of global connections" (Osterhammel and Petersson 2003, 10).

Social protest and critique provided a major stimulus for taking up and further developing this perspective. The obvious power of globalization in particular – which often appears to be irresistible, subordinating individuals or even governments – provoked questions on who or which power actually was behind all of this. Since Karl Marx, this was often answered by pointing at internationally operating business interests or trusts, to multinational banks or enterprises, which allegedly sought financial benefits and left behind their loyalty to home countries as well as social responsibilities (Barber 1995; Winseck and Pike 2007). I will not decide here whether or not that is true, but I will point out that there is a relatively new and telling differentiation being made often in recent historiography, which is the distinction between winners and losers of globalization. To highlight effects like these only makes sense, however, if one understands globalization as a process that is manmade, and that is neither natural nor irreversible.

Returning to myself again here: In my academic career, I have approached global questions like many others did, addressing a specific aspect of global entanglement in turning towards *colonial history* (van Laak 2004). For a long time, in Germany it had almost been forgotten that from 1884 to 1914 the German Empire had 'possessed' a range of colonies in Africa and in the Southern Pacific. Colonial history is a part of most national histories in Europe, even in countries like Belgium, Denmark, or Switzerland (Purtschert, Lüthi, and Falk 2012), but first of all and primarily it plays a very powerful role in the story of globalization. Colonialism developed from the exploration of foreign territories, and it aimed at appropriating and subduing those territories and the people living in them.

Seen from the perspective of governance, colonial policy was a failure in almost every respect, and during the era of decolonization after 1945, colonial rule could be abolished almost everywhere in the world. However, during the preceding era of imperialism, which was full of conflicts, *both* the colonizing and the colonized societies were deeply altered. This can also be recognized with respect to Germany, even though its involvement in colonial policy compared to France, Great Britain, or the Netherlands did not last for long (van Laak 2005).

In 2014, and together with students at Giessen University, I embarked upon tracing the imprints that German colonialism left in the history of Hesse. Among other stories, we found out that one of the most important schools for the education of colonial farmers and settlers was opened in 1898 quite close to Giessen, in the small city of Witzenhausen, north of Kassel. The school still exists today, of course not as a colonial training center, but as part of the University of Kassel; it instructs in tropical and subtropical farming, and among its students of today are many Africans and Asians (Linne 2017). We also found out that a lot of memorials, street names, or places that refer to colonialists or other colonial aspects still exist, as do museum collections generated in the era of German colonialism.[2]

Our research, which mixed together the global and the local, conveyed exactly what often makes the history of globalization surprising: It accounts for the fact that our present world is deeply permeated by global references even to the remotest places, and these references often document a long and complex prehistory. To give an example: For many centuries we have enjoyed foods that originate in other continents, yet we are rarely aware of the twisted roads these foods have taken to please us today: coffee and tea, spices, sugar and salt, fruits and cereals, tobacco and textiles, fish and meat, oil and timber, and many other products had to travel far and long to finally become an everyday matter of course for Europeans (Wendt 2007).

If we apply historical perspectives to these flows of people, ideas, and commodity chains, often very exciting stories emerge. Take, for instance, the triangle trade between Europe, Africa, and the Americas, in which money, goods, and slaves were exchanged, or the extended routes that many plants, crops, or animals have taken to become domesticated and acclimatized in very different places. Quite generally, the hunger for resources – be it gold or diamonds, foodstuffs or rare metal – was a kind of *leitmotif* for transcending existing borders, for appropriating different objects, for subduing foreign people, for influencing them to work for the Europeans, or for simply establishing trade. These actions are salient subjects of *transnational history*, which furthermore has been driven by interventions into the natural habitat, by the building up of facilities for travel and communication, by the transfer of goods and ideas, or by environmental disasters as generated by industrialization or global touristification.[3]

Many of these global and transnational aspects formerly were odd subjects for historical research because they were focused on national units, and because actors and agents therein were neither governmental nor confined within given borders. For a couple of years, historians have discussed the methodology of a 'histoire croisée' or *entangled history* (Werner and Zimmermann 2002). It tries to explain processes of interaction among actors of different localities and cultures, by which all involved actors are challenged. Often, something happens that has been called 'hybridization,' something that none of the involved actors actually has intended. It absorbs and transforms elements from all involved cultures, but

---

2 See <www.inst.uni-giessen.de/hessen-postkolonial> [accessed: 5 June 2019].

3 A great example of 'new' stories to be told from this perspective are Iriye and Saunier 2009 and Espagne, Geyer, and Middell 2010.

at the same time represents something that has a quality of being new and sovereign (Fischer-Tiné 2013).

This said, it is crucial to characterize *global history* as separate from the history of globalization. This approach can be understood as a more general invitation to add historical dimensions and narrations to the existing *one world* of today.[4] This offer, however, involves a lot of questions and problems. For instance: Do we have to imagine global history as a kind of history that converges towards an integrated and homogeneous world? And what will this unified world eventually look like?

Asking questions like these moves global history very close to the *universal history* approaches of the eighteenth and early nineteenth centuries, when world history was conceptualized and written with philosophical aspiration. Writers like August Ludwig von Schlözer, Friedrich Schiller, or Georg Friedrich Wilhelm Hegel assessed history as a gradual fulfillment of human rights, freedom, peace, democracy, and equality. Taking a similar attitude, some recent macro-sociologists have suggested space and time being categories of the past, not of the future, wherein everything would merge and become networked into a global system, which would eventually be defined by coexistence, connections, and synchrony (Castells 1996; Khannah 2016).

However, there is more evidence that global history does *not* evolve toward a synthetic or certain goal, but should instead be viewed as something that neither develops in a straight line nor with purpose. It displays many rupture zones, as nation states and empires rise and fall, territories seem to return to oblivion like in some parts of Africa, former centers again become provincialized, etc. Global history is especially considerate of activities that constantly change without arriving anywhere, which is why the transfer and migration of people, ideas, goods, and things are saliently represented in it to such an extent (Middell and Engel 2005).

Seen from this perspective, the problem arises as to whether 'natural borders' in fact do exist, or whether this concept is a fixed idea of a certain time or specific interests (Butlin 2009). The political geography of the late nineteenth century had stressed this notion of 'natural borders' to legitimize the expansion of nation states and the manipulation of existing borderlines, but in fact, borders can be justified almost exclusively by arguing historically. Globalization and the definition of national confines were reciprocal forces that intensified each other. The emphasis with which coherent nation states and integrated territories were

4 See the six volumes of *A History of the World*, edited by Akira Iriye and Jürgen Osterhammel (2012–2018).

conjured up since the early nineteenth century can only be interpreted by taking into account existing definitions becoming liquid, people increasing in mobility, and flows of commodity diffusing more and more (Conrad 2010). Consequently, global historians like to talk about processes of re- or deterritorialization, about imagined communities, mental maps, and other categories that paraphrase spatial concepts as hypothetical constructs (Anderson 1983; Schenk 2002).

In this view, the building of nation states was an act of defense: One that could create unities and establish transitional political control before transnational processes soaked these voluntary definitions again and provoked further modification. Global history mirrors the complex interplay between national sovereignty, on the one hand, and the ambition to cooperate internationally, on the other. One of the main advantages of global history is that in applying to regions and nations a view from above, almost nothing appears to be 'natural' or self-evident. Rather, it were compromises, migrations, and mixtures that constituted global history (as a good example see Fischer-Tiné 2007).

Whereas international history in its traditional style explored obvious questions of power, global history finds them concealed in factual issues like technical questions, the setting of international standards, the competition among political, economic, technological, or cultural systems, and their scaling and synchronization, e.g., in air traffic or financial transactions (Murphy and Yates 2008; Wenzlhuemer 2010). Global historians are less inclined to research large political conferences at which the fate of peoples or nations were decided. Rather, they frequently turn towards congresses of scientists or experts at which decisions were made about topics such as the implementation of the metrical system, the technological adjustment of radio or telegraphy, the sustainable management of timber, or the coordinated exploitation of the Arctic regions.

Likewise, the League of Nations and the United Nations have become subjects of an intensified historical research (Mazower 2009; Yearwood 2009; Housden 2012). And indeed, in analyzing what they and their sub-organizations have achieved – often in the background – their records are much more impressive than if one only takes into account official politics and diplomacy where lofty expectations often were not fulfilled (see www.lonsea.de; Löhr 2010; Pedersen 2015).

There is also a rising historical interest in the phenomenon of large empires. Seen from a global perspective, empires have been longer-lasting and in many respects even more successful than nation states. Their existence is questioned not only for the centuries of hegemony and dominance that their military power allowed, but also as a successful model of integrating migrants and tolerating minorities (Darwin 2010; Leonhard and von Hirschhausen 2012). In some aspects, empires, with their rather flexible frontiers and their social stratifications, appear even more modern than nation states do.

In the 1990s, US political scientist Samuel Huntington predicted that the world would usher in a 'clash of cultures.' This sparked attention in its assertion that the global networks of traffic, communication, goods, people, and ideas do not automatically create a unified world, but also provoke defense actions that legitimize themselves culturally or religiously (Huntington 1996). There seemed to be a dialectic process of cross-border opening on the one hand, and on the other, a need for man to define distinctive spaces of law, culture, religion and morality. This dialectic is still valid and is one of the major problems of our time.

Today, more than ever before, global history has to deal with a problem that traditional world history was also faced with from its beginning: Who is allowed to research and write history? Who is capable of adequately surveying different world regions, provinces, and cultures? Such is a difficult task for a single person. Consequently, it has often been scholars working outside of the guild of historians – like Oswald Spengler, Arnold J. Toynbee, or William H. McNeill – who dared to do so. Nonetheless, there are some encouraging examples of global syntheses with respect to the nineteenth century coming from Christopher Bayly (2005) or Jürgen Osterhammel (2009). Other global histories resort to assembling a number of specialized authors to portray certain aspects of world history, and this also works well.

In earlier times, writing a coherent history of one nation seemed the most prestigious job that a historian could seek, while writing world history was regarded a field of mavericks, and writing local history was seen as an enterprise mostly for amateurs. Today, it seems as if historians should be ready to adapt to *all* regional and spatial levels. A 'natural' hierarchy of competence and of allegedly more and less important fields of historical research is about to dissolve (Geyer and Bright 1995; Manning 2003). In their everyday work, of course, historians have to focus and specialize on certain aspects; usually they are well educated to do so. But as everything is connected to everything in the end, a historian should be able to keep her or his eye on all these levels.

Today, more than ever, historians must stress the tentativeness of their findings and assertions. It was a distinguishing mark of former research that historians tended to assess their own national histories to be the most important in the world, to claim their own positions as being the most advanced, and also to rank their own virtues as the most significant (Wimmer and Glick Schiller 2003). The perspective on world history that the writer presented was centralized and aligned to the writer's position in society. Historians generally spoke of the world that was familiar to them at a given moment, and they tended to divide that world between leading civilizations on the one hand, and barbarians and antediluvians on the other. Following this path, meaningful histories of salvation were written, which tended to legitimize one's own position while devaluing counter-positions (see Osterhammel 2002).

Today, regional histories or *area studies* are researched and written beyond such presumptions – or at least they should be (Schäbler 2007). One of the most salient and exciting aspects of global history is that it shows quite plainly how restricted our historical knowledge actually is, especially with respect to non-European affairs. The notion of Europeans to assess certain cultures of Latin America, Africa, or Asia as 'people without history' has been unmasked as a colonial attitude (Wolf 1982). Even more important is that it's not true. The more we know about other regions, the more we realize that Europe is just one of the world's provinces among others, the more Europe is 'provincialized' (Chakrabarty 2000).

To refer to two striking examples: what would have happened if, in 1485, Sultan Bayezid II had not prohibited the printing of books, a ban that remained in effect in the Ottoman Empire until the nineteenth century? How would world history have developed if, in the same century, the Chinese fleet had turned away from the African coast because Emperor Zeng-He's successors were too afraid of adventure expeditions? Chances and accidents are major forces of history, as are often-disregarded aspects like the weather or changes of climate. This has changed fundamentally (Grove 1997; Crosby 1986; Diamond 1997).

In recent book productions, there has been a revealing boom of histories telling exemplary stories of certain objects. Sometimes this is about crossroads, or supposed 'magic moments' of history (see Demandt 2004). Sometimes this is the story of things shaking the world like resources, inventions, or ideas, and their global impact, like the histories of spices, or tulip bulbs, or certain weapons like the notorious Kalaschnikow gun (see, for instance, Dash 1999). Alternatively, it is a story, sometimes referred to as *glocalization*, that describes or encompasses global changes as they are reflected in local settings (Epple 2007, 2013): For example, the protection of certain animals, like elephants, had far-reaching effects for the ivory industry in the Odenwald, a region located southwest of Frankfurt. This is another dialectic process, with the global and the local influencing each other. The ways in which global changes affect locals can most frequently be traced in the spheres of economy and industry (Giese, Mossig, and Schröder 2011). In this respect, telling the story of cocoa or cotton in a 'glocalized' way can be extremely illuminating (Beckert 2014).

There is yet another revealing approach that exemplifies global developments. *Biographies* enable historians to describe a multitude of activities and experiences, especially if the subjects in question traveled a lot and collected impressions about many places. A lot of them have lived 'transnational lives,' and the more closely we approach the present day, this kind of life, instead of being an exception, becomes a common phenomenon (Woollacott, Deacon, and

Russell 2010). This concept definitely applies to histories of human migration, irrespective of the motives that are responsible for the 'restlessness' of people moving (Bade 2003, Harzig and Hoerder 2009). There are many good reasons to stress 'nomadic' aspects of human life and to place them on an equal footing with 'settled' variations of human existence.[5] Thus, it is no coincidence that the very person who wrote one of the first comprehensive records of European migration, Eugene M. Kulischer, also coined the phrase of 'displaced persons' (1948; see also Schlögel 2005).

History and cultural sciences should contribute to a better or more appropriate analysis of, and also a justification for, people being on the move and migrating. Today, everything is expected to circulate: goods, ideas, energies – why should *people* of all things stay home and nourish their 'fixed identities'? It is certain that nothing as a 'fixed identity' exists.[6] In contrast to the domesticating tendencies of modern nationalism, people more and more were 'on the road' and understood this as their prime mode of existence.[7] Therefore, a 'heuristic of flows and circulation' should prevail over a heuristic of what is statically determinable. Today, *relations* are more relevant than causalities and comparisons, *movements* more relevant than conditions, *transitions* and *mixtures* more relevant than differences (Dommann 2016). I would add that this also applies to processes of synchronization, to perceived asynchronicities of simultaneous processes, and to (attempted) blockades of circulation as caused, for example, by terrorists. From current perspectives, this appears even more relevant than questions of war and peace.

2 Some Conclusions and Outlooks

To sum up some of the evidence here: Whereas in the 1980s, national histories were extended to many more agents and actors, since the 1990s, world history has been globalized, and also extended to allow even more actors to be involved. This can emphatically be assessed as a process of *democratizing* history. At the same time, it has also become much more difficult to identify distinctive tendencies of historical development, because the more factors one must regard, the more elaborate one's conclusion must be. The times are gone when the course of world history could be attributed to a chosen few actors, such as statesmen or

5 This aspect is methodologically discussed by Monika Dommann 2016.
6 For an archaeology of the concept, see Niethammer 2000; Assmann and Friese 1998.
7 The fatal effects of nationalism are portrayed by Philipp Ther 2014, 2017.

inventors. And the times have also passed when the history of a globalized world could be viewed from a 'Eurocentric' perspective, unless one were to consciously and intentionally choose that as a central theme (Diogo and van Laak 2016).

Global history, however, carries the risk of almost everything it approaches appearing convertible and constructed. To counterbalance this notion, historical research should collaborate with other sciences to describe the more stable aspects of life, like the natural environment or anthropological constants, from the more diversified aspects – much as the French historian Fernand Braudel did in the 1940s in his classic study on the Mediterranean Sea (Braudel 1949). Research results from biology, meteorology, or medicine can provide fresh perspectives and insights for our historical understanding. This has already been exemplified by another version of recent historiography that is called *Big History*. It discusses relationships between natural and human history, and puts man-made history into a much longer developmental perspective (Spier 1996; Christian 2005). In doing so, certain findings are given a dramatic twist, such as the insight into how deeply man has interfered with the natural environment during the last two or three hundred years, creating a manipulated environment called the 'Anthropocene' (Möllers, Schwägerl, and Trischler 2014).

Historians have once been labeled as 'prophets of the past' (see Brenner 2010). But it seems more appropriate to view them as experts of collected and evaluated experiences. We cannot contribute to the future, but we definitely can help others to come to terms with the past. Given the explosion of complexity in a world being networked on a global scale, this certainly is no trivial task. We realize broadening gaps between those people obviously profiting from 'globalization' and people who reckon they just pay the bills. Cultural scientists and historians should do more on explaining the term 'globalization' and its complexities, and they also should discuss and historically better explain 'diversity,' 'identity,' 'exile,' 'diaspora,' or 'home' – categories that desperately deserve to become historicized.

The recent past has experienced an incredible extension of phenomena being subjected to cultural historians' scrutiny: past emotions, space, time, animals, our built environment, pictures, sleep, the senses, and perceptions of all kind – for my part, I am quite determined that the next big thing will be a history of plants. This extension of phenomena has occurred because many traditions of restricting topics to certain disciplines have obviously eroded – and this is another liquidity being welcomed. What is needed, now, is a cultural history that is open and sensitive to different rhythms and paces of change – without preferring perceived 'avant-gardes' or depreciating purported 'backwardnesses.'

Using examples from my own generation I have aspired to show how the horizons of recent historical research have expanded with respect to contents, to space, and to methodology. Spaces have not been liquidated, but rather have been liquefied. The great privilege of historians remains that they can devote themselves to virtually *every* question, provided that it somehow addresses the *past*. Concerning the *future*, I am quite convinced that cultural and historical sciences that further promote globalization are about to enter into astonishing new spheres of knowledge.

References

Hessen (post)kolonial <www.inst.uni-giessen.de/hessen-postkolonial> [accessed: 5 June 2019].
www.lonsea.de [accessed: 5 June 2019].
Anderson, Benedict. *Imagined Communities: Reflections on the Origins and Spread of Nationalism*. London: Verso, 1983.
Assmann, Aleida, and Heidrun Friese, eds. *Identitäten*. Frankfurt a.M.: Suhrkamp, 1998.
Augstein, Rudolf, ed. *Historikerstreit. Die Dokumentation der Kontroverse um die Einzigartigkeit der nationalsozialistischen Judenvernichtung*. Munich: Piper, 1987.
Bade, Klaus J. *Migration in European History*. Oxford: Blackwell, 2003.
Badenoch, Alec, and Andreas Fickers, eds. *Materializing Europe: Transnational Infrastructures and the Project of Europe*. Basingstoke/New York: Palgrave Macmillan, 2010.
Barber, Benjamin. *Jihad vs. McWorld: How Globalism and Tribalism are Reshaping the World*. New York: Ballantine Books, 1995.
Bayly, Christopher. *The Birth of the Modern World, 1780–1914: Global Connections and Comparisons*. Oxford: Blackwell, 2005.
Beckert, Sven. *King Cotton. Eine Globalgeschichte des Kapitalismus*. Munich: Beck, 2014.
Berger, Stefan, ed. *Writing the Nation: A Global Perspective*. Basingstoke/New York: Palgrave Macmillan, 2007.
Braudel, Fernand. *La mediterranée et le monde méditerranéen à l'époque de Philippe II*. Paris: Colin, 1949.
Brenner, Michael. *Prophets of the Past: Interpreters of Jewish History*. Princeton: Princeton University Press, 2010.
Butlin, Robin A. *Geographies of Empire: European Empires and Colonies c. 1880–1960*. Cambridge: Cambridge University Press, 2009.
Carvalho, Susana, and François Gemenne, eds. *Nations and their Histories: Constructions and Representations*. Basingstoke/New York: Palgrave Macmillan, 2009.
Castells, Manuel. *The Rise of the Network Society*. Malden, MA: Blackwell, 1996.
Chakrabarty, Dipesh. *Provincializing Europe: Postcolonial Thought and Historical Difference*. Princeton: Princeton University Press, 2000.
Christian, David. *Maps of Time: An Introduction to Big History*. Berkeley: University of California Press, 2005.
Conrad, Sebastian. *Globalisation and the Nation in Imperial Germany*. Cambridge/New York: Cambridge University Press, 2010.
Conrad, Christoph, and Sebastian Conrad, eds. *Die Nation schreiben. Geschichtswissenschaft im internationalen Vergleich*. Göttingen: Vandenhoeck & Ruprecht, 2002.

Crosby, Alfred W. *Ecological Imperialism: The Biological Expansion of Europe, 900–1900*. New York/Cambridge: Cambridge University Press, 1986.
Darwin, John. *Der imperiale Traum. Die Globalgeschichte großer Reiche 1400–2000*. Frankfurt a.M./New York: Campus, 2010.
Dash, Mike. *Tulpenwahn. Die verrückteste Spekulation der Geschichte*. Munich: Claassen, 1999.
Demandt, Alexander. *Sternstunden der Geschichte*. Munich: Beck, 2004.
Diamond, Jared. *Guns, Germs, and Steel: The Fates of Human Societies*. New York: Norton, 1997.
Diogo, Maria Paula, and Dirk van Laak. *Europeans Globalizing: Mapping, Exploiting, Exchanging*. Basingstoke/New York: Palgrave Macmillan, 2016.
Dommann, Monika. "Alles fließt: Soll die Geschichte nomadischer werden?" *Geschichte und Gesellschaft* 42 (2016): 516–534.
Duchhardt, Heinz, ed. *Nationale Geschichtskulturen – Bilanz, Ausstrahlung, Europabezogenheit*. Stuttgart: Steiner, 2006.
Epple, Angelika. "'Global History' und 'Area History': Plädoyer für eine weltgeschichtliche Perspektivierung des Lokalen." *Area Studies und die Welt. Weltregionen und neue Globalisierung*. Ed. Birgit Schäbler. Vienna: Mandelbaum, 2007. 90–116.
Epple, Angelika. "Lokalität und die Dimensionen des Globalen. Eine Frage der Relationen." *Historische Anthropologie* 21.1 (2013): 4–25.
Espagne, Michel, Michael Geyer, and Matthias Middell. *European History in an Interconnected World: An Introduction to Transnational History*. Basingstoke/New York: Palgrave Macmillan, 2010.
Fischer-Tiné, Harald. "Global Civil Society and the Forces of Empire: The Salvation Army, British Imperialism and the 'Pre-History' of NGOs (ca. 1880–1920)." *Competing Visions of World Order: Global Moments and Movements, 1880s–1930s*. Eds. Sebastian Conrad and Dominic Sachsenmaier. Basingstoke/New York: Palgrave Macmillan, 2007. 29–67.
Fischer-Tiné, Harald. *Pidgin-Knowledge. Wissen und Kolonialismus*. Zurich/Berlin: Diaphanes, 2013.
Friedrichsmeyer, Sara, Sara Lennox, and Susanne Zantop, eds. *The Imperialist Imagination: German Colonialism and Its Legacy*. Ann Arbor: University of Michigan Press, 1998.
Gellner, Ernest. *Nations and Nationalism*. Oxford: Blackwell, 1983.
Geyer, Michael, and Charles Bright. "World History in a Global Age." *American Historical Review* 100.4 (1995): 1034–1060.
Geyer, Martin H., and Johannes Paulmann, eds. *The Mechanics of Internationalism: Culture, Society and Politics from the 1840s to the First World War*. Oxford: Oxford University Press, 2001.
Giese, Ernst, Ivo Mossig, and Heike Schröder. *Globalisierung der Wirtschaft. Eine wirtschaftsgeographische Einführung*. Stuttgart: Schöningh, 2011.
Gorman, Daniel. *The Emergence of International Society in the 1920s*. Cambridge: Cambridge University Press, 2012.
Grove, Richard H. *Ecology, Climate and Empire: Colonialism and Global Environment History, 1400–1940*. Cambridge: White Horse Press, 1997.
Harzig, Christiane, and Dirk Hoerder. *What is Migration History?* Cambridge: Polity, 2009.
Herren, Madeleine. *Internationale Organisationen seit 1865. Eine Globalgeschichte der internationalen Ordnung*. Darmstadt: Wissenschaftliche Buchgesellschaft, 2009.
Hobsbawm, Eric. *Nations and Nationalism since 1780: Programme, Myth, Reality*. Cambridge: Cambridge University Press, 1990.

Hobsbawm, Eric. *The Age of Extremes: The Short Twentieth Century 1914–1991*. London: Abacus, 1995.
Hobsbawm, Eric, and Terence Ranger. *The Invention of Tradition*. Cambridge: Cambridge University Press, 1992.
Housden, Martyn. *The League of Nations and the Organization of Peace*. London: Pearson Education, 2012.
Huntington, Samuel P. *The Clash of Civilizations*. New York: Simon & Schuster, 1996.
Iriye, Akira. *Global Community: The Role of International Organizations in the Making of the Contemporary World*. Berkeley: University of California Press, 2002.
Iriye, Akira, and Pierre-Yves Saunier, eds. *The Palgrave Dictionary of Transnational History*. Basingstoke/New York: Palgrave Macmillan, 2009.
Iriye, Akria, and Jürgen Osterhammel, eds. *A History of the World*. 6 vols. Munich: Beck/ Cambridge, MA: Harvard University Press, 2012–2018.
Kaelble, Hartmut. *Der historische Vergleich. Eine Einführung zum 19. und 20. Jahrhundert*. Frankfurt a.M./New York: Campus, 1999.
Khannah, Parag. *Connectography: Mapping the Future of Global Civilization*. New York: Random House, 2016.
Kleßmann, Christoph, Hans Misselwitz, and Günter Wichert, eds. *Deutsche Vergangenheiten – eine gemeinsame Herausforderung. Der schwierige Umgang mit der doppelten Nachkriegsgeschichte*. Berlin: Links, 1999.
Kulischer, Eugene M. *Europe on the Move: War and Population Changes, 1917–1947*. New York: Columbia University Press, 1948.
van Laak, Dirk. *Gespräche in der Sicherheit des Schweigens. Carl Schmitt in der politischen Geistesgeschichte der frühen Bundesrepublik*. Berlin: Akademie-Verlag, 1993.
van Laak, Dirk. "Alltagsgeschichte." *Aufriss der Historischen Wissenschaften*. Ed. Michael Maurer. Vol. VII: *Neue Themen und Methoden der Geschichtswissenschaft*. Stuttgart: Reclam, 2003. 14–80.
van Laak, Dirk. *Imperiale Infrastruktur. Deutsche Planungen für eine Erschließung Afrikas 1880 bis 1960*. Paderborn et al.: Schöningh, 2004.
van Laak, Dirk. *Über alles in der Welt. Deutscher Imperialismus im 19. und 20. Jahrhundert*. Munich: Beck, 2005.
Leffler, Melvyn P., and Odd Arne Westad, eds. *The Cambridge History of the Cold War*. 3 vols. Cambridge: Cambridge University Press, 2010.
Leonhard, Jörn, and Ulrike von Hirschhausen, eds. *Comparing Empires: Encounters and Transfers in the Long Nineteenth Century*. Göttingen: Vandenhoeck & Ruprecht, 2012.
Linne, Karsten. *Von Witzenhausen in die Welt. Ausbildung und Arbeit von Tropenlandwirten 1898 bis 1971*. Göttingen: Wallstein, 2017.
Löhr, Isabella. *Die Globalisierung geistiger Eigentumsrechte. Neue Strukturen internationaler Zusammenarbeit 1886–1952*. Göttingen: Vandenhoeck & Ruprecht, 2010.
Löhr, Isabella, and Roland Wenzlhuemer, eds. *The Nation State and Beyond: Governing Globalization Processes in the Nineteenth and Early Twentieth Centuries*. Berlin: Springer, 2013.
Loth, Wilfried. *Europas Einigung. Eine unvollendete Geschichte*. Frankfurt a.M.: Campus, 2014.
Loth, Wilfried, and Jürgen Osterhammel, eds. *Internationale Geschichte. Themen – Ergebnisse – Aussichten*. Munich: Oldenbourg, 2000.
MacKenzie, David. *A World Beyond Borders: An Introduction to the History of International Organizations*. Toronto: University of Toronto Press, 2010.

Manela, Erez. *The Wilsonian Moment: Self-Determination and the International Origins of Anticolonial Nationalism*. Oxford/New York: Oxford University Press, 2007.
Manning, Patrick. *Navigating World History: Historians Create a Global Past*. Basingstoke/New York: Palgrave Macmillan, 2003.
Mattelard, Armand. *Networking the World 1794–2000*. Minneapolis/London: University of Minnesota Press, 2000.
Mazower, Mark. *No Enchanted Palace: The End of Empire and the Ideological Origins of the United Nations*. Princeton: Princeton University Press, 2009.
McLuhan, Marshall. *Understanding Media: The Extensions of Man*. New York: Signet Books, 1964.
Middell, Matthias, and Ulf Engel, eds. *Comparativ* 15.5/6 (2005) (special issue *Bruchzonen der Globalisierung*).
Möllers, Nina, Christian Schwägerl, and Hellmuth Trischler, eds. *Welcome to the Anthropocene: The Earth in Our Hands*. Munich: Deutsches Museum, 2014.
Murphy, Craig N., and JoAnne Yates. *The International Organization for Standardization: Global Governance through Voluntary Consensus*. London: Routledge, 2008.
Niethammer, Lutz, ed. *Der 'gesäuberte' Antifaschismus. Die SED und die roten Kapos von Buchenwald*. Berlin: Akademie Verlag, 1995.
Niethammer, Lutz. *Kollektive Identität. Heimliche Quellen einer unheimlichen Konjunktur*. Reinbek: Rowohlt, 2000.
Osterhammel, Jürgen. "Weltgeschichte." *Lexikon Geschichtswissenschaft. Hundert Grundbegriffe*. Ed. Stefan Jordan. Stuttgart: Reclam, 2002. 320–324.
Osterhammel, Jürgen. *Die Verwandlung der Welt. Eine Geschichte des 19. Jahrhunderts*. Munich: Beck, 2009 (Engl. transl. Princeton: Princeton University Press, 2014).
Osterhammel, Jürgen, and Niels P. Petersson. *Geschichte der Globalisierung. Dimensionen – Prozesse – Epochen*. Munich: Beck, 2003.
Pedersen, Susan. *The Guardians: The League of Nations and the Crisis of Empire*. Oxford: Oxford University Press, 2015.
Prashad, Vijay. *The Darker Nations: A People's History of the Third World*. New York: New Press, 2008.
Pratt, Mary Louise. *Imperial Eyes: Travel Writing and Transculturation*. London/New York: Routledge, 1992.
Purtschert, Patricia, Barbara Lüthi, and Francesca Falk, eds. *Postkoloniale Schweiz. Formen und Folgen eines Kolonialismus ohne Kolonien*. 2nd ed. Bielefeld: Transcript, 2012.
Raj, Kapil. *Relocating Modern Science: Circulation and the Construction of Knowledge in South Asia and Europe, 1650–1900*. Basingstoke/New York: Palgrave Macmillan, 2007.
Said, Edward W. *Orientalism*. New York: Vintage, 1978.
Said, Edward W. *Culture and Imperialism*. London: Vintage, 1993.
Schäbler, Birgit, ed. *Area Studies und die Welt. Weltregionen und die neue Globalgeschichte*. Vienna: Mandelbaum, 2007.
Schenk, Frithjof Benjamin. "Mental Maps. Die Konstruktion von geographischen Räumen in Europa seit der Aufklärung." *Geschichte und Gesellschaft* 28 (2002): 493–514.
Schlögel, Karl. "Verschiebebahnhof Europa. Joseph B. Schechtmans und Eugene M. Kulischers Pionierarbeiten." *Zeithistorische Forschungen/Studies in Contemporary History* 2.3 (2005): 468–472.

Schot, Johan, and Vincent Lagendijk. “Technocratic Internationalism in the Interwar Years: Building Europe on Motorways and Electricity Networks.” *Journal of Modern European History* 6.2 (2008): 196–216.
Sluga, Glenda. *Internationalism in the Age of Nationalism*. Philadelphia: University of Pennsylvania Press, 2013.
Smith, Anthony D. *Nationalism: Theory, Ideology, History*. Cambridge: Polity, 2001.
Spier, Fred. *The Structure of Big History from the Big Bang until Today*. Amsterdam: Amsterdam University Press. 1996.
Stuchtey, Benedikt, ed. *Science Across the European Empires, 1800–1950*. Oxford: Oxford University Press, 2004.
Ther, Philipp. *The Dark Side of Nation-States: Ethnic Cleansing in Modern Europe*. New York/Oxford: Berghahn, 2014.
Ther, Philipp. *Die Außenseiter. Flucht, Flüchtlinge und Integration im modernen Europa*. Berlin: Suhrkamp, 2017.
Wehler, Hans-Ulrich. *Nationalismus. Geschichte, Formen, Folgen*. Munich: Beck, 2011.
Wendt, Reinhard. *Vom Kolonialismus zur Globalisierung. Europa und die Welt seit 1500*. Paderborn: Schöningh, 2007.
Wengst, Udo, and Hermann Wentker, eds. *Das doppelte Deutschland. 40 Jahre Systemkonkurrenz*. Bonn: Links, 2008.
Wenzlhuemer, Roland. “The History of Standardisation in Europe.” *European History Online* (EGO), published by the Institute of European History (IEG), Mainz 2010. http://www.ieg-ego.eu/wenzlhuemerr-2010-en [accessed: 5 June 2019].
Werner, Michael, and Bénédicte Zimmermann. “Der Ansatz der Histoire croisée und die Herausforderung des Transnationalen.“ *Geschichte und Gesellschaft* 28 (2002): 607–636.
Westad, Odd Arne. *The Global Cold War: Third World Interventions and the Making of Our Times*. Cambridge: Cambridge University Press, 2007.
Wimmer, Andreas, and Nina Glick Schiller. “Methodological Nationalism, the Social Sciences, and the Study of Migration: An Essay in Historical Epistemology.” *The International Migration Review* 37.2 (2003): 576–610.
Winseck, Dwayne R., and Robert M. Pike, eds. *Communication and Empire: Media, Markets, and Globalization, 1860–1930*. Durham, NC/London: Duke University Press, 2007.
Wolf, Eric. *Europe and the People without History*. Los Angeles: University of California Press, 1982.
Woollacott, Angela, Desley Deacon, and Penny Russell, eds. *Transnational Lives: Biographies of Global Modernity, 1700–Present*. Basingstoke/New York: Palgrave Macmillan, 2010.
Yearwood, Peter J. *Guarantee of Peace: The League of Nations in British Policy 1914–1925*. Oxford: Oxford University Press, 2009.

IV Future Connectivities: Economy, Natural Sciences, Ecology

Tom Clucas

Culture in the Marketplace

1 The Hypostasis of the Market

In one of his final publications, Stuart Hall noted that "Marketing and selling metaphors now threaten to swamp public discourse. The market is hypostacized: it 'thinks' this, 'does' that, 'feels' the other, 'gets panicky', 'loses confidence', 'believes'" (Hall 2011, 722). In the light of this hypostatization of the marketplace, it is important to consider possible futures of the study of culture in the age of the mass market and mass consumerism. This is a formidable task, since market thinking now permeates almost every aspect of society. Market forces help to determine not only which cultural artefacts get produced and consumed, but also the conditions under which people produce and consume them and the technologies available for their dissemination. The market also helps to determine how the study of culture is funded and conducted, in and beyond the universities, as well as which studies are published and how they are received. Given such complex entanglements, this short essay can only provide a few programmatic observations, combined with personal reflections drawn from the experience of that highly marketized being, the 'early career academic.' In the process, it offers some thoughts about how those working within the study of culture might deal with the fact that not only their objects of study, but also their critical languages, and their own existences as students and teachers, are increasingly determined by the market logic that Hall describes. It also suggests that the 'consumer turn' might eventually become the turn that subsumes all others in the study of twenty-first-century cultures.

Since Marx's original formulation of the dynamic between base and superstructure, generations of writers have considered how relations of production affect cultural, and specifically artistic, activity. In one of the classic texts of Marxist criticism, Terry Eagleton argued that:

> Each element of a society's superstructure – art, law, politics, religion – has its own tempo of development, its own internal evolution, which is not reducible to a mere expression of the class struggle or state of the economy. Art, as Trotsky comments, has 'a very high degree of autonomy'. (Eagleton 2002, 13)

However, this 'degree of autonomy' is precisely what Hall and others saw being progressively eroded by the continual hypostatization of the market. In his seminal analysis of postmodernism, Fredric Jameson asked "whether it is not precisely this semiautonomy of the cultural sphere which has been destroyed

https://doi.org/10.1515/9783110669398-014

by the logic of late capitalism" (Jameson 1991, 48). Since the appearance of this text, the 2008 financial crisis has given the market and its metaphors even more prominence in public life. As early as 1959, Reinhart Koselleck analyzed how the state of crisis came to be "practiced as the referent from which history is both apprehended and comprehended" in modern societies (Koselleck 1988, 69). Since 2008, however, this "ubiquitous, apparently 'self-evident', characterization of contemporary life in terms of crisis" (Crosthwaite, Knight, and Marsh 2015, 129) has been anchored ever more securely in the language and logic of modern financial markets. Although relations of production have always influenced cultural activity, the new trend is for marketing and selling metaphors to be applied self-consciously in every facet of life. Johannes Angermuller has shown that this is no less true in universities, where academic discourse has become a "multileveled positioning practice" in which the "driving force[s] for researchers are power structures in academic organisations and markets" (Angermuller 2013, 265, and 2017). One challenge for those in the study of culture is how to remain objective when the critical metalanguages that govern not only their research, but also its reception and management are influenced by the market. First, however, it is necessary to consider how the concept of culture itself has been affected by the rise of consumerism.

2 The Widening Remit of the Study of Culture

One of the most notable trends within the study of culture in recent years has been its widening remit. Doris Bachmann-Medick has noted that the emergence of "multifaceted reorientations in the study of culture is by no means attributable only to a postmodern fragmentation. They also have a clear material-economic and social foundation" (Bachmann-Medick 2016, 8). On the one hand, this pluralism of approaches may be interpreted as the result of a marketized "competition between theories" (Bachmann-Medick 2016, 8). From this perspective, the academic marketplace described above drives a perpetual but possibly illusory quest for innovation, as scholars are encouraged to anticipate 'trends' in the market.[1] On the other hand, Bachmann-Medick shows that this new pluralism within the

1 I say "possibly illusory" because what may at first seem like a free choice between topics and approaches is actually governed by the logic of the academic marketplace: If (untenured) academics do not follow the current market trend, then their work is proportionally less likely to be published. To this extent, academic research may be seen to replicate the principle known as 'reputational herding' in financial markets; see Roider and Voskort 2016.

study of culture has the beneficial effect of breaking down what Bourdieu called the "profound conformisms" of the intellectual world (Bourdieu 2007, 106). This corresponds to Jameson's prediction that:

> [T]he dissolution of an autonomous sphere of culture is rather to be imagined in terms of an explosion: a prodigious expansion of culture throughout the social realm, to the point at which everything in our social life – from economic value and state power to practices and to the very structure of the psyche itself – can be said to have become "cultural" in some original and yet untheorized sense. (Jameson 1991, 48)

The breaking down of academic disciplines and cultural canons has been welcomed as liberation on many grounds. One must be mindful, however, that there is always a principle of selection at work when audiences and academics choose where to focus their attention. Arguably, the new logic of selection at work in what Thomas H. Davenport famously termed the "attention economy" is the logic of the marketplace (Davenport 2001). In the process, not only the field known as the study of culture but also the concept of culture itself has been profoundly transformed.

As Jameson predicted, the concept of culture has expanded in recent years to encompass every aspect of social life. Raymond Williams famously claimed that culture became available during the nineteenth century "as the court of appeal in which real values were determined, usually in opposition to the 'factitious' values thrown up by the market and similar operations of society" (Williams 1960, 37). However, scholars like Williams and E. P. Thompson inevitably recognized the influence of the market, as the cultures they studied became increasingly commercialized throughout the twentieth century (Williams 1960, 319–320). Accordingly, Stuart Hall describes how the concept of 'high culture' gave way to that of 'mass culture' or 'popular culture' and then to an 'anthropological' definition, which emphasizes "participants interpreting meaningfully what is happening around them, and 'making sense' of the world, in broadly similar ways" (Hall 1997, 2). In recent years, this anthropological definition has once again been challenged by the development of 'corporate culture' and 'organizational culture.' These terms were popularized by Terrence Deal and Allan Kennedy in their 1982 book *Corporate Cultures*, which defined organizational culture matter-of-factly as "the way we do things around here" (Deal and Kennedy 2000, 4). In business schools, the sociological and anthropological tools of cultural studies departments have been successfully adapted for the purposes of "maximizing the value of human capital" and ensuring "organizational success" (Baker 2002). Although these definitions have not yet migrated from management to cultural studies departments, it is clear that the concept of culture has undergone a major transformation, if not a reversal, in the past fifty years. In the era of bestsellers, blockbusters, and downloads charts, it is no longer possible to define culture in opposition to the marketplace.

A corollary of this redefinition of culture is that the types of cultural practices and artefacts being studied has also altered dramatically. For example, a recent Facebook post by the Bodleian Libraries in Oxford noted that "Bodleian readers have access right now to over 165 different dissertations that name Harry Potter in their titles, and over 4,000 more that reference the Potter books or films as part of their arguments" (Bodleian Libraries 2017). Naturally, there are reasons to study a global phenomenon like Harry Potter, but this concentration of effort may prove detrimental if researchers (particularly in the early stages of their careers) feel compelled to follow the market. In contrast, while the pioneers of cultural studies never excluded mass-market and commercial products from their studies, they were often critical of their value. Richard Hoggart's ground-breaking study of *The Uses of Literacy* concluded that "[m]ost mass-entertainments are in the end what D. H. Lawrence described as 'anti-life'. They are full of a corrupt brightness, of improper appeals and moral evasions" (Hoggart 1992, 340). Nowadays, this statement might seem dated for two reasons: The first is its overt and perhaps antiquated value judgments ('corrupt,' 'improper,' 'moral evasions') and the second is its dismissive attitude towards the more commercial forms of popular culture.

The first objection is symptomatic of what Helen Small terms a care to "eschew the language of moralism for critical reason" and "not to be seen to assert that the activities of the humanities are necessarily ethically driven" (Small 2013, 144–145). Though the renewed focus on critical reason over moralism is surely to be welcomed, it is worth considering the implications if academics within the humanities renounce what Williams called the focus on 'real values.' Recently, commentators like Stefan Collini and Martha Nussbaum have argued that, faced with the increasing marketization of higher education, those in the humanities must continue to emphasize values other than those of the market or risk "becoming door-to-door salesmen for vulgarized versions of their increasingly market-oriented 'products'" (Collini 2009, 18–19, qtd. in Nussbaum 2012, 130). Ultimately, a value-neutral version of the study of culture is unachievable, since attempts to realize this vision of neutrality replicate the logic of the free market.

The second possible objection to Hoggart's statement above – namely, its dismissal of commercial products – raises questions about what stances those within the study of culture can adopt towards the marketplace. Arguably, this is more of a concern within the British tradition, where the model of *cultural criticism/critique* implied a more politicized stance than the German tradition of *Kulturwissenschaft*. The German tradition, founded on the works of thinkers like Max Weber and Georg Simmel, was quicker to examine the effects of the marketplace on modern cultures. Following Walter Benjamin's seminal analysis of the transformative effects of mechanical reproduction, Theodor W. Adorno and Max Horkheimer described the development of a 'culture industry' that replicated the logic of the market:

> The less the culture industry has to promise and the less it can offer a meaningful explanation of life, the emptier the ideology it disseminates necessarily becomes. Even the abstract ideals of the harmony and benevolence of society are too concrete in the age of the universal advertisement. (Horkheimer and Adorno 2002, 118)

In a similar vein, Jürgen Habermas argued that: "Along the path from a public critically reflecting on its culture to one that merely consumes it, the public sphere in the world of letter [...] has lost its specific character" (Habermas 1991, 175). This permeation of culture by the logic of the market raises questions for both the German and the Anglo-American traditions. Should those within the study of culture celebrate the market for its achievements? Should they attempt to describe its cultural effects objectively (which, as discussed above, brings challenges of its own)? Or should they attempt to maintain a position of criticism or critique (in which case, the question arises: In what ways and to what extent is this possible)?

It seems inevitable that the trend is towards greater engagement with the marketplace. Arguably, anyone studying contemporary cultures cannot overlook its influence if they wish to explain them comprehensively. Furthermore, there seems to be a move away from earlier oppositional approaches to the market towards greater objectivity. The introduction to a recent volume on *Cultural Studies and Anti-Consumerism* recognizes that:

> Inquiries into consumption as a cultural process have emerged from a range of fields [...] many bearing the influential stamp of cultural studies' early inquiries into consumption as a rich semantic domain [...]. (Binkley and Littler 2011, 3)

Increasingly, these studies have dropped the 'anti-' from their titles to investigate the relationship between culture and consumerism on a more objective footing. However, the fields of culture and economics, culture and finance, and culture and consumption are relatively new and the complexities of their supposed objectivity regarding the marketplace remain to be fully explored. As those within the study of culture embark on this exploration, it is important for researchers to acknowledge their situatedness within the economy. More than ever, those working in the study of culture need to reflect on how their critical languages and methods are influenced by market thinking. Equally, there is a need for tenured and untenured academics alike to recognize the constraints of their positions within the academic market. It is sometimes tempting for those who teach culture to present the academy as an autonomous sphere that is hermetically sealed from the marketplace. Naturally, the values of scholarship and diligence need to be upheld, but it is also important for those who teach culture to educate their students about the knowledge economy and to give them a realistic understanding of the challenges they face as they seek to enter it.

3 New Methods in the Study of Literature and Economics

Many working within the study of culture have attempted to meet the challenges of explaining the mutual influence of markets and cultures. This section focuses on the field of literary studies to investigate the development of new methods for studying the marketplace. It also examines the surge in projects dealing with literature and economics since the 2008 financial crisis. In the process, the chapter considers the extent to which the fields of literature and economics can and should remain autonomous, as well as how those undertaking interdisciplinary work between these fields might transfer methods from economics and finance and attempt to describe the effects of the market objectively.

Until recently, there was a relative dearth of critical methods in the field of literature and economics. From the 1960s to the 2000s, the pace of technological and financial innovation in the marketplace arguably outstripped the development of literary methods for studying the marketplace. Though Marxist criticism became increasingly sophisticated through the development of cultural materialism and new historicism, some perceived a need for alternative, more politically 'neutral' methods for the study of literature and economics (Osteen and Woodmansee 1999, 12).[2] Inevitably, these new methods needed to explain why their worldview was more 'neutral' or 'objective' than earlier approaches. Rather than importing market thinking uncritically into literary studies, they needed to devise methods to analyze the interdependence of markets and cultures. Within literary studies, this need was initially met by Mark Osteen and Martha Woodmansee's collection *The New Economic Criticism*. Osteen and Woodmansee noted that the first wave of economic criticism primarily comprised historicist studies, which attended to "contextual discursive formations – law, banking, art history, etc. – as they impinge upon literary texts" (Osteen and Woodmansee 1999, 12). Such studies tended to focus on the market as a context or theme, rather than attempting to translate methods from economics and finance into literary studies. Although not interdisciplinary at the level of method, the historicist study of literature and the marketplace continues to provide a rich field for exploration. Raymond Williams's model of the opposition between culture and marketplace has been complicated by later scholars like Philip Connell, who noted the "extent to which early nineteenth-century political economy [...] played a formative role in the emergence of the idea of 'culture' itself" (Connell 2001, 7). In a similar vein,

2 The term 'neutral' is applied in this sense by Mark Osteen and Martha Woodmansee to the work of Marc Shell.

Mary Poovey has shown how profoundly the developing credit economy in the nineteenth century influenced the development of literary genres and the definition of literature itself (Poovey 2008). Such studies have proved important in breaking down the traditional, and often historically questionable, dividing lines between disciplines.

However, one challenge facing studies of literature and economics is that their themes and methods are extremely disparate. For this reason, some have questioned whether economic criticism represents a new school of literary criticism, with distinct methods for studying the market, or whether it constitutes a series of discrete studies that use existing methods and take aspects of the market as their theme. In an often-cited contribution to the collection *Money and the Age of Shakespeare*, Douglas Bruster referred to new economic criticism as an "open unity, an emergent mode of criticism defined by its willingness to treat the economic basis of social interaction both in and out of literary texts, and supporting the production of literature itself" (Bruster 2003, 69). This openness certainly offers the potential for innovation, but it also suggests that the development of new methods for studying the marketplace is still in its infancy. Given the absence of established methods, there remains great potential and a need for new methods that enable literary scholars to study the market objectively, disentangling the (historical) economic world of the text from that of the critic. A recent survey of new economic criticism noted that works professing this approach are often characterized as 'apolitical':

> The analyses explored herein largely appear to have no ideological axe to grind; freed from the dialectics that underpinned previous modes of economic criticism, these works instead seem to assume that the economic was an intrinsic part of early modern life. (Grav 2012, 133)

As described above, however, it is not possible to be apolitical when writing about culture in the marketplace. Elizabeth Hewitt remarks that "for all our savvy about the intersection between art and commerce, we nonetheless seem nostalgically to hold out hope that our scholarship will offer some loophole from the tyranny of commerce – even as we describe its all-consuming embrace" (Hewitt 2009, 621). Those studies that set traditional dialectics aside are often the ones confronted most urgently with the need to explain the vexed relations between literature and economics, culture and the marketplace. Here, the hypostatization of the market described by Hall becomes highly problematic. The field of culture and economics is perhaps uniquely challenging, because scholars in this area must seek – possibly in vain – to prevent their studies from being colored by the market thinking and relations of academic production in their own time.

In addition to pre-existing historicist approaches, Osteen and Woodmansee perceived a new avenue of exploration in the work of Marc Shell and Kurt

Heinzelman. Beginning in the 1980s, these scholars investigated the relationship between language and money, arguing that "all metaphors are in a sense economic, since the etymology of 'metaphor' contains within it the concept of transfer or exchange" (Osteen and Woodmansee 1999, 18). In a seminal passage, Marc Shell suggests that "money, which refers to a system of tropes, is also an 'internal' participant in the logical or semiological organization of language, which itself refers to a system of tropes" (Shell 1982, 3). As a result, a distinct approach has developed within new economic criticism that examines how literary texts "produce and respond to reformulations of the nature of representation and credit embodied in money and in the economic system in general" (Osteen and Woodmansee 1999, 4). This approach is rich in possibilities and provides the opportunity for a rigorous rethinking of the relation between language and money, as well as how the relations of production influence semantic systems and the production of meaning. However, this approach faces two key challenges. Firstly, the metaphor of language as money upon which it is based is fraught with complexities. To what extent can one speak of metaphor as 'exchange' or of money as a 'system of tropes' without distorting the object of study? More than ever, it is necessary to ensure that the implications and imperfections of these metaphors are foregrounded before any false critical assumptions are made. Secondly, although these approaches are important in turning the focus onto language as the medium of both literary and economic texts, the inevitable result is that the language of academic criticism also becomes embroiled in the metaphor.

The last few years have seen a burgeoning interest in the relation between language and money. With its emphasis on semiosis, the field of literary studies is often seen to be in a strong position for "developing a theoretical value for understanding money's mimetic and textual implications" (Crosthwaite, Knight, and Marsh 2015, 118). Many have seen similarities between poststructuralist theories of language, which emphasize the inexhaustibility of meaning, and money's virtue of being infinitely exchangeable. However, there is a danger that equating language with money, especially if this is done in a reductive or simplistic way, may damage our understanding of language as a system capable of producing types of knowledge and discussing types of value other than those of the market. In the context of the financial crisis, Ansgar Nünning has shown the detrimental effects of too easily accepting the metaphors of crisis and illness:

> In short: metaphors of money and financial crises serve to narrativize and naturalize complex cultural, economic and political transformations, projecting ideologically charged plots onto the developments they purport merely to represent or to illustrate.
>
> (Nünning 2015, 63)

The same is true of metaphors that attempt to compare language with money. Such metaphors can yield enlightening results about the similarities and differences between the organization of meaning in markets and texts. Yet the condition for this is that critics must consider any ways in which the metaphor may be imperfect, 'ideologically charged,' or misleading.

Georg Simmel famously argued that: "Money is not, by its nature, a valuable object whose parts happen to have the same proportion to each other or to the whole that other values have to each other. The significance of money is only to express the value relations between other objects" (Simmel 2004, 145). Yet the linguistic turn in the humanities has shown that (abstract) language does not merely express value relations between other objects, but that it constitutes those values and shapes people's perception of reality. To equate language with money therefore has profound implications. On the one hand, treating language as a currency foregrounds the fact that values are determined by a kind of market consensus among language users. On the other hand, denying the role of language in producing and constituting value risks undermining the objectivity that academic writing tends to assume. As language approaches the status of money, its critical purchase upon the marketplace is reduced. It is no longer capable of conveying meanings whose value can be independently assessed, but becomes an inherently valueless medium of exchange whose worth is determined solely by the market of readers. Instead of being descriptive and analytical, the critic's language merely reproduces the market relations it attempts to explain.

The view of language as money exists in an ambivalent relation to the principles of poststructuralist criticism. Recently, several critics have suggested that post-structural and deconstructionist models of language may be more indebted to economic ideas than was previously assumed. In the *Course in General Linguistics*, a founding text of both structural and post-structural linguistics, Ferdinand de Saussure argued that:

> Unlike language, other human institutions – customs, laws, etc. – are all based in varying degrees on the natural relations of things [...] Language is limited by nothing in the choice of means, for apparently nothing would prevent the associating of any idea whatsoever with just any sequence of sounds. (De Saussure 1966, 75–76)

Having rejected these institutions as analogies for language, Saussure suggests that a better comparison might be made with economics, arguing that in linguistics "as in political economy we are confronted with the notion of value; both sciences are concerned with a system for equating things of different orders – labour and wages in one and a signified and signifier in the other" (De Saussure 1966, 79). For Saussure, linguistics and economics are alike in requiring both a static (synchronic) approach, which studies the structure of the system in a moment of time, and an evolutionary (diachronic) approach, which examines how the system changes over time. However, this equation of semantic and monetary values has led some to question the economic basis of Saussure's synchronic model of language.

As David Holdcroft points out, Saussure's synchronic approach assumes that "it is the present state of the language which determines for the speaker [which signs are] to be considered, not the history of the signs themselves, of which most speakers can be presumed to know nothing" (Holdcroft 1991, 74). With history stripped away, the arbitrary series of relations between a system of signs in the present becomes reified as a self-determining structure. This parallels Simmel's description of how the "function of exchange, as a direct interaction between individuals, becomes crystallized in the form of money as an independent structure" (Simmel 2004, 174). In both cases, the cultural context of human actors negotiating values and reaching a consensus over time is overlooked. In a recent discussion of securities markets, Andreas Langenohl argues that "securitization is part of a discursive structure that 'becomes self-referential. It does not refer to an external, objective reality but establishes a security situation by itself'" (Langenohl 2017, 136). The synchronic approach to language envisages a similarly self-referential structure, in which the 'direct interaction between individuals' becomes 'crystallized' as a series of arbitrary relations between signs. As Roy Harris observes, once the "notion of a synchronic system applies both to language and to economics", there is no reason why it should not "apply to *all* human activities where signs and values are involved" (Harris 2001, 199). Saussure's analogy between linguistics and economics thus gives rise to a "philosophical thesis of far deeper import," which proves "very subversive of the limits of historical explanation" (Harris 2001, 199). Following this line of thinking, there is a chance that language and culture itself will be 'hypostacized' as another market.

4 Conclusion: Futures in the Study of Culture

The study of culture in the marketplace is an extremely rich and complex field, which urgently invites the development of new methods. In the process, it may be necessary to revisit fundamental assumptions about the study of culture from the bottom up, beginning with the definition of culture and with the prevailing theories of language and metaphor. Given the hypostatization of the market, it is important for those working in this field to develop critical languages to describe the marketplace without simply replicating its logic or applying its assumptions unquestioningly. There is also a need to move beyond reifying the market as a series of 'forces,' 'movements,' and 'trends' towards understanding it as a complex series of human interactions within social and technological networks. In this way, the market can be analyzed in cultural terms as an arena within which individual and collective decisions are made about fundamental human needs, desires, values, and priorities. There exists a huge potential for interdisciplinary exchange between cultural studies and economics departments. Along the way, however, it will be necessary for those working in the study of culture to be open about their position within the attention economy and academic funding structures. The futures of the study of culture may be greatly curtailed if cultural studies departments attempt to increase their visibility and relevance by simply following the market and adopting economics as a theme. Instead, what is required is the further development of rigorous methods – building on the great number of studies already published in this area – which consider how one can gain critical purchase on the market after it has been hypostacized to incorporate every facet of human life.

References

Angermuller, Johannes. "How to Become an Academic Philosopher: Academic Discourse as a Multileveled Positioning Practice." *Sociología Histórica* 2 (2013): 263–289.

Angermuller, Johannes. "The New Political Economy of Higher Education: Between Distributional Conflicts and Discursive Stratification." *Higher Education* 73.6 (2017): 795–812.

Bachmann-Medick, Doris. *Cultural Turns: New Orientations in the Study of Culture*. Berlin/Boston: De Gruyter, 2016.

Baker, Kathryn A. "Organizational Culture." *Management Benchmark Study*. United States Air Force Air War College, Office of Planning and Analysis. http://www.au.af.mil/au/awc/awcgate/doe/benchmark/ch11.pdf. 2002. (9 July 2017).

Binkley, Sam, and Jo Littler. "Introduction." *Cultural Studies and Anti-Consumerism: A Critical Encounter*. Eds. Sam Binkley and Jo Littler. London: Routledge, 2011. 1–12.

Bodleian Libraries. "It's the 20th Birthday of Harry Potter and the Philosopher's Stone." Facebook post. 26 June 2017.

Bourdieu, Pierre. *Sketch for a Self-Analysis*. Cambridge: Polity, 2007.

Bruster, Douglas. "On a Certain Tendency in Economic Criticism of Shakespeare." *Money and the Age of Shakespeare: Essays in New Economic Criticism*. Ed. Linda Woodbridge. Basingstoke: Palgrave Macmillan, 2003. 67–77.

Collini, Stefan. "Impact on Humanities: Researchers Must Take a Stand Now or be Judged and Rewarded as Salesmen." *Times Literary Supplement* (2009): 18–19. Quoted in Nussbaum, Martha C. *Not for Profit: Why Democracy Needs the Humanities*. Princeton: Princeton University Press, 2012.

Connell, Philip. *Romanticism, Economics and the Question of 'Culture.'* Oxford: Oxford University Press, 2001.

Crosthwaite, Paul, Peter Knight, and Nicky Marsh. "Economic Criticism." *The Year's Work in Critical and Cultural Theory* 23 (2015): 108–133. DOI: 10.1093/ywcct/mbv005 [accessed: 30 June 2019].
Davenport, Thomas H. *The Attention Economy: Understanding the New Currency of Business*. Boston: Harvard Business School Press, 2001.
Deal, Terrence E., and Allan A. Kennedy. *Corporate Cultures: The Rites and Rituals of Corporate Life*. Reissued with a new introduction. New York: Perseus Books, 2000.
De Saussure, Ferdinand. *Course in General Linguistics*. Eds. Charles Bally and Albert Sechehaye. New York: McGraw-Hill, 1966.
Eagleton, Terry. *Marxism and Literary Criticism*. London: Routledge, 2002.
Grav, Peter F. "Taking Stock of Shakespeare and the New Economic Criticism." *Shakespeare* 8.1 (2012): 111–136.
Habermas, Jürgen. *The Structural Transformation of the Public Sphere: An Inquiry into a Category of Bourgeois Society*. Transl. Thomas Burger. Cambridge, MA: MIT Press, 1991.
Hall, Stuart. "Introduction." *Representation: Cultural Representations and Signifying Practices*. Ed. Stuart Hall. London: Sage, 1997. 1–11.
Hall, Stuart. "The Neo-Liberal Revolution." *Cultural Studies* 25.6 (2011): 705–728.
Harris, Roy. *Saussure and His Interpreters*. Edinburgh: Edinburgh University Press, 2001.
Hewitt, Elizabeth. "Review: The Vexed Story of Economic Criticism." *American Literary History* 21.3 (2009): 618–632.
Hoggart, Richard. *The Uses of Literacy: Aspects of Working-Class Life with Special Reference to Publications and Entertainments*. Harmondsworth: Penguin, 1992.
Holdcroft, David. *Saussure: Signs, System, and Arbitrariness*. Cambridge: Cambridge University Press, 1991.
Horkheimer, Max, and Theodor W. Adorno. "The Culture Industry: Enlightenment as Mass Deception." *Dialectic of Enlightenment: Philosophical Fragments*. Ed. Gunzelin Schmid Noerr. Stanford: Stanford University Press, 2002. 94–136.
Jameson, Fredric. *Postmodernism, Or the Cultural Logic of Late Capitalism*. Durham, NC: Duke University Press, 1991.
Koselleck, Reinhart. *Critique and Crisis: Enlightenment and the Pathogenesis of Modern Society*. Oxford: Berg, 1988.
Langenohl, Andreas. "Securities Markets and Political Securitization: The Case of the Sovereign Debt Crisis in the Eurozone." *Security Dialogue* 48.2 (2017): 131–148.
Nünning, Ansgar. "Metaphors We Pay For, or: Metaphors of the 'Financial Crisis' Shaping the Cultural Life of Money." *The Cultural Life of Money*. Eds. Isabel Gil and Helena Gonçalves da Silva. Berlin/Boston: De Gruyter, 2015. 47–67.
Nussbaum, Martha C. *Not for Profit: Why Democracy Needs the Humanities*. Princeton: Princeton University Press, 2012.
Osteen, Mark, and Martha Woodmansee. "Taking Account of the New Economic Criticism: An Historical Introduction." *The New Economic Criticism: Studies at the Intersection of Literature and Economics*. Eds. Mark Osteen and Martha Woodmansee. London: Routledge, 1999. 2–42.
Poovey, Mary. *Genres of the Credit Economy: Mediating Value in Eighteenth- and Nineteenth-Century Britain*. Chicago: University of Chicago Press, 2008.
Roider, Andreas, and Andrea Voskort. "Reputational Herding in Financial Markets: A Laboratory Experiment." *Journal of Behavioral Finance* 17.3 (2016): 244–266.

Shell, Marc. *Money, Language, and Thought: Literary and Philosophical Economies from the Medieval to the Modern Era.* Berkeley/Los Angeles/London: University of California Press, 1982.
Simmel, Georg. *The Philosophy of Money.* Ed. David Frisby. 3rd enlarged ed. London/New York: Routledge, 2004.
Small, Helen. *The Value of the Humanities.* Oxford: Oxford University Press, 2013.
Williams, Raymond. *Culture and Society 1780–1950.* New York: Anchor Books, 1960.

Laura Meneghello

Cultural History, Science Studies, and Global Economy: New and Future Approaches

While the concept of a 'global economy' becomes more and more central in current debates, the history of such a conception of economy has hardly been investigated. Since when has the economy been perceived as a global issue? What ethical values were related to it, and what "epistemic virtues" (Daston and Galison 2007) were related to economics as a discipline dealing with global issues? What institutions and networks contributed to shaping the conception of the global economy? These and other questions should become the subject of a cultural history of the global economy. With a few exceptions, the perspectives of cultural history, science studies, and the history of knowledge are largely missing or at least underrepresented in economic history, where scholars have mostly taken the categories of economics for granted, basing their research upon these categories instead of examining their historicity (Dejung et al. 2011). However, other approaches are possible and, in part, they are already an emerging reality. As Michel Callon poignantly expressed it in an interview with Michel Ferrary: "Si l'économie-chose existe, c'est parce qu'elle a été performée par l'économie-discipline: 'no economy without economics!'" (Callon and Ferrary 2006). In the following pages, I will attempt to sketch these new approaches, and point to possible foci for future research and new ways of understanding the history of the economy and economics. At the same time, I will exemplify one possible approach through an analysis of discourses on the global economy between the second half of the nineteenth century and the first half of the twentieth century.[1]

An entangled history of knowledge of the global economy should approach economics as a scientific discipline not through its own categories, but through a discourse-analytical approach that allows taking a critical distance from the concepts of classical economic theory, and questioning their universal validity, thus historicizing economics as a science in the same way that classical categories within the humanities are increasingly historicized.[2] In recent years, some schol-

1 I would like to thank my colleague in Siegen and former GCSC member, Katharina Kreuder-Sonnen, for critically reading my paper.

2 On questioning the universality of the categories of the humanities, see Bachmann-Medick 2014; for discourse analysis applied to economy and economics, see the contributions in Diaz-Bone 2015; on the history of discourses on the relationship between economy and politics in the twentieth century, see Scholl 2015.

https://doi.org/10.1515/9783110669398-015

ars in the fields of history, sociology, science studies, and cultural studies have not only unveiled practices and conceptions in the economic sphere as being culturally determined (Berghoff and Vogel 2004), but also, and most importantly, have argued that the categories and methods of the discipline of economics play a key role in shaping economic processes themselves (Callon 1998, 2007; Knorr-Cetina and Preda 2005; Speich Chassé 2013). Among historians, Dejung, Dommann, and Speich Chassé (2014) have attempted to question the assumption that 'the economy' be a given rather than the product of historical transformations, negotiations, and shifts of meaning. They pled for an analysis of the relations between economy and economics on the one hand, and ethics, religion, and politics on the other hand, which should uncover their reciprocal intertwinement. Other historians have done research on the practices and techniques of economic knowledge production (e.g. Tooze 2001; Schneider 2013).

In this paper I would like to highlight emerging approaches to the history of economy and economics, as well as to suggest new ones. As I explained above, I will concretize one of these possible new approaches through a short analysis of discourses on the global economy. Thus, my aim is to uncover the historicity of the 'global economy' not by studying the development of economic processes, but by questioning the very understanding of these processes and the development of economics as a discipline being concerned with global economic processes. Like Monika Dommann argues with regard to legal history in her book on the history of copyright, *Autoren und Apparate* (2014), I maintain that economic history should also be approached with the methods of cultural history and the history of knowledge (see Speich Chassé and Gugerli 2012). The suggested approach is inspired by research in the humanities (Bod et al. 2014; in particular Daston 2014); by what Michel Foucault has initiated through his archeology of knowledge especially in the fields of medicine, psychiatry, and systems of punishment (Foucault 1969); as well as by what the history of science and the actor-network-theory have achieved in the history of the natural sciences and technology (Lenoir 1997; Mitchell 2005; Latour 2005).

The approach I exemplify seeks both to deconstruct the discourses on global economy and to reconstruct the networks of experts who contributed in spreading conceptions of world economy. "Travelling concepts" (Neumann and Nünning 2012; Bachmann-Medick 2014) between both disciplinary and national borders constitute a key theoretical approach for such an entangled history of economics, which should be both interdisciplinary and transnational. This approach implies a broadening of the future study of culture to a field, economics, whose laws are often considered to be universal and independent from cultural variances. At the same time, it potentially allows for interactions between historical disciplines on the one hand, and sociological and economic disciplines on the other hand, joining their efforts in the examination of the self-understanding of economics as a discipline and finally questioning its very categories.

The sources I will analyze are not limited to scientific publications, but include non-scientific publications that were influenced by, and themselves possibly influenced, scientific discourse.[3] I do not adhere to an internalist conception of science, and therefore I do not think that considering economics as an abstract sphere isolated from other disciplines or from popularizing and political discourses would do justice to the breadth of the dynamics of knowledge circulation.

The time between 1850 and 1945 is of particular interest for my questions, which include: how did the term 'global economy' originate and spread? What contexts and constellations favored its multiplication and stabilization? What shifts in its meaning and what processes of translation were involved? To address this, I will indicate a few case studies as well as possible answers to these questions.

1 The Emergence of the Concept of Global Economy

In the following pages, I will describe a concrete example of a discourse-analytical approach to the emergence of discourses on world economy. That is: applying discourse analysis to some sources on the 'global economy' dating approximately from 1850 to 1945, I argue that discourses on the global economy (not 'global trade' or 'global commerce') emerged in this period of time. In fact, a bibliographic search based on library catalogues or databases will show that the term 'global economy,' in its variants of 'world economy,' '*Weltwirtschaft*' or '*Weltökonomie,*' '*wereld economie,*' '*économie mondiale*' and '*economia mondiale,*' is nearly impossible to find before 1850.[4] There are isolated examples from the 1830s, a few more since the 1850s, and the frequency always increases until the twentieth century, when the term becomes omnipresent after the 1940s.

The following examples show that, around 1870, discourses on world economy started to spread among economists. In a book on *American Political Economy* dating from 1870, the author, the American Francis Bowen (1811–1890),

3 On the circulation of scientific knowledge even beyond the scientific or academic sphere, and back, see Fleck 1980.

4 See Karlsruher Virtueller Katalog, https://kvk.bibliothek.kit.edu; Anno, Volltextsuche in Zeitungen und Zeitschriften, Österreichische Nationalbibliothek, Suche: '*Weltwirtschaft,*' http://anno.onb.ac.at/anno-suche#searchMode=simple&query=weltwirtschaft&resultMode=list&from=1: the search in ANNO gives the following results for each of the following time periods: 1867–1881 (7 results); 1882–1897 (46); 1898–1912 (455); 1913–1928 (3.259); 1929–1945 (2.988) [both last accessed: June 2016].

Professor at Harvard, criticizes Malthusian theories appealing to a greater dimension in the economic space, namely "world's economy":

> We can now see with sufficient distinctiveness the two great facts which afford a complete refutation of Malthusianism. The *first* is, that *the limit of Population, in any country whatever, is not the number of people which the soil of that country alone will supply with food, but the number which the surface of the whole earth is capable of feeding;* and it is a matter of demonstration, that *this* limit cannot even be approached for many centuries. The inability of England alone, or of Ireland alone, to supply her teeming population with food, is a fact of no more importance in the **world's economy**, than the inability of the city of London alone to supply her two millions of people with farm-produce from her own soil. London taxes all the counties of England for her sustenance; England taxes all the countries of the world for her sustenance; – I cannot see any difference between the two cases.
> (Bowen 1870, 140, emphasis mine, italics in the original)

As we can see from this justification of colonial politics, expansionism and economical exploitation of the resources of other continents entailed thinking in a global scale, and it is in this context that talking about a national economy was not sufficient any longer.

In a German source from 1873, an article by the political economist Karl Knies (1821–1898), a representative of the historical school, the "global economy" ("die *Weltwirtschaft*, als ein großes Ganzes betrachtet") is opposed to isolated national economy ("eine von allen übrigen Völkern *isolierte Volkswirtschaft*"), thus also implying that the global economy represents a higher dimension, to which national economies are subordinated (Knies 1873, 307, italics in the original). However, 'world economy' was already present also beyond the scientific discourse. In the same year, the international exhibition in Vienna also spoke of a global economy as opposed to national economy, and its journal pled for a transition from a national to an all-encompassing global economy:

> However, we should not lose sight of at least one characteristic of our times at the World Exhibition in Vienna: the transformation of national economies [*Volkswirtschaften*] into the global economy [*Weltwirtschaft*]. International conventions and boards should be established as basis for later diplomatic or political action. (Anonymous 1873, 2; my translation)[5]

From this example we can induce that international exhibitions, such as the one that was organized in Vienna in 1873, were places where the world economy was actively performed, and its very conception was actively shaped and spread. In

5 "Zumal aber einen Charakterzug der Zeit dürfen wir bei der Wiener Weltausstellung nicht aus dem Auge verlieren: den Uebergang der Volkswirtschaften in die Weltwirtschaft. Internationale Einigungen und Berathungen sollen als Grundlage für eine spätere diplomatische oder politische Action geschaffen werden."

the following decades, the higher dimension of a world economy acquired an increasing important role in economic theory, and already in 1905 we can find that a congress was held in Belgium on the "world's economic development" ("Congrès international d'expansion économique mondiale") (Anonymous 1905). A search in the abovementioned databases lets us also see that the frequency of the term increased from 1919, after World War I, and especially in the 1930's and 40's. In particular, we can observe a stabilization of the term in scientific language, as well as in handbooks: in 1919 we can find, for instance, a publication called *Die Weltwirtschaft der Fettstoffe* (Fitzner 1919), in 1916 one called *U-Boot und Weltwirtschaft* (Thielemann 1916); some books asked about "the place" of particular nations in "world economy" (Arndt 1908; Rathgen 1911; Schumacher 1917), other books opposed the global economy to national economy *("Volkswirtschaft"* or *"Nationalwirtschaft")* (Dietzel 1900; Oppenheimer 1915).

Besides that, the proceedings of conferences relating to economic and political matters, organized by supranational entities such as the League of Nations, show that the term was becoming more and more present in the political sphere between the 1920s and 1930s (see League of Nations 1927). This is confirmed by publications such as *L'économie mondiale et la Société des Nations* (Hantos 1930). A further important step in the institutionalization of the study of the 'global economy' was the foundation of the journal *Weltwirtschaftliches Archiv* in Kiel in 1913 (still existing as *Review of World Economics*), edited by Bernhard Harms. The *Weltwirtschaftliches Archiv* presented itself as a scientific journal with the aim of studying the history and the functioning of world economy, an issue that, as the editor maintained, was highly important but had not received the due attention until that moment (Harms 1913, 1–36). Bernhard Harms founded the "Königliches Institut für Seeverkehr und Weltwirtschaft" at the Christian-Albrechts-Universität in Kiel in 1914, another step in the institutionalization of the 'world economy.'[6] As we will see later on, its importance went beyond national borders and contributed in spreading the concept in the international scientific discourse; in fact, the German concept of '*Weltwirtschaft*' and, in particular, the translation of (the titles of) some works by German scholars publishing in the journal *Weltwirtschaftliches Archiv*, played a significant role in the European scenario. Another example is the "Hamburgisches Welt-Wirtschafts-Archiv," founded in 1908 and being part of the Hamburgisches Kolonialinstitut (Thilenius 1920); as we can see, even in this case the idea of global economy went hand-in-hand with colonialism.

6 On scientific disciplines and institutions, in particular on the natural sciences, see Lenoir 1992 and 1997, as well as Stichweh 1994. On economics as a scientific discipline: Schneider 2013, Hesse 2010, and Mitchell 2005.

2 Socialism and the Concept of World Economy

In the 1920s, socialist and communist economic theories were an important factor of the increased popularity of concepts of global economy. Above all, they brought the term outside the scientific discourse, as we can see from the communist newspaper *Die Rote Fahne*, published in socialist Vienna of the 1920s. Here, 'world economy' was conceived in opposition to 'national economy':

> For years, the principle of global economy has been competing with the one of national economy. The longer the struggle went on, the clearer it was that the idea of the global economy, of the increasingly more intimate economic connection of all countries, is the stronger one. (Anonymous 1920, 2; my translation)[7]

According to these lines, the global economy, where all countries are economically bound to each other, had proved to be "stronger" than the principle of a national economy, and had triumphed over it. As we can see from the following passage, capitalist economy was perceived as being bound to older, national economic models, and the war was conceived as a break in global economic relations:

> Short before the beginning of the war, capitalist economy was showing its weakest side. Almost everywhere there was a quite heavy crisis, unemployment was remarkable. At this point, the war administered the heaviest blow to the principle of the global economy. Great nations broke all relations with one another and wanted to support only the resources and the laws of development [economic development, but the analogy with Darwinism and the natural sciences is obvious, L.M.] of their own country. From then on there was no global economy for the Central Powers anymore, but only an economy, which did not overcome the national borders. For years, the capitalists considered the decline of the global economy as a fruitful result. Instead of the economic relations on the world market, military [relations] on the battlefield emerged. (Anonymous 1920, 2; my translation)[8]

---

7 "Jahrzehntelang kämpfte das Prinzip der Weltwirtschaft mit dem der nationalen Wirtschaft. Je länger der Kampf währte, desto offenbarer wurde es, daß der Gedanke der Weltwirtschaft, der immer innigeren ökonomischen Verknüpfung aller Länder, der stärkere sei."

8 "Kurz vor Ausbruch des Krieges zeigte die kapitalistische Weltwirtschaft ihre schwächste Seite. Beinahe überall herrschte eine ziemlich schwere Krise, die Arbeitslosigkeit war bedeutend. In diesem Augenblicke wurde dem Prinzip der Weltwirtschaft der schwerste Schlag durch den Krieg versetzt. Große Staaten lösten alle wirtschaftlichen Beziehungen zueinander und wollten nur die Hilfsquellen und die Entwicklungsgesetze des eigenen Landes zur Geltung kommen lassen. Von nun an gab es für die Mittelmächte keine Weltwirtschaft mehr, sondern nur eine Wirtschaft, die über die Staatsgrenzen nicht hinausreichte. Jahrelang galt für die Kapitalisten der Verfall der Weltwirtschaft als fruchtbringendes Ergebnis. An Stelle der wirtschaftlichen Beziehungen auf dem Weltmarkte traten die militärischen auf dem Schlachtfelde."

Capitalism took advantage from the crisis of world economy, it is stated, substituting trade relations with military relations. Of course, communist and socialist papers were not the only that interpreted the war as interruption of global economic relations: in the Habsburg Monarchy of 1915, the *Fremden Blatt* encouraged the development of national economy, since the global economy had been destroyed by the war:

> Since the global economy is destroyed, there are no international relations anymore, and we must strive with all our forces to nationalize our economy again [...].
>
> (Wurm-Arnkreuz 1915, 4; my translation)[9]

The concept of the global economy can be found in the socialist discourse well before the 1920s, and constituted a topic of socialist congresses already in 1894: for instance, it was mentioned in the *Protokoll des Internationalen Sozialistischen Arbeiterkongresses* held in Zurich in 1894.[10] In this case, the fact that the world's economy was accompanied by a condition of constant warfare was identified as one of the contradictions of capitalism.

The importance of the 'world economy' for communist and socialist discourses was well known, and the term was recognized as being directly connected with these currents of thought. For instance, an Italian scientific publication already described 'world economy' as an economic category typical for communism as early as 1876 (Ciccone 1876, 178).[11] At that time, the expression was rare to find in other Italian-language contexts. *Socialism from Genesis to Revelation* (1892), by Franklin M. Sprague, also mentioned "world economy" in relation to socialist economy (Sprague 1892, 404). Socialist newspapers thus possibly played a key role in spreading the discourse on world economy beyond the scientific domain of economics. As Ludwik Fleck maintained with regard to knowledge circulation in the natural sciences (Werner and Zittel 2011, 23), economic knowledge was transferred beyond the scientific sphere and became exoteric knowledge (which then eventually might have come back into the esoteric sphere of economics, having been transformed and adapted).

---

9 "Weil die Weltwirtschaft zusammengebrochen ist, gibt es keine internationalen Beziehungen mehr, und wir müssen mit allen Kräften bemüht sein, unsere Wirtschaft wieder zu nationalisieren [...]."

10 *Protokoll des Internationalen Sozialistischen Arbeiterkongresses in der Tonhalle Zürich vom 6. bis 12. August 1893*: "Nur das Proletariat hat ein Lebensinteresse, den Widersinn der heutigen Gesellschaft zu beseitigen: Weltwirtschaft – und ständige Kriegsgefahr [...]," 1894, 2.

11 Ciccone: "[...] economia pubblica, economia politica, cui aggiungono i comunisti una economia mondiale," 1876, 178.

3 Technology and Communication: Changing Representations of Global Space and Economy

Focusing on the conception of space, we can ask how the idea of a global economic space was constructed, but also which economic processes were seen as influencing economy on a global scale. For instance, the first issue of the *Weltwirtschaftliches Archiv* contains an article on the Panama Canal (Stubmann 1913), as well as on the global significance of railways (Thieß 1913). It would also be interesting to explore the connection between technical innovations – especially infrastructures – and conceptions of global economy.

In fact, means of communication such as the telegraph and the telephone were understood as being deeply bound to the development of global economic relationships, as we can see from volume fifteen of the British *Board of Trade Journal*, which at the end of the nineteenth century stated: "The Americans were the first to realise the importance of telephony and to appreciate the part it was likely to play in the world's economy" (Commercial Department of the Board of Trade 1893, 264).

At the same time, means of transport and communication routes – and thus technical innovations in general – were also conceived as a way of fostering the 'world economy.' In the 1893 handbook *Descriptive Economics: An Introduction to Economic Science*, we can read:

> Fast lines of steamships bring all parts of the globe into communication. They have facilities for transporting all products. The Suez canal went a long way toward furthering the world economy. A Nicaragua canal may sometime do even more. (Bly 1893, 233)

Thus, we can infer that the growth of communication and transport routes, as well as technical innovations that allowed faster exchange of goods and opened up new trade relations, signified a change in the perception of space, and that these changes favored thinking of economy not as a national or colonial but as a global issue.

For this reason, the spatial turn (Bachmann-Medick 2016, 211–243) and the history of technology must also be considered central theoretical approaches for an entangled history of economy and economics, whereas questions of mobility (of persons as well as of objects) should be as important as the questions of knowledge transfer that I examine in the next section.

4 Travelling Concepts in World Economy

Finally, a possible research focus in the history of the global economy lies on transnational and interdisciplinary transfer: on knowledge transfer on the international level, on the one hand, and on conceptual transfer between disciplinary and social fields, on the other. Knowledge transfer can be analyzed through a comparison of economic journals from different countries, and by asking how their terminology and conception of world economy show reciprocal influences. Conceptual transfer could be highlighted through the analysis of metaphors and imageries from fields other than classical economic theory. In both cases, we can conceptualize these phenomena as 'travelling concepts' in the history of economics.

Interestingly, the translation of German book titles (though not necessarily of their contents) in the 1920s seems to have been important for the diffusion of the term *'économie mondiale'* in French.[12] However, even well before the founding of the *Weltwirtschaftliches Archiv* in Kiel, discourses on global economy were triggered by the translation of the German word *'Weltwirtschaft.'* For example, in 1855 the *Annali universali di statistica*, published in Milan, displayed the term 'world economy' in Italian (*'economia mondiale'*), which at that time was still very rarely used, in the translation of a German title, *System der Weltökonomie* (Anonymous, 1855, 117–118).[13] The use of the term in English was also influenced by German terminology: for instance, Kaufmann's *Socialism* (1874), declaredly "based on" Schäffle's book *Kapitalismus und Socialismus* (1870), uses the words "universal economy" with the meaning of world economy. It relates the concept to a series of negative values:

> Over-little as well as over-much state help is undesirable. The individual citizen now-a-days is so intimately connected with and dependent on the movements of the great social body, and with the progressive extension of the economic circle all over the world (comprehending not only individuals but whole nations), every isolated domestic economy is drawn so completely into the general vortex of universal economy, or at least the economy of nations, that it becomes the undoubted duty of the society to protect its component members against those contingencies which arise from such a complicated system of economy.
>
> (Kaufmann 1874, 284)

As we can see, the text reflects the German opposition of *"Weltwirtschaft"* and *"isolierte Volkswirtschaft"* discussed above. In this quotation, however, the global economy was presented as dangerous and complicated. Moreover, in this

12 Brocard 1929, 25, quoted Bernhard Harms, *Économie nationale et économie mondiale*, 1924; von Tyrka, *Problèmes d'économie mondiale des états industriels modernes*, 1916; Nachimsen, *Économie mondiale avant et après la guerre*, 1926. The expression *'économie mondiale'* is rare to find in French titles in the same period.

13 The title referred to is Winkelblech 1853, translated as: *Ricerche sull'organizzazione del lavoro, o Sistema d'economia mondiale.*

publication, dating from the 1870s, the concept was projected onto other epochs, and it was understood as a typical feature of capitalism:

> From times immemorial capitalism has maintained its place among other systems in the world's economy. It was the ruling principle in the commerce of the Phoenicians, the Italians, the German towns, long before slavery and feudalism ceased to dispute its pretension as the exclusive motor of society. (Kaufmann 1874, 63)

In fact, communism and capitalism were identified as two different "modes of world's economy":

> To return then to a public state communism, pure and simple, as the best mode of the world's economy, instead of simply purifying and further developing our present capitalistic forms, would be a lamentable step backwards. (Kaufmann 1874, 247)

In this passage, the author suggested to modify, improve, and "purify" capitalism, opposing "public state communism."

The projection of the concept of global economy onto other historical periods can also be conceived as a form of conceptual transfer or translation. This began around 1915, and developed especially after the 1930s and 1940s, when the concept was stabilized in scientific language and enjoyed increasing usage also beyond the scientific field. English imperial and colonial economy, for example, was now described as *"Englische Weltwirtschaft,"* an expression that was not used before the twentieth century (Predöhl 1949, 65–66). Thus, discourses on the global economy, whose beginnings I have briefly sketched, can be better understood through the analytical category of "travelling concepts" or "concepts in translation" (Bachmann-Medick 2014, 119–136). In this way, we can relativize and historicize concepts that have often been considered as being obvious and universally valid – their universality being granted by economic 'laws.'

5 Conclusion: Approaches to Deconstructing Economics

In this article, I have pointed to a new way of looking at economics and its concepts and categories. Furthermore, I have tried to exemplify this through a brief analysis of discourses on global economy and how they were translated from and into different languages and contexts. This is, of course, only a short and exemplary case study, and there is still much work to do in order to uncover the making of the global economy and to deconstruct the categories of economic science. Such a process of deconstruction of economics implies several challenges, but at the same time new and interesting results. First, it means questioning the universality of economic laws, as has been done with the categories of the humanities and the natural sciences, thereby also paving the way for a "cross-categorical translation" (Chakrabarty) of even too "obvious" economic terms (see Bachmann-Medick 2012, 33–35). Second, it allows a deeper understanding of the ethical and epistemic values attributed to economic concepts and ideas, following the methods that historians of science have applied to the history of the natural sciences (Daston and Galison 2007) and of the humanities (Daston 2014). Third, it results in studying the multiple processes of translation and adaptation of the categories of economic science. Last, but not least, science studies applied to the history of economy and economics entail analyzing the construction of scientificity, of 'the market' and of 'the economy' in and through the discipline of economics, which can be both visual (e.g. through diagrams or other visualizations of statistical data) and textual (Latour 2006 [1986], 279).

In order to point to how one could go even further than discourse analysis, I have suggested studying the perception of global economic processes in its interaction with the perception of new media, means of transport and communication, and mobility (see Callon and Ferrary 2006). In this sense, a cultural history of the global economy cannot be conceived without history of science and history of knowledge, science and technology studies (STS), and media studies. However, it would only then be truly interdisciplinary when engaging in a dialogue with literary studies (see the contribution by Tom Clucas in this volume as well as Joseph Vogl 2002). Its dimensions should go beyond Eurocentric approaches and consider the translational, transnational, global, and (post-) colonial aspects of the understanding of economy and the development of economics as a discipline. Since this implies that researchers can understand the languages and have a deep knowledge of the history of geographic areas other than those traditionally belonging to the 'West,' area studies are indeed much needed for such new approaches to an entangled history of economics. Finally, those who dare to go beyond the traditionally established borders attributed not only to written sources, but also to the textual as a form of publishing research results, should follow Peter Galison's (2014) and Andressa Schröder's (in this volume) hints and allow filmmaking and art to enter the domain of scientific research. As Peter Galison has argued in his 2014 article and shown through movies he has (co-)directed, documentaries can represent a complementary way of doing history of science. I would suggest that the same is true for economic processes and the understanding of economics, where historians and sociologists could only profit from cooperation with visual artists and filmmakers. It is a question of materiality, too: indeed, what could be a better means than film to make the voices of the (subaltern) people heard, the technological and natural soundscape listened to, and the interplay of nature, humans, animals, technology, and economic interests seen? For instance, in the documentary *Sonic Sea* (2016), dealing with man-made ocean noise and related environmental problems, the film directors Michelle Dougherty and Daniel Hinerfeld succeed in this aim. I thus conclude encouraging more openness to non-traditional, non-textual forms of expression from the side of historians, not only as sources and as means of knowledge dissemination, but also as epistemic tools, as ways of doing research and producing knowledge they should seriously confront.

References

Anonymous. "Bibliografia: Rassegna di opere straniere." *Annali universali di statistica* 5.14 (1855): 116–120.

Anonymous. *Internationale Ausstellungszeitung*. Beilage der *Neuen Freien Presse* zu Nr. 3140 (21 May 1873).

Anonymous. *Protokoll des Internationalen Sozialistischen Arbeiterkongresses in der Tonhalle Zürich vom 6. bis 12. August 1893*. Zurich: Buchhandlung des Schweiz. Grütlivereins, 1894.

Anonymous. *Congrès international d'expansion économique mondiale tenu à Mons du 24 au 28 septembre 1905 sous le haut patronage Sa Majesté le roi Leopold II. et du gouvernement Belge*. 8 vols. Bruxelles: Hayez, 1905.

Anonymous. "Weltwirtschaft und Revolution." *Die Rote Fahne* 3.279 (14 April 1920).

Arndt, Paul. *Deutschlands Stellung in der Weltwirtschaft*. Leipzig: Teubner, 1908.

Bachmann-Medick, Doris. "Translation: A Concept and Model for the Study of Culture." *Travelling Concepts for the Study of Culture*. Eds. Birgit Neumann and Ansgar Nünning. Berlin/New York: De Gruyter, 2012. 23–43.

Bachmann-Medick, Doris. "From Hybridity to Translation: Reflections on Travelling Concepts." *The Trans/National Study of Culture: A Translational Perspective*. Ed. Doris Bachmann-Medick. Berlin/Boston: De Gruyter, 2014. 119–136.

Bachmann-Medick, Doris. *Cultural Turns: New Orientations in the Study of Culture*. Berlin/Boston: De Gruyter, 2016.

Berghoff, Hartmut, and Jakob Vogel, eds. *Wirtschaftsgeschichte als Kulturgeschichte. Dimensionen eines Perspektivenwechsels*. Frankfurt a.M./New York: Campus, 2004.

Bly, Myron T. *Descriptive Economics: An Introduction to Economic Science for Use in Academies, High and Normal Schools, and Business Colleges*. New York/Cincinnati: American Book Co., 1893.

Bod, Rens, Jaap Maat, and Thijs Weststeijn, eds. *The Making of the Humanities*. Vol. 3: *The Modern Humanities*. Amsterdam: Amsterdam University Press, 2014.

Bowen, Francis. *American Political Economy: Including Strictures on the Management of the Currency and the Finances Since 1861, with a Chart Showing the Fluctuations in the Price of Gold*. New York: Scribner & Co., 1870.

Brocard, Lucien. *Principes d'économie nationale et international*. Vol. 1. Paris: Sirey, 1929.

Callon, Michel, ed. *The Laws of the Markets*. Oxford/Malden, MA: Blackwell, 1998.

Callon, Michel, ed. *Market Devices*. Oxford/Malden, MA: Blackwell, 2007.
Callon, Michel, and Michel Ferrary. "Les réseaux sociaux à l'aune de la théorie de l'acteur-réseau." *Sociologies pratiques* 13.2 (2006): 37–44. DOI 10.3917/sopr.013.0037 [accessed: 25 November 2019].
Ciccone, Antonio. "La nuova scuola economica tedesca detta socialismo della cattedra e la sua introduzione in Italia." *Atti dell'Accademia di scienze morali e politiche di Napoli* 14 (1876): 1–256.
Commercial Department of the Board of Trade. *Board of Trade Journal* 15 (1893).
Daston, Lorraine. "Objectivity and Impartiality: Epistemic Virtues in the Humanities." *The Modern Humanities: The Making of the Humanities*. Vol. 3. Eds. Rens Bod, Jaap Maat, and Thijs Weststeijn. Amsterdam: Amsterdam University Press, 2014. 27–42.
Daston, Lorraine, and Peter L. Galison. *Objectivity*. New York: Zone Books, 2007.
Dejung, Christof, Monika Dommann, and Daniel Speich Chassé. "Wissen und Wirtschaften." *WerkstattGeschichte* 58 (2011): 3–7.
Dejung, Christof, Monika Dommann, and Daniel Speich Chassé, eds. *Auf der Suche nach der Ökonomie. Historische Annäherungen*. Tübingen: Mohr Siebeck, 2014.
Diaz-Bone, Rainer, ed. *Diskurs und Ökonomie. Diskursanalytische Perspektiven auf Märkte und Organisationen*. 2nd ed. Wiesbaden: Springer VS, 2015.
Dietzel, Heinrich. *Weltwirtschaft und Volkswirtschaft*. Dresden: Zahn und Jaensch, 1900.
Dommann, Monika. *Autoren und Apparate. Die Geschichte des Copyrights im Medienwandel*. Frankfurt a.M.: Fischer, 2014.
Dougherty, Michelle, and Daniel Hinerfeld (film directors). *Sonic Sea*. 2016.
Fitzner, Rudolf. *Die Weltwirtschaft der Fettstoffe*. 8 vols. Berlin: Heymann, 1919.
Fleck, Ludwik. *Entstehung und Entwicklung einer wissenschaftlichen Tatsache. Einführung in die Lehre vom Denkstil und Denkkollektiv*. Eds. Lothar Schäfer and Thomas Schnelle. Frankfurt a.M.: Suhrkamp, 1980.
Foucault, Michel. *L'Archéologie du Savoir*. Paris: Gallimard, 1969.
Galison, Peter L. "Meanings of Scientific Unity: The Law, the Orchestra, the Pyramid, Quilt, and Ring." *Pursuing the Unity of Science: Ideology and Scientific Practice from the Great War to the Cold War*. Eds. Harmke Kamminga and Geert Somsen. Burlington, VT: Ashgate, 2016. 12–29.
Hantos, Elemér. *L'économie mondiale et la Société des Nations*. Paris: Marcel Giard, 1930.
Harms, Bernhard. "Weltwirtschaft und Weltwirtschaftslehre." *Weltwirtschaftliches Archiv* 1.1 (1913): 1–36.
Hesse, Jan-Otmar. *Die Wirtschaft als Wissenschaft. Bundesdeutsche Volkswirtschaftslehre zwischen Weltkrieg und Ölpreiskrise*. Frankfurt a.M.: Campus Verlag, 2010.
Kaufmann, Moritz. *Socialism: Its Nature, Its Dangers, and Its Remedies Considered*, based on the German work *Kapitalismus und Socialismus*, by Dr. A.E.F. Schäffle. London: Henry S. King & Co., 1874.
Knies, Karl. *Geld und Kredit*. Vol. 1. Berlin: Weidmann, 1873.
Knorr-Cetina, Karin, and Alex Preda, eds. *The Sociology of Financial Markets*. Oxford: Oxford University Press, 2005.
Latour, Bruno. *Reassembling the Social: An Introduction to Actor-Network-Theory*. Oxford: Oxford University Press, 2005.
Latour, Bruno. "Drawing Things Together. Die Macht der unveränderlich mobilen Elemente." *ANThology. Ein einführendes Handbuch zur Akteur-Netzwerk-Theorie*. Eds. Andréa Bellinger and David J. Krieger. Bielefeld: transcript, 2006 [1986]. 259–307.

League of Nations. *Journal de la Conférence économique internationale, Genève 1927*. Genève: A. Kundig, 1927.
Lenoir, Timothy. *Politik im Tempel der Wissenschaft. Forschung und Machtausübung im deutschen Kaiserreich*. Frankfurt a.M.: Campus, 1992.
Lenoir, Timothy. *Instituting Science: The Cultural Production of Scientific Disciplines*. Stanford: Stanford University Press, 1997.
Mitchell, Timothy. "Economists and the Economy in the Twentieth Century." *The Politics of Method in the Human Sciences: Positivism and Its Epistemological Others*. Ed. George Steinmetz. Durham, NC: Duke University Press, 2005. 126–142.
Neumann, Birgit, and Ansgar Nünning, eds. *Travelling Concepts for the Study of Culture*. Berlin/New York: De Gruyter, 2012.
Oppenheimer, Franz. *Weltwirtschaft und Nationalwirtschaft*. Berlin: Fischer, 1915.
Predöhl, Andreas. *Aussenwirtschaft. Weltwirtschaft, Handelspolitik und Währungspolitik*. Göttingen: Vandenhoeck & Ruprecht, 1949.
Rathgen, Karl. *Die Japaner in der Weltwirtschaft*. 2nd ed. Leipzig: Teubner, 1911.
Schäffle, Albert. *Kapitalismus und Socialismus*. Tübingen: Laupp, 1870.
Schneider, Michael C. *Wissensproduktion im Staat: Das königlich preußische statistische Bureau 1860–1914*. Frankfurt a.M.: Campus Verlag, 2013.
Scholl, Stefan. *Begrenzte Abhängigkeit. 'Wirtschaft' und 'Politik' im 20. Jahrhundert*. Frankfurt a.M./New York: Campus, 2015.
Schumacher, Hermann Albert. *Belgiens Stellung in der Weltwirtschaft*. Leipzig: Hirzel, 1917.
Speich Chassé, Daniel, and David Gugerli. "Wissensgeschichte. Eine Standortbestimmung." *Traverse. Zeitschrift für Geschichte/Revue d'histoire* 19.1 (2012): 85–100.
Speich Chassé, Daniel. *Die Erfindung des Bruttosozialprodukts. Globale Ungleichheit in der Wissensgeschichte der Ökonomie*. Göttingen: Vandenhoek & Ruprecht, 2013.
Sprague, Franklin Monroe. *Socialism from Genesis to Revelation*. Boston: Lee and Shepard, 1892.
Stichweh, Rudolf. *Wissenschaft, Universität, Professionen. Soziologische Analysen*. Frankfurt a.M.: Suhrkamp, 1994.
Stubmann, Peter. "Panamakanal und Weltwirtschaft." *Weltwirtschaftliches Archiv* 1.1 (1913): 286–309.
Thielemann, Walter. *U-Boot und Weltwirtschaft. Das Wirtschaftsleben im Gesamtunterricht der Arbeitsschule*. Leipzig/Prague/Vienna: Haase, 1916.
Thieß, Karl. "Die Weltspur der Eisenbahnen." *Weltwirtschaftliches Archiv* 1.1 (1913): 325–341.
Thilenius, Georg Christian. "Hamburgisches Kolonialinstitut." *Deutsches Kolonial-Lexikon*. Vol. 2. Ed. Heinrich Schnee. Leipzig: Quelle & Meyer, 1920. 12.
Tooze, J. Adam. *Statistics and the German State, 1900–1945: The Making of Modern Economic Knowledge*. Cambridge/New York: Cambridge University Press, 2001.
Vogl, Joseph. *Kalkül und Leidenschaft. Poetik des ökonomischen Menschen*. Zurich/Berlin: Diaphanes, 2002.
Werner, Sylwia, and Claus Zittel, eds. *Ludwik Fleck. Denkstile und Tatsachen. Gesammelte Schriften und Zeugnisse*. Berlin: Suhrkamp, 2011. 9–38.
Winkelblech, Karl Georg. *Untersuchungen über die Organisation der Arbeit oder System der Weltökonomie*. Kassel: Appel, 1853.
Wurm-Arnkreuz, k.k. Oberbaurat v. "Der Krieg und die Kulturträger." *Fremden-Blatt* 69.17 (1915): 4.

Silke Schicktanz

Normativity and Culture in the Context of Modern Medicine: A Prospective Vision of an Elective Affinity

> Cursed, cursed creator! Why did I live? Why, in that instant, did I not extinguish the spark of existence which you had so wantonly bestowed?
> – The 'monster' in Mary Shelley's *Frankenstein*; ch. 15

Mary Shelley's romantic novel *Frankenstein* (1818) symbolizes a particular position towards the means and achievements of modern life sciences: It is marked by hopes and fears related to the outcomes of such research as well as an ambivalent admiration for the courage of those inventors' desire to try out what is at least thinkable. Apart from the rich ambivalence of this novel, it is a kind of irony that 'Frankenstein' as term still appears in many public debates on genetic modification, transplantation medicine, or synthetic biology as a label to classify such innovations as repugnant and dismissible. Hereby, the creator and the creation are almost equated.

Almost two centuries later, pop-cultural phenomena such as the 2011 Hollywood movie *Rise of the Planet of the Apes*, directed by Rupert Watt, show, in a similar vein, the subtle ways in which science fiction serves as platform to address various ethical issues and concerns. This film tells the story of the application of a new somatic gene therapy targeting dementia in a near future. When applied in human primates it significantly boosts the primates' social and cognitive intelligence. Feeling inspired by the outstanding intelligence performances of the tested chimpanzees, the young leading researcher (here rather a failed hero than a mad scientist) is tempted to apply the new drug to his demented father. He does so by bypassing the common professional ethics of informed consent of his father or any review board assessment, though his actions seem justified as a case of 'ultimate ratio,' and by his passionate love for his father. While the movie overall presents more fiction than science, it touches upon many ethical concerns related to modern science: From animal testing and the underlying arbitrariness of a human-ape distinction, to the ethics of inserting artificially genes into the genome of a species, and finally experimental drug testing, the movie seems to leave nothing out.

Considering both, *Frankenstein* and *Rise of the Planet of the Apes*, what is most relevant is the observation that novels and other types of cultural narratives

https://doi.org/10.1515/9783110669398-016

allow the expression of concerns, feelings, or worldviews otherwise easily neglected by the rational language of scientific enthusiasm. These examples mark a period of almost two hundred years of cultural reflection on science and medicine, each embedded within a framework of understanding of science and narrative techniques typical to its own era.

It is exactly this interplay between the socio-cultural dimension and the norms expressed that is worth reconsideration. In the following, I use the terms 'socio-cultural' and 'cultural' synonymously as umbrella terms, to signify 'culture' in a wide sense. By this, despite the difficulties in defining such a highly loaded term, I mean meanings, understandings, and practices 'presenting' or 'symbolizing' our interactions with other beings or the environment by verbal or non-verbal forms. According to my definition, 'culture' stands here for the robust and historically (partly) stable system of ideas and practices within a collective, expressed in various forms of mediality.[1] By this definition, culture differs significantly from psychological attempts or spontaneous social inter-individual interactions.

Culture overlaps with economics, politics, and law, which are also collective, public endeavors, but is less explicit than those orders and rules. It is – to use Clifford Geertz's metaphor – a "web" (1973) in which we collectively feel cocooned without often being aware of it. Culture exists only when we give it a cultural meaning – and we tend to do so, when others ask for explanations of things that are not self-evident. However, I am aware – and it might be particular methodologically relevant – that social interactions and cultural forms can strongly differ with regard to their material, spatial, and temporal validity. The study of culture – as I here understand it – can therefore range from ethnographic, anthropological, sociological, or empirical-ethical approaches to the study of literature, media, or art.

For instance, a cultural study of German science fiction novels of twentieth century and an empirical study of social interactions by non-participatory observation of doctor-patient-communication during cancer care differ significantly with regard to their methodological accounts and, perhaps, theoretical assumptions. However, what these studies share (or can share, according to the idea I defend here) is that they allow us insights in the often-hidden meta-structure of providing 'meaning' to the way how norms or values and medical practices or ideas regarding the human body are mutually shaped in a particular setting (see Schicktanz 2007). The reservoir of cultural attempts can help us to reflect upon

1 In a broad sense, I am interested in the parallels of different media (including literature, movies, artistic performances, or even images and paintings), but of course I am aware that each media has its historical meanings and methodological constraints.

both, the hidden as well as the explicit meanings that the body 'has' and that inform so even our normative reflections of what one ought to do with the body.

The relationship between applied ethics and cultural studies is not an easy one to determine. The juxtaposition of culture and normativity can provoke objection, when norms and values are seen already as part of 'culture.' However, such an almighty concept of culture seems almost impenetrable and therefore unproductive for any reflective study. Still, there exist many ways to conceptualize the relationship between culture and normativity. At one end of this spectrum, as in post-modern cultural studies, there is a trend to see things from a social-constructivist point of view (Clifford and Marcus 1986; Bauman 1993). In such a view, all explicit and hidden values and norms are construed as arbitrary, local, or contingent. Any underlying moral message is seen as just one of many possible messages, and the body is understood as a text that can be rewritten and reinterpreted in various terms. At the other end of the spectrum, where analytical (Searle 1995) or structuralist approaches (Douglas 1970, 1992; Lévi-Strauss 1961, 1963) are located, the cultural narrative provides a vehicle to infect its listener with 'true' propositions of what is and what ought to be. However, whether one of these extreme positions really does justice to the normativity embedded and entangled with cultural interpretations of our bodily practices and images must be critically questioned.

In the following, I suggest a third, alternative way to conceptualize and analyze the productive joints and links between current approaches of applied ethics and socio-cultural studies. The aim of the approach I propose is to open up future cultural studies for an 'ethical turn,' but not in the naïve sense of ethics, which conflates it with a pre-fixed set of norms and values (whether western or non-western morality does not matter here). As I will illustrate later on, the ethical perspective that can propel cultural studies further requires a reflective, participatory, and theoretically informed take.

Given the various meanings that 'culture' has gained within the broad field of cultural studies, it seems almost impossible to provide one simple working definition of the term.[2] In the following, therefore, I will use various concepts of 'body' as analytical lenses to illustrate how culture and normativity can be fruitfully brought together at the intersection of medicine and bioethics. My restriction on this particular intersection has its historical and pragmatic reasons: science and medicine have reconstructed the (late-) modern worldview with regard to

2 This is an almost unavoidable problem that various ethicists and political philosophers struggle with when reflecting on 'culture' from a normative point of view, as Seyla Benhabib 2002 illustrates.

ontology and epistemology and almost replaced the former religious hegemony. They have also fundamentally restructured our everyday life (at least in highly-industrialized societies). Even without any explicit reference to high-tech medicine, human life is from its beginning now structured by modern practices of hygiene, birth control, or prenatal care. The body serves as a locus of all interventions, projects, and expectations. Life expectancy, still an average likelihood, is a mutually shaped result of modern medicine and social conditions, including the composition of expectations and life plans regarding education, family planning, working career, or retirement. That said, the massive impact of other factors such as capitalism, communication technologies, or political orders must be recognized as signifiers of late-modern culture. All of these culminate in the field of medicine. No current debate on medical advancements such as embryonic stem cell research, uterus transplantation, or robotics in health care can restrain the economical, communicative, or legal frameworks in which such debates *and* research practices are embedded.

It seems almost impossible to escape modern medicine's influence from the minute 'we' were created in a pre-birth stage. In a similar vein to the question that Frankenstein's monster rhetorically asks its creator, should we be anxious or thankful about medicine as structural creator of our lives? To escape the emotional stalemate of such a question, I propose to address both the ethical *and* cultural dimensions attached to this matter.

For this purpose, I want to suggest the term of 'elective affinity' (*Wahlverwandtschaft*). It is a productive concept for describing the relationship between normative ethics and socio-cultural studies of medicine that I suggest here. I use the term much as it was used by the German sociologist Max Weber, to describe the fact that two social systems or mentalities are related to or gravitate to each other, even though there is no simple causality or natural logic for such a relationship (see also Swedberg 2005, 83).[3] To construe this relationship as elective affinity is an attractive alternative to the idea that morality and culture are bound by natural kinship. Both the naturalization of cultural values as well as the universal justification attached to social norms neglect the structural differences between the study of norms (the 'ought') with the study of social facts (the 'is'). The elective decision to relate normative judgements to a social practice of morality, embedded in cultural practice, allows for a critical distance to the facts as well as to commonly made claims about how people should behave. The affinity, however, stresses the compatibility of understanding and interpreting social norms and

---

3 The term *Wahlverwandtschaft* itself stems from early chemistry and was culturally made popular by Johann Wolfgang Goethe's novel *Die Wahlverwandtschaften* 1808.

cultural practice as an expression of moral judgements how things should be. By this, a pragmatic assumption is shining through, meaning that social practice is coined by and therefore expresses moral convictions.

The programmatic shift I suggest by bringing cultural studies and ethical analysis into a productive interplay – intended to allow for an important future turn in cultural studies – intersects on four different dimensions:

First, a mutual critical reflection upon underlying basic assumptions within each area – bioethics as well as cultural studies – is needed. The cultural assumption within applied ethics – here understood as a theoretical reflection on moral practice and everyday norms – often includes limited descriptive conceptions of the self, society, or hegemonic structures. On the other side, the normative premises often buried under the attempt of a 'critical' analysis in cultural studies should be made transparent, visible, and explicit.

Second, on the descriptive-analytical dimension, we need a better, more detailed understanding of the dynamics between biomedicine, lived morality, and socio-cultural factors and how they interact in specific time-space constellations. Here, we are interested on the one side in the processes of negotiations between the somatic, material body defended in biomedicine, and the understanding of the body as locus of cultural inscriptions on the other. Examining this dynamic requires theoretical openness and detailed descriptions of global, local, or glocal developments in the area to enrich our understanding of the complexity.

Third, on a methodological dimension, we should involve lay and patients' moral perspectives beyond the scholars' view. By this, we may appreciate the complexity of the sensing body as promoted by phenomenological or some feminist approaches. Until now, bioethical expertise as well as cultural scholarliness methodologically prioritize the scholar's view on problems, outcomes, and norms. While this, as such is legitimate, it limits our epistemology as well as the range of justifiable claims for generalization. Methodological experimentation and diverse models of inclusiveness need to be addressed as important innovations for the future studies.

Fourth, on a normative dimension, we need to consider integrated approaches to addressing commonalities and parallels in the ethical and cultural space. I suggest the concept of responsibility to increase our analytical sensibility for the political, social body. The 'social body' refers to the power relations defining and ascribing vulnerability, personhood, or injustice related to medical practices. The language and concepts of bioethics need to bring in such concepts for practical and social reasons to overcome the still-unquestioned paradigms of individualism and (neo)liberalism prevailing in bioethics. This opening up of a political-ethical space allows us to rejoin attempts from both sides, from the cultural and the ethical perspectives.

In the following, I will enfold each of these four dimensions. Hereby, I understand each dimension in itself as a field worthy of future research, while the combination of more than one dimension is also welcomed. My programmatic approach takes its self-reflective starting point from bioethics. Regarding the potential of future cultural studies, I do not promote a concrete way to ethicize cultural studies, but I suggest to use the approach I propose here as an analytical lens for disciplinary self-reflection and inspiration. This implies rethinking the underlying assumptions regarding political impacts, the conceptualization of the body, the tendency toward expertocracy and scholarly elitism, as well as issues of responsibility (as scholar, citizen, or society) within cultural studies.

1 Culture and Bioethics: Where to Start

Bioethics is a wide field. On one end of its spectrum, it covers political activities undertaken to implement expert advice (e.g. in form of council or committee). In this context, 'ethics' or 'bioethics' does not mean one clear-cut scholarly way of moral philosophical reasoning; instead, it extends to a broad range of social roles and practical functions. Experts involved are rarely philosophers or ethicists, but can be any kind of academics or legal scholars. Ethicization herein aims at setting up so-called 'soft-law,' often bypassing democratic structures such as parliament or civil society. It presents a governance solution to regulate new social and technological trends (Nowotny, Scott, and Gibbons 2001; Jasanoff 2003). This part of bioethical practice suffers from an underdevelopment of political-ethical theorization as well as from a lack of deliberative and participatory methodologies (Schicktanz, Schweda, and Wynne 2012).

On the other end of its spectrum, bioethics describes a purely scholar activity, based on analytical or sometimes hermeneutic approaches, for developing theories, arguments, or concepts to address ethical problems related to medical practice or life sciences research. As Stephen Toulmin (1982) once put it, this 'applied' context has saved the life of ethics within twentieth-century philosophy. Before then, the area was generally preoccupied with theoretical debates over meta-ethics and formalistic analytical approaches; and moral philosophy fell victim to this priority. The approach of bioethics as academic endeavor can be characterized by a strong analytical methodology (i.e., considering the moral status of an embryonic stem cell in comparison to a living animal, etc.) or by a narrow focus on very practical questions (i.e., solving clinical ethical conflicts).

A third alternative aims at a cultural and empirically informed bioethics. Apart from all of the challenges and limitations inherent to such an interdisciplinary enterprise, its real advantages and strengths lie in the integration of intersubjective approaches into a wider context of political-ethical considerations. While conventional bioethics has a strong focus on doctor-patient relationships and on ethical conflicts arising for patients or citizens facing modern science, the cultural and political context (e.g., consumerism and capitalism, Western values and medical ethos, expertocracy and health illiteracy, etc.) in which such a relationship is already embedded is otherwise neglected or disregarded.

Re-contextualizing bioethics as an intellectual activity that acknowledges the political needs and requirements for the public as well as for the academic means to bring back the political-ethical argument. But why was academic bioethics stripped of political-ethical considerations?

This can be explained by at least three factors. First, as a close political-institutional perspective reveals, medicine and life sciences operate mainly outside of parliamentary political structures in many western democracies. While other areas of social life such as trade, work, or education have been highly politicized and heavily regulated since their beginnings, medicine and health care often operate in a rather loose web of political structures. The number of state laws regulating medicine and life science research is rather specific (and are often only a result of public ethical controversies, i.e., as it was the case for embryonic stem cell research, abortion, or organ transplantation). In most western democracies, it is an expertocracy that self-regulates the dos and don'ts by soft-law.[4] Differences between countries exist and it is therefore crucial to study and reflect on the medico-legal culture when examining any particular medical practice and its ethical framework.

A second perspective, in line with a more Foucauldian understanding of 'biopolitics,' acknowledges that there are strong state or institutional interests directly implemented in modern medicine and health care (Lemke 2006). However, they remain implicit and are hidden in the rational language of needs, diagnoses, or treatments. They are already internalized by modern citizens or patients looking after their healthy lifestyles, reproductive behaviors, or end-of-life planning. Such a biopolitical perspective in governance risks eliminating the individual's perspective. The political is all and everywhere, and therefore the productive, analytical tension of the political vs. the non-political is suspended (Bishop and Jotterand 2006). This supra-political perspective might be relevant to understand hegemonic grammar and hegemonic position but it underrates and

4 This argument is supported by the immense impact not only of national academies of science and medicine but also of international organizations such as the World Medical Association or the World Health Organization etc. on the health policy regulation.

oversees practices of resistance or renitence by affected persons (Fraser 1989). Another worry about the domination of biopolitical power as discursive power is that it hides biosociality, unutterable sensation, or embodiment as human factors (Hazan 2015, 27).

A third explanation acknowledges that dominant bioethical approaches such as utilitarian, deontological, or principle-oriented approaches are always embedded in political-ethical assumptions of modern liberalism. However, this political framing became almost invisible because of its presumption of priority given to individualism, to the moral principle of individual choice, and to respect for individual autonomy. This 'naked' version of liberal bioethics suffers from the fact that political assumptions about liberalism entail much more than just this triumvirate. Political-ethical assumptions of liberalism should always include in-depth analyses of the relationship between state, expert, and the citizen; questions of tolerance and its limits; the meaning of collectivity for self-understanding and understanding other's citizens interest; and so on. All these questions surface from time to time in conventional bioethics, but are yet insufficiently addressed. The alternative would be to enlarge the bioethical analysis from the bedside beyond the doors of the hospital: to explore how inter-individual decisions, expectations, and negotiations of lays and professionals are embedded in a broader context of state-market-citizen relationships. Of course, such a zoom is methodologically challenging and limited. However, focusing, for example, on central actors or new political institutions such as patient organizations and patient collectives, allows for such an expanded perspective, which brings together the socio-cultural practices of such collectives (Brown et al. 2004), their political-ethical claims and legitimacy, as well as their impact on bioethical controversies (Beier et al. 2016; Raz, Jordan, and Schicktanz 2014; Schicktanz 2015). Such a normative perspective would complement the cultural study of the collective body – in its explicit as well as more implicit versions – reflecting on the gendered, the disabled, and the colored body, as those bodies are always collectivized.

2 The Body as Local Inscription or as Global Soma: The Dynamics of Medicine, Morality, and Culture

Cultural studies and STS (science and technology studies) have revealed many astonishing facts regarding the dynamics of medicine, cultural practice, and norms. On the one hand, there are areas that can be characterized by strong local differences or even local resistance against global standards. One example here is the non-acceptance of postmortem organ donation and brain death in Japan and in many other countries of the Asian or Arabic world, while the western world seems to see this as self-evident and taken for granted (Lock 2002; for limits within the western world, see Schicktanz and Wöhlke 2017). On the other hand, there are cases of strong global uniformity and global conformance in medicine, based on assumptions of the body as purely materialistic soma, detached from any interpretation or value (see also Joralemon and Cox 2003).

An example for the global spread of new body technologies is the genetic selection of in-vitro fertilized eggs before they are implanted into a woman's uterus, called as pre-implantation diagnosis (PGD). It is now a commonly accepted practice in most regions of the world across the western/eastern division, if such expensive reproductive medical technologies are affordable. Given the extreme concerns expressed in the early 2000s when PGD was established, its triumphal procession since then is quite impressive and a result of an international active community of scientists and ethicists defending the idea that the fertilized eggs are not yet morally relevant as the 'adult' human body and its attached personhood.

However, there are also more complex examples, of how modern medicine is both globally spreading *and* locally adapted to fit into the respective cultural context. This process can be understood as 'glocalization.' Here, the concept of glocalization is understood to analyze the process of negotiation, refraction, and mimicry between globalization and localization (see Bauman, 1998; Roudometof 2016, 1–42). In contrast to globalization – here understood as the modern version of a market-driven soft-colonialism – glocalization as a conceptual approach sensitizes for a detailed analysis of how the global and the local are negotiated case by case in medical and health care practice. The local-global relationship of various medical practices might differ with regard to their legal-ethical frameworks (for example in the case of organ donation and its different regulations worldwide: Shepherd, O'Carroll, and Ferguson 2014; Lopp 2013; Randhawa and Schicktanz 2013). Economical aspects, regarding when and how much a new medical technology is covered by public health insurance, are also an obvious striving force for global spread. While some public health systems cover all costs for in-vitro fertilization for every woman, others cover a limited number of treatments only for heterosexual couples (Brigham, Cadier, and Chevreul 2013). Even the scientific practices might also differ, for example, in which gynecological examinations are conducted in the US, France, or Germany, as once observed by Lynn Payer (1989). More often it remains globally robust, because common medical diagnoses or treatments are now conducted along international standards to satisfy the quality criteria of the 'gold standard' of the World Medical Association.

To provide a more detailed picture of what I mean by 'glocalization,' I will expand upon the current practice of *surrogacy* as an illustrative example of the way that medical practice, culture, and morality are mutually negotiated. Surrogacy is an artificial reproductive treatment where a so-called surrogate, the gestational mother, is implanted with a genetically often non-related embryo, then, after birth, hands the baby over to the so-called social parents. The surrogate and the intended parents are bound via a contract, and the intended parents normally adopt the child after birth or are legally acknowledged because the embryo is genetically related to them.[5]

The idea to implant a fertilized egg into a womb of a woman not genetically related became technically possible after the introduction of in vitro fertilization, from the end of the 1970s on. Two decades later this practice has spread worldwide (see Mitra, Schicktanz, and Patel 2018, 3–6). While some South Asian regions are often portrayed in the media as hot spots for surrogacy markets, surrogacy is also now practiced in all other continents of the world. However, the concrete practice varies extremely with regard to the selection process of surrogates as well as access for potential parents. The surrogate can include a close relative acting out of 'altruistic' reasons, or an almost unknown person selected from an internet databank. Defense of a commercialized practice of surrogacy sees the surrogate as a 'womb to rent,' and the delivery of a baby as bio-labor, which needs to be reimbursed in 'fair' prices. Alternately, the proponents of 'altruistic' practice assume an emotional bond between surrogate and baby via physical unity, and therefore want to avoid any commercialization or allow bonds between the child and the surrogate. The legal justifications for eligible intended parent(s) differ strongly, too: in India, currently, only heterosexual couples are allowed for medical reasons, while in Israel religious reasons determine who can be a surrogate in relation to the intended parents (e.g., only a Jewish surrogate for Jewish-intended parents). In California, homosexual couples or single (male or female) parents can approach a surrogate as intended parents. This is for conservative reasons in many countries not possible because their sexuality is seen as 'unnatural' or 'immoral.' Moreover, the scientific practice differs among countries according to the selection procedure of fertilized eggs or the absolute number of embryos to be implanted into the surrogate's uterus. In most European countries, one, two, or a maximum of three embryos are permitted for implantation, while in the US or India more are possible, despite the significant increase of medical risks associated with multiple

5 In some cases, sperm or egg or both stem from the social parents, but there are also cases where both, eggs or sperms, are donated by another third party.

pregnancies for the surrogate and for the fetuses. In India, selective abortion is practiced to reduce again the number of fetuses if the intended parents want this (see Mitra and Schicktanz 2016).

These variations are inevitably linked to different ethical and social debates regarding the problem of exploitive market conditions for surrogates in low- and middle-income countries (such as India or Thailand), the right of reproductive freedom for intended parents or surrogates, the question of agency of surrogates under unequal social conditions, and the right of intended parents to select or to not come for the baby. The social concerns might be even more general regarding the impact of such a medical practice on the mundane understanding of kinship and motherhood, gender, or ethnicity – always attached to the body.

While an international overview of the debate offers a broad or even balanced picture of the 'pros' and 'cons,' the question needs to be posed whether national, local discourses are also so broad. They seem rather be dominated by few or selective arguments. Such a cultural taming of the ethical debate must be understood as a result of negotiating between the local and global context in which such debates are evolving and – at least for a particular moment in time – are fixable as culturally 'significant.' From a distant, comparative view there gleam some peculiarities: for instance, Indian sociologists have pointed out, despite critiques of the large economic and caste disparities, that it would be important to see the agency and opportunities for self-determination for surrogates, even in situations of commercial surrogacy (Tanderup et al. 2015). In Germany, the agency of surrogates is rarely considered as leading point but ethicists have emphasized rather the 'best interest of the child' as a criterion of legitimization. Whether this argument results in a permission or moral veto is dependent on how the 'best interest of the child' is then concretely interpreted (see also Wiesemann 2016, 133; Beier and Wiesemann 2013). In the US, various scholars have focused on the social risks of commercialization as it might increase social disparities between races or classes and could lead to the exploitation of poor women or to a racist practice of dismissing non-white women as surrogates or egg donors (Thompson 2005, 66). Again, such points to consider are yet rarely addressed in the German context, although might perhaps in the same way be relevant once the practice is implemented.

While none of these points is made exclusively in any of these three different national discourses, it is striking how some main lines of argument prevail in each context. We need more detailed studies to understand how the bioethical discourse depends on the culture in which it is embedded. Such a descriptive-analytical reflection, however, does not solve the quest for a more rational or universal understanding of moral norms – a project still worth to be defended as an ideal orientation, not as a simple solution.

But indeed, such cultural framing of differences in the discourse[6] has led to some misunderstanding within normative ethics: This misunderstanding assumes that the socio-cultural study of differences in bioethical positions results in "normative relativism" (Schicktanz 2018, 117–119). Normative relativism means that we are not allowed to question each other about the local forms of norm validation and norm hierarchy, because all moral considerations are only locally valid. In a more pervasive form of neoliberal argumentation, such relativism is even used to justify any permissive stance towards new technologies: because nation A, B, or C (e.g. the US, India, or the UK) is doing X (e.g. surrogacy), it would be also acceptable in D (e.g. Germany) to do X. The underlying premise is then that there exists no universally valid argument to forbid it.

However, the here-defended idea of 'post-conventional' bioethics' interest in cultural and social studies of medical practice means nothing more or less than contextualizing the leading moral justification by taking into account the 'real perceived' social conditions in which the respective agents (e.g., patients, doctors, citizens) live, as well as the interpretation given to these living conditions. 'Post' because conventional bioethics neglects any cultural embedding or social factors such as gender, ethnicity, class, etc. to influence bioethical positions. Such a distinction between conventional and post-conventional bioethics might provoke objections, because the generalization does injustice to individual scholars who are already open to interdisciplinary exchanges with sociology and cultural studies. However, it is used here to mark a more general shift in the field without discrediting any of those former approaches. Furthermore, it is important not to conflate post-conventional bioethics with postmodern approaches, as the analytical focus on non-relativistic normative traditions is still viable. In a same vein, it would be productive to reflect on normativity within cultural studies by leaving the conventional paths of anti-normativity or radical constructivism. In such a sense, future cultural studies could try to embed their analysis in the lived experiences of moral reflections – not just considering moral standards and values as taken for granted, but to put more emphasis on the human practices of doubts, concerns, sensing dilemmas, seeking deliberation, and how this is culturally mediated.

6 This might be explained by national law and local regulation, though the law also depends on cultural accounts of what is seen as ethical acceptable or not, see Hansen 2012. According to most philosophers the proper way should be that law follows ethics and not the other way around. However, in political practice, legal regulation is sometimes quicker implemented than a thorough ethical deliberation takes place. Therefore, it is important, from a cultural point of view, to assume a rather complex interplay between law and ethics.

The production of hopes, expectations, or fears is accessible through the study of cultural media by understanding the body as text or as narrative (see Dekkers 1998; Frank 1995; Squier 2004, 20–21). Visual and performative art as well as literature are media that allow access to the emotional dimensions in form of a bodily narrative. Examples for this can be found in the work of the French artist Orlan. She examined in her own body art the relationship of pain, medical surgery, beauty, gender stereotypes, and norms. A very different example is Philip Roth's novel *Everyman* (2006), where he examines aging, dying, and end-of-life-planning, and how they impact the relationship of body, personhood, and narration. Both Orlan and Roth share the attempt to display pain and fear of death by pointing to scars, fragility, dependency, and by narrating a lived body in its particular social, moral, and cultural embedding. And both provide a narration of moral doubts and concerns: where are the advantages of modern medicine, where are limits?

However, it is also necessary to contextualize the moral claims made in a historical course of the discourse. The search for reasons for differences or similarities in arguments and norms – and by this, the transgression of geopolitical boundaries – serves as part of a rationalization of each claim made. This is a main condition for the 'elective affinity' of applied ethics and cultural studies and can be seen as productive future for both disciplines.

The challenge of such an approach is not to lead to 'factual fallacy.' Such a factual fallacy would mean jumping directly from empirical or descriptive findings of how people actually think or how practice currently works to the normative conclusion about how it *should* be. Such a normative positivism must be avoided. Instead, we need a critical assessment of how any moral claim or argument brought forward is culturally embedded in a hegemonic presentation. A transparent strategy for a comprehensible, proper making of a *practical-moral judgment* refers to an uncontroversial understanding of practical-moral judgment as *mixed* judgment. The mix consists of a prescriptive (normative) *and* descriptive (empirical, factual) statement combined, but avoids any crypto normativity.

Let us consider for a moment the above-mentioned example of surrogacy. Consider that somebody states in a public debate that surrogacy should be allowed in Germany, because it allows women a good income and women want this. This claim is a conclusion as practical-moral judgment and built on normative premises (A) *and* on descriptive premises (B). The normative premises can be summarized as the following: A1) Surrogacy is as such morally not wrong and A2) all women should have the right to a good income. Descriptive assumptions that underlie such a conclusion are: B1) Good income is the main interest for women, which presents an empirical question as to whether this is true and women would not value other opportunities to gain more money elsewhere; and B2) It is culturally uncontroversial what defines 'good income'; or B3) Women have no other chance to get a good income than by surrogacy. What we see by this is that even if we would agree all on A1) or A2), the moral conclusion depends in a paretic version from the descriptive part. This descriptive part requires therefore socially robust knowledge (Nowotny, Scott, and Gibbons 2001, 116–119) about social practice, effects, opinions, etc. Hence, practical-moral judgments do not only depend on common, shared reflections about what is ethically right or wrong, but in a similar way on shared robust interpretations of the world.

3 The Sensing Body as Situated and Affected: Enlarging the Experts' View

Conventional bioethics has a one-sided tendency for the expert discourses. Such a tendency has its historical roots in the analytical tradition of ethics as well as in the close orientation towards the legal discourse. While the analytical approach is not necessarily expertocratic, its formalistic methods and abstract language often hinders non-experts in participation. The legal discourse definitively has an expertocratic manner, given the fact that public education never ever touches upon it and we mainly leave it to specialists, apart from some areas where lay judges are involved. For post-conventional bioethics, the critical assessment of expertocracy is a central element (Schicktanz, Schweda, and Wynne 2012). Cultural insights similarly foster skepticism towards the idea that those not directly affected by or outside of the messiness of everyday life struggles (such as physicians, academic ethicists, or lawyers) can anticipate hypothetically and properly such a complexity in its ambivalence.

Whether this intellectual representation works for the perspectives of persons who are socially marginalized or excluded must be problematized, though, for these persons, as social and political inequality hampers their opportunity to be represented in exclusive circles of academia or other elite groups. Marginalization is here mainly based upon involuntarily, non-mutual membership such as belonging to a particular gender, ethnicity, or nationality. Such a group membership was not voluntarily chosen by these persons, but assigned to them from outside. Marginalization only takes place if a particular group identity is seen as 'negative' (Williams 1998, 15–18). As Melissa Williams has convincingly shown, typical examples of such marginalization have concerned women, people of color, or people with disabilities, depending on particular historical or political-cultural conditions. In relation to such a social exclusion from many intellectual resources or access to socio-political decision making, there is a serious risk that social stereotypes related to such a negative group assignment hinder those in power to decide to trust testimonials of those from marginalized groups. By this, many public and legal discourses suffer from "epistemic injustice" (Fricker 2007) due to unbalanced power relations in the presentation of knowledge. Countermeasurements include awareness increasing participation or representation by members of one's own group. These very general considerations are particular relevant in the field of medicine (Schicktanz 2015). This is because persons with chronic illness or with a disability have very specific perspectives and insights in the challenges of pain/bodily experiences or social and spatial exclusion when it comes to bioethical issues related to their condition. Being marginalized and affected (meaning that decisions will have a causal effect on them, see Schicktanz, Schweda, and Franzen 2008; Schicktanz 2015 for a detailed definition of 'affected persons') justifies a significant ethical priority of such voices. The embodied or affected experiences as well as the illness identity are legitimate and valuable sources for a privileged understanding of the everyday complexity relevant to the bioethical issue at stake. People in the fourth age, with dementia, or with autism, are too quickly excluded because of the non-translational content of their experiences (Hazan 2015, 47).

Experts, in contrast, cannot phenomenologically rely on such experiences. Of course, they can indirectly reconstruct such experiences by referring to social, cultural, or psychological studies. But finding the right language, the right translation, to transform these special experiences into a social, publicly shared space is not trivial. Hence, the direct involvement of affected persons – in one way or another – is a necessary element for any future bioethics. Because limited resources and basic needs of persons affected might restrict their interest or factual opportunity to take actively part in such discourses and debates, new, joint methods in the cultural and socio-empirical studies can bridge the need for such a reconstruction. Here I see a particular area for future cultural studies to explore various means and methods to bring the affects, interests, vulnerabilities, and needs of those excluded into the broader discourse. In terms of exploration, more anthropological or ethnographic studies of people in the fourth age (Hazan 2015, 46–47, 71) or with dementia are needed to challenge stereotypical and often discriminating views of them as "almost dead" or "cognitive zombies," to enlighten their untypical, but yet human nature. Experimental designs are required to explore the social and ethical issues of biotechnical innovations and their impact on our understanding of humanity. Following the course of cloning novels served Solveig Hansen in her dissertation (2016) as an orientation to examine the historical practice of social othering. By her joint cultural and ethical analysis of how clones have been anticipated and depicted in cultural discourse, she provides a thorough and complex picture of how our moral relationships are built upon assumptions of sameness in quality (and not in quantity as the clones imply) but also independence and self-reliance as a basis for mutual respect. The limited socio-empirical perspective access to these future scenarios can be productively complemented with such cultural studies of novels and anticipations.

However, this does not mean to incorporate any view of an affected person in an uncritical way. All perspectives shall be reflected with regard to their moral and epistemic claims. Assuming that affected persons are neither able nor willing to transcend their own personal interest into the social sphere is, however, problematic. Emotional as well as biased views are similarly common in experts' debates about patients, persons with disability, or others who are particularly affected. Therefore, any particular position or moral perspective needs to be understood as "situated knowledge" (Haraway 1988) or "situated ethics," but this does not hinder the opportunity to enter a mutual discourse for finding the better argument.

Transferring these thoughts to a future of cultural studies means, for example, to radically revisit the divide of high/classical and pop culture. Especially for any work on medicine and literature, pop culture, such as 'trashy' science fiction or medical thrillers, provide deep insights in common moralities and understandings of modern medicine and biotechnology (see, for instance, Pethes 2005).

4 Body and Responsibility: The Certainty of Moral Tensions as Conjunctions of Deliberation

To illustrate my understanding of post-conventional bioethics as a continuous challenge between practice and theory, between descriptive and normative claims, I want to refer to the performance *Zerreißprobe* (tensile test) of the Austrian artist Günter Brus from 1970. As a performance artist he shocked the public by making his body to the subject of artistic performances. He injured himself by cutting his head and thigh with a razor blade and arranging his vulnerable, naked body half stretched and half hanging within a web of strings crossing a room. Hereby, the vulnerability of the flesh was shown by means of the extreme display of a body disfigured by pain and by interventions from the outside. By being thus displayed, the body itself becomes both the medium of the artistic work, and the scene in which it takes place. It is this mutual meaning and interaction that symbolizes the performance of bioethics by focusing on particular events or single bodies but being aware of the embedding of such entities in a broader context. In a second line of thought, the work of Brus also marks in an abstract sense the particular meaning the 'body' has as intersectional space between bioethics and cultural studies (see Barkhaus and Fleig 2002, 9–23, 27–36). Although the cultural irritations such artistic presentations of the body produce on their own is not the point here, I assume that almost every viewer of such a performance feels moved, touched, or disturbed. This common sense of vulnerability of the human as body *and* as person is a certainty that provokes the claims of relativism and arbitrariness – Brus's tensile test serves here as a litmus test for the tensions built into modern medicine and biotechnology where they produce, via their innovations, such anticipations of bodily vulnerability and personhood.

So it is precisely the field of body modification and related bioethics where we observe a clash of perspectives in two ways, but which can also serve us as a productive intersection for an elective affinity.

In the first place, there is a serious distinction in the normative ways of ethically judging how we assess the right of self-determination towards our bodies; secondly, there are solid variations in how the body and embodiment are theoretically addressed. I have suggested earlier a methodological approach of making the tension between different meanings of body and autonomy explicit by setting out a dialectical method for heuristic use to be made of the recent dichotomies in bioethics (Schicktanz 2007). By this we cannot easily resolve moral dilemmas, but we can proceed in a dialogical way for addressing theoretically the various descriptive and normative claims. At least, we will overcome simplistic pro- and contra- debates and we are opening up instead of closing down debates for various theoretical relationships between autonomy (and other relevant normative concepts) and body/embodiment. This provides a central interface for the ethical reflection about who can decide what, when, and how about one's own body. What elements of a person can be regarded as available or unavailable at which points in time during the process of this person's life or dying? Whether the 'body boom' in ethics is something avoidable can be questioned (see also Shildrick and Mykitiuk 2005). Even supporters of the liberal conception of self-determination, who primarily recognize the principle of non-maleficence (the general rule not to harm) as morally equivalent, need to clarify the idea of socio-cultural dimensions of embodiment and the framework for the meaning of bodily unavailability within social interaction. Who or what is the other entity that must not suffer damage, and what constitutes damage (Schicktanz 2007)?

Having said this, I need to propel my own focus on the relationship of the normative principle of 'autonomy' on one hand, and the conceptualization of the body/embodiment a bit further. The political-ethical sphere of social interaction requires a constant concern for more than individual autonomy. The most important concepts are then justice and responsibility. Starting from cultural observations and political practice, the bioethical enterprise is not only to set out ideal theories of justice or responsibility but to address witnessed forms of injustice and irresponsibility. For sure, our sense for such immoralities is neither independent nor free of theoretical presumptions of what justice or responsibilities are. The *ex negativo* start is often more robust regarding our intuitions and knowledge, but it does not free us from a reflective approach to clarify such presumptions.

Considering an opening-up of cultural studies for ethical thinking might be facilitated by art or performative acts that confront us with the limits of textual analysis, rising issues of affects. However, it would be a great self-restriction to limit the ethical perspective to the sphere of aesthetics. Critical sociology and cultural studies studying the presentation or performances of the liminal, excluded, or resistant human existences share a long-standing tradition with concerns about injustice along class, gender, ethnicity, national belonging, or injustice regarding the exclusion of disabled or sick persons. Their arsenal to address injustice is manifold, be it a dense description of exclusion mechanisms or a quantitative summary of the suffering of discriminated parties. They can bring often-unheard voices into the discourse and highlight the agency of parties often neglected or denied: women, children, the ill, or others often overseen (de Beauvoir 2000). This sociological practice, according to Wayne Brekhus (1998), devotes greater epistemological attention to "politically salient" and "ontologically uncommon" features of social life. Addressing women, the elderly, homosexuals, etc., means "marking" those excluded entities, but this practice unreflects or even repeats the hegemonic grammar and leaves the "unmarked" (whites, heterosexuals, men, etc.) unrevoked. Brekhus' critique of the epistemic practice of identity labels and problematic singling-out is important and highly relevant to overcoming simplistic, unreflective assumptions of the good and bad guys. We need to acknowledge that this epistemic practice within sociology is already embedded in a normative theory of justice and fair treatment of which ethics can help to unmask them in future co-operations.

I have suggested somewhere else (Schicktanz and Schweda 2012) that the concept of 'responsibility' is particularly helpful in linking everyday languages of morality and ethical-normative reflection. Providing a theoretical formula for what the concept of responsibility entails offers a way to explicate moral claims of self-responsibility, social or professional responsibility, or family responsibility ubiquitous in medical practice, health policy, or health communication. While of course only working in limitations, the concept of responsibility is not just a moral idea among others, but as a meta-ethical concept, it provides a meaning of how ethical properties are formulated, logically expressed, and epistemically assessed. Therefore, using terms of responsibility means that we are explaining normative claims embedded in social presumptions about relationships. This helps to translate ordinary folk language into a more abstract form to proof for consistency or to detect contradictions.

This need for translation – from the everyday moral grammar to the theoretical-analytical level and back – is an endeavor that goes beyond the conventional understanding of 'education.' It requires theoretical sensitivity for what constitutes a responsible social relationship: It is always embedded in space and time and the relationship is enriched by cultural assumptions (i.e., in the case of the earlier mentioned surrogacy: what constitutes good parenthood; who counts as morally relevant actor: only the intended parents, or also the mother, the doctors, the government, etc.)?

Responsibility, however, in current sociology – especially medical sociology – has been narrowed down to a critical notion of moral imperatives, synonymous with the social practice of blaming and shaming (Rose 2006, 4; Arribas-Ayllon, Featherstone, and Atkinson 2011), and applied to criticize biopolitical strategies. As such, the sociological notion of 'responsibilization' emphasizes a very special application of the term 'responsibility' focusing on the individual or the family as both the moral agent and the moral object in biopolitics. As an alternative, the productive junction with ethical theory alludes that this application has a strong tendency to reduce the understanding and practical usage of responsibility and that there are better, more refined ways to address responsibility in its many dimensions by using a detailed, transparent description of normative complexity (Schicktanz 2016).

5 Summing Up: An Elective Affinity between Bioethics and Socio-Cultural Studies of Medicine and Life Sciences

I have suggested in the beginning the concept of 'elective affinity' to bring forward a new relationship between bioethics and cultural studies – and would mean somehow a double turn-over: a cultural turn for bioethics and an ethical turn for cultural studies. Whether cultural studies have already adopted such an ethical gaze, I am not sure. However, a current trend to differentiate 'critical' cultural studies can be read as tendency to explicitly address issues of marginalization, discrimination, and exclusion. Being informed by various strands of critical theory might, however, not be the only future direction for cultural studies. Other approaches stemming from applied bioethics to address various ethical and social concerns can be innovative and helpful, as suggested here.

Normative studies and moral languages provide access to moral practice and help to signify the consistency as well as inconsistency in moral practice and ethical thinking. There exists no absolute demarcation between moral practice in everyday life and theoretical ethical reflection, rather it is a continuum with smooth transitions. The theoretical concept of "reflective equilibrium" serves as dialectical model of normative judgment between theory and practice to describe and reflect on this continuum as philosophical method (Daniels 1996).

It is, however, seen as a legacy of Max Weber's idea of value-free sociology and economical sciences that until today lead to an unrealistic or even wrong ideal of value-free sociology or cultural studies. As the dispute over 'value-free' (*Wertfreiheit*) vs. 'value judgment-free' (*Werturteilsfreiheit*) revealed already almost a half century ago (Albert and Topitsch 1971), it is not only a strange mythos of modern social and cultural studies to be value-free, as the study of values as well as the explication of values is part of any scientific activity – in social and cultural sciences, as well as life science areas such as medicine or agriculture aiming for 'saving life,' 'curing disease,' or providing 'better living conditions' or 'sustainability.' However, such scholars should be explicit and transparent when making value judgments instead of allowing crypto-normativity in scientific terms or scholarly language. Terms and concepts such as 'critique,' 'power,' 'social inequality,' 'vulnerability,' and 'colonialism' always reflect a pejorative, moral meaning that we cannot escape as either speaker nor listener (Fraser 1989, 17–20). However, not any value judgment can claim to count as well-considered judgment. Applied ethics and moral philosophy provide the methodological arsenal to win this battle over crypto-normativity and hidden values in scholarly work.

A flourishing, productive elective affinity between bioethics and cultural studies requires a crucial clarification about all own normative premises on the why, the how, and the what of ongoing research. The 'why'-question focuses on the motivations and programmatic reasons behind a study and for singling out a problem to being relevant for in-depth examination. The 'how'-question follows the lines of a chosen methodology and asks how far normative premises are already embedded in the research program (Merton and Storer 1973, 229–250). For example, does the selection of qualitative vs. quantitative methods only refer to epistemic assumptions of generalization or depth, or might it also include who should be in the focus of examination (the lay public, the experts, the media, etc.). The 'what'-question critically reflects which underlying assumptions of injustice, responsibility, or vulnerability are already attached to the selection of a particular topic (the topic of terror in Europe, the topic of dementia in India, the topic of education in Africa, etc.). How does the spot on this topic risk shading other topics, and is the priority well-justified?

From a bioethicist's point of view, there are many good arguments as to why and how socio-cultural studies are important or even indispensable for a well-defined and well-argued problem definition (what is the moral problem we want to solve) as well as the practical recommendations often following bioethical inquiries 'how to solve' the conflict in the future. The concrete function of socio-cultural empirical insights for norm justification is instead very controversial and perhaps for the purpose of the here-proposed collaboration modus not needed.

Cultural studies provide not only, but still importantly, a challenge to monological or one-sided perspectives on bioethics. From a theoretical point of view, the solution to the problem of legitimacy lies not in simple forms of public participation in research and policy making, but in a conceptual analysis of the kind of perspectives needed. I am here assuming that there is no single, ultimate perspective. Only a combination and pluralization of different perspectives can offer us an approximation of the 'whole picture.' This requires a systematic adoption of other perspectives (Schicktanz 2015, 251–252). With this increased complexity, we enhance our understanding of the dependence of morality on affects and social dimensions of power. Thinking with stories, narratives, or images as cultural studies provides the arsenal and methodology that help us to test for consistency, for the wideness of the chosen perspective, or for the peculiarity of it. However, there are new risks such as more hidden morality, exclusivity, and ambiguity awaiting such concerted efforts. It would be worth going still further in this direction in the future.

Acknowledgement: I am particularly grateful to Haim Hazan, Tel Aviv, and Solveig L. Hansen, Göttingen, for critical-constructive comments to an earlier version, and would also like to thank Doris Bachmann-Medick and Jens Kugele, Giessen, for their editorial feedback.

References

Albert, Hans, and Ernst Topitsch, eds. *Werturteilsstreit.* Darmstadt: Wissenschaftliche Buchgesellschaft, 1971.

Arribas-Ayllon, Michael, Katie Featherstone, and Paul Atkinson. "The Practical Ethics of Genetic Responsibility: Non-Disclosure and the Autonomy of Affect." *Social Theory & Health* 9.1 (2011): 3–23.

Barkhaus, Annette, and Anne Fleig, eds. *Grenzverläufe. Der Körper als Schnittstelle.* Munich: Fink, 2002.

Bauman, Zygmunt. *Postmodern Ethics.* Hoboken, NJ: Wiley, 1993.

Bauman, Zygmunt. "On Glocalization: or Globalization for some, Localization for some Others." *Thesis Eleven* 54.1 (1998): 37–49.

Beier, Katharina, and Claudia Wiesemann. "Reproduktive Autonomie in der liberalen Demokratie – Eine ethische Analyse." *Patientenautonomie. Theoretische Grundlagen –*

Praktische Anwendungen. Eds. Claudia Wiesemann and Alfred Simon. Münster: Mentis, 2013. 199–216.
Beier, Katharina, Isabella Jordan, Claudia Wiesemann, and Silke Schicktanz. "Understanding Collective Agency in Bioethics." *Medicine, Healthcare and Philosophy* 19.3 (2016): 411–422.
Benhabib, Seyla. *The Claims of Culture: Equality and Diversity in the Global Era*. Princeton: Princeton University Press, 2002.
Bishop, Jeffrey P., and Fabrice F. Jotterand. "Bioethics as Biopolitics." *Journal of Medicine and Philosophy* 31.3 (2006): 205–212.
Brekhus, Wayne. "A Sociology of the Unmarked: Redirecting Our Focus." *Sociological Theory* 16.1 (1998): 34–51.
Brigham, Karen, Benjamin Cadier, and Karine Chevreul. "The Diversity of Regulation and Public Financing of IVF in Europe and its Impact on Utilization." *Human Reproduction* 28.3 (2013): 666–675.
Brown, Phil, Stephen Zavestoski, Sabrina McCormick, Brian Mayer, Rachel Morello-Frosch, and Rebecca Gasior Altman. "Embodied Health Movements: New Approaches to Social Movements in Health." *Sociology of Health & Illness* 26.1 (2004): 50–80.
Brus, Günter. *Performance Art: Zerreißprobe [Tensile Test]*. https://www.youtube.com/watch?v=Gmtm_A5fvbc. Munich, 1970 [accessed: 1 July 2019].
Clifford, James, and George E. Marcus, eds. *Writing Culture: The Poetics and Politics of Ethnography.* Berkeley/Los Angeles/London: University of California Press, 1986.
Daniels, Norman. *Justice and Justification: Reflective Equilibrium in Theory and Practice.* Cambridge: Cambridge University Press, 1996.
De Beauvoir, Simone. *Das andere Geschlecht. Sitte und Sexus der Frau*. Reinbek: Rowohlt, 2000.
Dekkers, Wim. "The Body as a Text: The Interpretive Tradition." *Advances in Bioethics*. Eds. Rem B. Edwards, Edward E. Bittar, and Martyn Evans. Stamford, CT: JAI Press Inc., 1998. 209–228.
Douglas, Mary. *Natural Symbols: Explorations in Cosmology.* London: Barrie & Rockliff, 1970.
Douglas, Mary. *Risk and Blame: Essays in Cultural Theory.* London/New York: Routledge, 1992.
Frank, Arthur W. "Illness as a Call for Stories." *The Wounded Storyteller: Body, Illness, and Ethics.* Ed. Arthur W. Frank. Chicago/London: University of Chicago Press, 1995. 53–74.
Fraser, Nancy. *Unruly Practices: Power, Discourse and Gender in Contemporary Social Theory.* Minneapolis: University of Minnesota Press, 1989.
Fricker, Miranda. *Epistemic Injustice: Power and the Ethics of Knowing*. Oxford/New York: Oxford University Press, 2007.
Geertz, Clifford. *The Interpretation of Cultures: Selected Essays*. London: Hutchinson & Co, 1973.
Goethe, Johann Wolfgang. *Die Wahlverwandtschaften. Ein Roman.* Berlin: Insel (2012 [1808]).
Hansen, Solveig L. "Utopische Dynamiken von Macht, Reproduktion und Gesetz? Ein interdisziplinärer Beitrag zur Law-and Literature-Debatte." *Gehört mein Körper noch mir?* Ed. Susanne Beck. Würzburg: Nomos, 2012. 193–221.
Hansen, Solveig L. *Alterität als kulturelle Herausforderung des Klonens. Eine Rekonstruktion bioethischer und literarischer Verhandlungen.* Münster: Mentis, 2016.
Haraway, Donna. "Situated Knowledges: The Science Question in Feminism and the Privilege of Partial Perspective." *Feminist Studies* 14.3 (1988): 575–599.
Hazan, Haim. *Against Hybridity: Social Impasses in a Globalizing World.* Cambridge/Malden, MA: Polity, 2015.

Jasanoff, Sheila. "Technologies of Humility: Citizen Participation in Governing Science." *Minerva* 41.3 (2003): 223–244.
Joralemon, Donald, and Phil Cox. "Body Values: The Case Against Compensating for Transplant Organs." *The Hasting Center Report* 33.1 (2003): 27–33.
Lemke, Thomas. "Genetic Responsibility and Neo-Liberal Governmentality: Medical Diagnosis as Moral Technology." *Michel Foucault and Power Today: International Multidisciplinary Studies in the History of the Present.* Eds. Alain Beaulieu and David Gabbard. Lanham, MD: Lexington Books, 2006. 83–91.
Lévi-Strauss, Claude. *Tristes Tropiques.* London: Hutchinson & Co, 1961 [1955].
Lévi-Strauss, Claude. *Structural Anthropology*. New York: Basic Books, 1963 [*Anthropologie Structural*, 1958].
Lock, Margaret. *Twice Dead: Organ Transplants and the Reinvention of Death.* Berkeley/Los Angeles/London: University of California Press, 2002.
Lopp, Leonie. *Regulations Regarding Living Organ Donation in Europe: Possibilities of Harmonisation.* Berlin/Heidelberg: Springer, 2013.
Merton, Robert K., and Norman W. Storer. *The Sociology of Science: Theoretical and Empirical Investigations.* Chicago: University of Chicago Press, 1973.
Mitra, Sayani, and Silke Schicktanz. "Failed Surrogate Conceptions: Social and Ethical Aspects of Preconception Disruptions During Commercial Surrogacy in India." *Philosophy, Ethics, and Humanities in Medicine* 11.9 (2016) https://peh-med.biomedcentral.com/articles/10.1186/s13010-016-0040-6 [accessed: 1 July 2019].
Mitra, Sayani, Silke Schicktanz, and Tulsi Patel. *Cross-Cultural Comparisons on Surrogacy and Egg Donation: Interdisciplinary Perspectives from India, Germany and Israel.* London: Palgrave Macmillan, 2018.
Nowotny, Helga, Peter Scott, and Michael Gibbons. *Re-Thinking Science: Knowledge and the Public in an Age of Uncertainty.* London: Polity, 2001.
Payer, Lynn. *Culture and Medicine*. London: Penguin, 1989.
Pethes, Nicolas. "Terminal Men: Biotechnological Experimentation and the Reshaping of 'the Human' in Medical Thrillers." *New Literary History* 36.2 (2005): 161–185.
Randhawa, Gurch, and Silke Schicktanz, eds. *Public Engagement in Organ Donation and Transplantation*. Lengerich: Pabst Science Publishers, 2013.
Raz, Avid, Isabella Jordan, and Silke Schicktanz. "Exploring the Positions of German and Israeli Patient Organizations in the Bioethical Context of End-of-Life Policies." *Health Care Analysis* 22.2 (2014): 143–159.
Rise of the Planet of the Apes. Dir. Rupert Wyatt. Narr. Rick Jaffa and Amanda Silver. 20th Century Fox, 2011.
Rose, Nikolas. *The Politics of Life Itself: Biomedicine, Power, and Subjectivity in the Twenty-First Century*. Princeton: University Press. 2006.
Roth, Philip. *Everyman.* New York: Houghton Mifflin, 2006.
Roudometof, Victor. *Glocalization: A Critical Introduction*. London/New York: Routledge, 2016.
Schicktanz, Silke. "Why the Way We Consider the Body Matters: Reflections on Four Bioethical Perspectives on the Human Body." *Philosophy, Ethics, and Humanities in Medicine* 2.30 (2007): 1–12.
Schicktanz, Silke. "The Ethical Legitimacy of Patients Organizations' Involvement in Politics and Knowledge Production: Epistemic Justice as Conceptual Basis." *The Public Shaping of Medical Research: Patient Associations, Health Movements and Biomedicine*. Eds. Peter Wehling, Willy Viehöver, and Sophia Koenen. London/New York: Routledge, 2015. 246–265.

Schicktanz, Silke. “Genetic Risk and Responsibility: Reflections on a Complex Relationship.” *Journal of Risk Research* (2016): 1–23.

Schicktanz, Silke. “Beyond Relativism: Comparing Practices and Norms of Surrogacy in India, Israel and Germany.” *Cross-Cultural Comparisons on Surrogacy and Egg Donation: Interdisciplinary Perspectives from India, Germany and Israel.* Eds. Sayani Mitra, Silke Schicktanz, and Tulsi Patel. London: Palgrave Macmillan, 2018.103–125.

Schicktanz, Silke, Mark Schweda, and Martina Franzen. “‘In a Completely Different Light?’ The Role of ‘Being Affected’ for the Epistemic Perspectives and Moral Attitudes of Patients, Relatives and Lay People.” *Medicine, Health Care and Philosophy* 11.1 (2008): 57–72.

Schicktanz, Silke, and Mark Schweda. ”The Diversity of Responsibility: The Value of Explication and Pluralization.” *Medicine Studies* 3.3 (2012): 131–145.

Schicktanz, Silke, Mark Schweda, and Brian Wynne. “The Ethics of ‘Public Understanding of Ethics’: Why and How Bioethics Expertise Should Include Public and Patients’ Voices.” *Medicine, Health Care and Philosophy* 15.2 (2012): 129–139.

Schicktanz, Silke, and Sabine Wöhlke. “The Utterable and Unutterable Anthropological Meaning of the Body in the Context of Organ Transplantation.” *Dilemata: International Journal of Applied Ethics* 9.23 (2017): 107–127.

Searle, John. *The Construction of Social Reality.* New York: Free Press, 1995.

Shelley, Mary. *Frankenstein.* New York: Dover Publications, 1994 [1818].

Shepherd, Lee, Ronan E O’Carroll, and Eamonn Ferguson. “An International Comparison of Deceased and Living Organ Donation/Transplant Rates in Opt-in and Opt-out Systems: A Panel Study.” *BMC Medicine* 12 (2014): 131. doi:10.1186/s12916-014-0131-4 [accessed: 30 June 2019].

Shildrick, Margrit, and Roxanne Mykitiuk, eds. *Ethics of the Body: Postconventional Challenges.* Cambridge, MA/London: MIT Press, 2005.

Squier, Susan M. *Liminal Lives: Imagining the Human at the Frontiers of Biomedicine.* Durham, NC: Duke University Press, 2004.

Swedberg, Richard. *Max Weber Dictionary: Key Words and Central Concepts.* Stanford, CA: Stanford University Press, 2005.

Tanderup, Malene, Sunita Reddy, Tulsi Patel, and Brigitte Bruun Nielsen. “Reproductive Ethics in Commercial Surrogacy: Decision-Making in IVF Clinics in New Delhi, India.” *Journal of Bioethical Inquiry* 12.3 (2015): 491–501.

Thompson, Charis. *Making Parents: The Ontological Choreography of Reproductive Technologies.* Cambridge, MA: MIT Press, 2005.

Toulmin, Stephen. “How Medicine Saved the Life of Ethics.” *Perspectives in Biology and Medicine* 25.4 (1982): 736–750.

Wiesemann, Claudia. *Moral Equality, Bioethics, and the Child.* Cham: Springer, 2016.

Williams, Melissa. *Voice, Trust, and Memory: Marginalized Groups and the Failings of Liberal Representation.* Princeton: Princeton University Press, 1998.

Ursula K. Heise

Multispecies Futures and the Study of Culture

1 Cultural Study and the Shape of Change

When I went to graduate school in the mid to late 1980s, first in Germany and then in the United States, it was hard to keep up with the pace of change. New theoretical paradigms in literary and cultural studies kept emerging, mingling, and inflecting each other: New Left Marxism, feminism, postcolonialism, poststructuralism, deconstruction, New Historicism, Lacanian psychoanalysis, gender studies, queer studies, and critical race theory all demanded attention to core theoretical texts, central concepts, distinctive methodologies, and often particular styles of argument. Cultural studies itself became another one in the march of paradigms. As the cross-disciplinary US offspring of the Birmingham School, cultural studies blended Marxist theories of revolution and the Foucauldian rhetoric of "circulation of power" with elements of gender and race theory in the late 1980s and early 1990s. In its case studies, it relentlessly attacked and sought to do away with, once and for all, the distinction between high culture and popular or mass culture.

Conceived in this way, cultural studies imported methods from anthropology and broadened the literary canon of study immensely. At the same time, it paid much more direct attention to the users and consumers of texts rather than just to their producers, with Janice Radway's study of romance and its readers and Constance Penley's analyses of technoculture and fan fiction standing as early models of this transformation (Radway 1984; Penley 1991). While this altered perspective arguably transformed literary studies for good, cultural studies itself ran out of steam over the course of the 1990s. The theoretical contradictions between Marxist and poststructuralist assumptions about social change, the fear of "totalization" and the consequent narrowing of analytical focus, and the detailed attention lavished on cultural phenomena that were either forgotten or became trivial in just a few short years all contributed to the fizzling-out of cultural studies as a dominant theoretical paradigm, even as some of its theoretical and methodological innovations endured.

Sometime in the 1990s, the shape of change itself changed in a good deal of humanistic research. Studies of memory and trauma moved to the forefront at about the same time as the focus on varieties of postmodernism in literature and the arts gave way to concepts such as globalization, hybridity, cosmopolitanism,

https://doi.org/10.1515/9783110669398-017

and border cultures, to name just a few. But while these concepts worked as powerful organizing categories of research across disciplines in the humanities and qualitative social sciences, they did not impose one dominant theoretical paradigm. Rather, they gave rise to clusters of approaches from different theoretical perspectives. This tendency has continued in the disciplinary innovations of the last two decades, which, in the United States at least, have mostly taken the shape of research areas organized around a central theme that does not in and of itself demand specific theoretical assumptions. Typically, these areas have come with labels such as 'x studies' or 'y humanities': disability studies, food studies, and human-animal studies, for example, or digital humanities, medical humanities, environmental humanities, and urban humanities. Many of them have catalyzed innovations in research and teaching that have crystallized in centers, programs, majors and minors; few of them, to date, have resulted in the creation of new departments.

Many, though not all, of these new areas of study have emerged from or sought to create new connections to particular areas of science and technology. Disability studies, narrative medicine, and the medical humanities, for example, use methods of analysis from anthropology, history, and literary and cultural studies to explore historically and culturally varying ideas about "normal" bodies, health, and disease; about the roles of doctors, patients, and their means of communication; about childhood, maturity, and old age; and about the representation of organs, illnesses, and cures in texts and images, including new technologies of medical imaging. Food studies focus on the complex interface of agriculture, economy, and culture in the production, distribution, consumption, and representation of food. In the process, the field draws on the practical knowledge of farmers and gastronomers as well as on the academic expertise of agronomists, anthropologists, ecologists, sociologists, and researchers of literature and film. The environmental humanities bring together anthropology, ecocriticism, cultural geography, history, and philosophy to analyze assumptions about ecologies, natures, landscapes, and nonhuman species that inform past and present environmental discourses. The digital humanities connect to the electronic landscape of computers, the Internet, and social media in two distinctive ways: either by analyzing digitally generated texts and images with existing methods of historical and literary research, or by applying new digital tools to established canons of texts (Fitzpatrick 2011).

The study of culture or, more precisely, of cultures (emphatically in the plural) is crucial to all of these fields even when the concept itself is not foregrounded. How gender, sexuality, race, and age inflect the practice of medicine varies by historical period and region (a question for the medical humanities). What foodstuffs are considered edible or inedible, good or bad for one's physical health, or spiritually acceptable or unacceptable in different communities often has deep historical, ecological, and cultural roots, and shapes current practices of growing, harvesting, and cooking that food studies explore. Human-animal studies engage with recent scientific insights into the cognition, perception, and skills of individual animals and the cultures and politics of animal communities, often so as to question the exceptionality of the human subject. In the process, the field draws on historical and cultural comparisons to explore the different ways in which the boundary between humans and animals has been drawn in the past and present. The urban humanities explore the historical memories, spatial sensibilities, social inequalities, and cultural frameworks that shape architecture, design, landscape architecture, and urban planning. A great deal of the intellectual energy in many of these recent fields, in fact, emerges precisely from their cultural, historical, linguistic, and media-theoretical reframing of questions that were earlier thought to be the unique purview of biology, ecology, engineering, medicine, computer science, public policy and planning, or other disciplines.

The study of cultures today demonstrates its relevance and urgency precisely at these intersections. In turn, cultural studies are being reshaped by the central research questions around which these new fields revolve. In this article, I will briefly highlight two areas in which culture is currently being re-envisioned: one, in the tension between discussions about the Anthropocene and the emergence of various strands of posthumanism; and two, in the expansion of culture beyond the human sphere. As I will show in the last section, this more-than-human reconceptualization of culture is also transforming literature and the arts.

2 Culture between the Anthropocene and Posthumanism

In my field of research, the environmental humanities, literary and cultural studies over the last decade have increasingly been re-envisioned under the dual influences of debates about the Anthropocene and posthumanist theories. The concept of the Anthropocene, proposed casually at a conference by the ecologist Eugene Stoermer in the 1980s and formalized in a series of publications spearheaded by the atmospheric chemist Paul Crutzen since 2000, is now ubiquitous in discussions of the environmental present and future. Crutzen and Stoermer proposed in their original publications that we no longer inhabit the Holocene, geologists' designation of the last 12,000 years, but a new epoch, the Anthropocene. The name "Age of Man," they argued, was justified because of humans' pervasive impacts on global ecosystems and the likelihood that traces of these impacts will be visible in the Earth's geological strata for millennia to come (Crutzen and Stoermer 2000; Crutzen 2002). As of this writing, professional associations of geologists have yet to issue their stamp of approval for this change of nomenclature. But in the meantime, the idea of the Anthropocene has developed a cultural life of its own in a wide range of academic and popular publications as well as exhibitions, conferences, and seminars in Australia, North America, and Western Europe (less so in other regions).

In the process, different narratives have accreted around the concept.[1] For Crutzen as well as many journalists, writers, and activists, the neologism of the Anthropocene re-emphasizes a narrative that has long shaped environmentalist thought and writing: that of the deterioration and destruction of nature under the impact of modern societies. In other words, it is a new term for an old story that portrays humans' interactions with nature as a process of nature's decline. But this narrative has not gone uncontested. Diane Ackerman's *The Human Age: The World Shaped by Us* (2014) takes the opposite tack, interpreting the Anthropocene not as an age of destruction but of unprecedented human ingenuity and creativity – qualities that in her view will let humankind overcome the environmental challenges it currently faces. In Ackerman's approach, the Anthropocene becomes a shorthand for the Enlightenment narrative of technological and social progress that environmental thinking has persistently criticized over the last two hundred years.[2] In between these extreme positions, environmentalists such as the biologists Peter Kareiva and Joseph Mascaro, the geographer Erle Ellis, and the science writers Emma Marris and Christian Schwägerl have sought to map out a more moderate landscape of hope. They take the recognition that no part of the Earth's atmosphere and biosphere remains untouched by human impacts – climate change alters even terrestrial and marine regions that humans have not visited in their own bodies – as a point of departure for envisioning a new environmentalism that is less beholden to conceptions of nature and environmental cultures of the past, and that does not so much seek to restore the ecosystems of the past as to design ecosystems for the future that will allow both humans and nonhumans to flourish. In the process, all of them move away from pristine nature and wilderness as yardsticks for environmental activism in the future, emphasizing instead the complexity and value of the mixed and cultivated landscapes to

1 For a more detailed discussion of the divergent Anthropocene narratives, see Heise 2016, ch. 6.

2 Over the last two decades, some strands of the environmental movement have turned against this critique of modernization and embraced environmental modernization instead; the most prominent advocates of this position in the United States have been Breakthrough Institute founders Ted Nordhaus and Michael Shellenberger 2004, 2007.

which human intervention has given rise (see Ellis and Ramankutty 2008; Kareiva et al. 2007; Marris 2011; Marris et al. 2011; Schwägerl 2010).

At stake in these more cautiously optimistic perspectives on the Anthropocene is the attempt to rethink human cultures as part of the nature that environmentalism seeks to conserve and sustain. The idea that climate change, in particular, puts conventional boundaries between nature and culture into question is not, of course, new: The environmental writer and activist Bill McKibben had already argued in his book *The End of Nature* (1989), one of the first nonfiction books to ring the alarm about climate change, that for modern societies at least, nature is defined by its separation from culture, and that climate change therefore implies the end of this type of nature and the experiences it enables. But if this change in nature presented itself as a relentlessly melancholy prospect to McKibben, writers such as Marris and Schwägerl highlight the opportunities and joys that a different understanding of the natural might bring, precisely because it does not interpret human interventions as by definition detrimental to what is most valuable about the nonhuman world.

From a different theoretical purview, the historians Dipesh Chakrabarty and Julia Adeney Thomas as well as the philosopher Dale Jamieson have also questioned the boundary between nature and culture. Climate change as an outcome of human agency though not of human intention, Chakrabarty argues in his by now classic essay "The Climate of History" (2009), turns humanity into a geological force. Other writers (the paleoanthropologist Richard Leakey, for example) had already suggested that humankind's current environmental impact could be compared to the impact of the meteorite sixty-five million years ago that triggered the extinction of the dinosaurs along with more than eighty percent of the other species then in existence – another way of suggesting that humans collectively have achieved geological or even cosmological force. But Chakrabarty is particularly interested in the challenges this power poses for historical thinking: "Humans have become geological agents very recently in human history. In that sense, we can say that it is only very recently that the distinction between human and natural histories – much of which had been preserved even in environmental histories that saw the two entities in interaction – has begun to collapse" (Chakrabarty 2009, 207). As a consequence, Dale Jamieson has pointed out, the sense of humankind's enormous power as a species goes along with a sense of utter powerlessness on the part of individuals, both equally symptomatic of the Anthropocene (Jamieson 2017, 15).

Julia Adeney Thomas sees questions about the distinction of human identity and culture from nature arising not only from the large scales of time and space that climate change forces us now to consider as part of humans' cultural history. She also points to the challenges that have come from other confrontations of culture with biology. Microbiology, she argues, has recently shown that microbes are constituent parts of the human body, "inseparably 'us,' more responsible than 'we' are for 'our' existence by most calculations" (Thomas 2014, 1594), and that they contribute more genes to human survival than humans themselves contribute. In this view, "a person is not an individual but a congregation" of such nonhuman organisms (Thomas 2014, 1594). By the same token, the hundreds of thousands of chemicals that twentieth-century societies have introduced into the natural environment are now imbricated into the human organism to the point where body and environment cannot be categorically distinguished – not only in the case of victims of industrial accidents, but quite ordinary humans as well (Thomas 2014, 1596–1602). These alterations, all part and parcel of the Anthropocene, not only challenge conventional definitions of the human, but also imply divergent understandings of human bodies and practices that are not compatible with each other: "in paleobiology, 'we' are an increasingly domineering species operating over vast eons of time; in microbiology, 'we' are a coral reef of many species spreading out in awkward archipelagos of co-dependent beings; and in biochemistry, 'we' are a semi-industrialized product of the last, brief half-century," Thomas argues (Thomas 2014, 1603). In this context, human agency as well as human cultures clearly need to be envisioned in very different terms, depending on which perspective on human nature one privileges.

In her multiscalar survey of the different types of human subjects that the Anthropocene implies, Thomas already maps some of the territory that theories of posthumanism have traversed over the last few decades (though she herself does not use the term). Typically, posthumanist approaches envision human existence, intentionality, and agency as neither singular nor exceptional, but as part of networks that also include other modes of being and agency. Posthumanisms, though, differ fundamentally in how they envision these networks. The "heterogeneous" social networks of actor-network theory, made up of human and nonhuman, animate and inanimate agents that relate to each other in material as well as semiotic ways, were proposed by Bruno Latour, Michel Callon, and John Law in the 1980s. The German sociologist Niklas Luhmann developed a systems theory whose central tenet is that individuals do not "form part" of societies, but that individuals and societies constitute one another's environments in a cybernetic model of communication. "New materialisms" such as those more recently formulated by Karen Barad, Stacy Alaimo, Serenella Iovino, and Serpil Oppermann, among others, have theorized human minds and bodies as "transcorporeal" vectors (Alaimo 2010) in material relations that constitute the human subject through ecological networks. Jane Bennett's new vitalism explores the vibrant agency of matter and its assemblages. Object-oriented ontology as proposed by Graham Harman, Levy Bryant, Quentin Meillassoux, and Timothy Morton seeks to free objects from their "correlationism" to human agency and to explore them on their own terms, even though object-oriented ontology also emphasizes that objects will ultimately always remain withdrawn from human knowledge. Human-animal studies, elaborated by Peter Singer, Tom Regan, Jacques Derrida, Giorgio Agamben, Roberto Esposito, Donna Haraway, and Cary Wolfe, has focused on the philosophical underpinnings and political consequences of historically varying distinctions between human and animal. Recent work in plant studies by anthropologists Matthew Hall and Edward Kohn has begun to expand this argument into the domain of plants.

Some posthumanist theories focus on systems, some on machines, others on objects, and yet others on animals, and their foundational assumptions are in quite a few cases incompatible with each other. But in the discourse of the humanities and qualitative social sciences, including the study of culture from various disciplinary perspectives, they have collectively tended to exert a conceptual pull that contravenes the Anthropocene debates. Discussions about the Anthropocene have over the last decade often revolved around the question of human agency, specifically the question as to whether the emphasis on the human species as a whole masks continuing social and economic inequalities that distinguish those human populations who mostly cause climate change from those populations who suffer most of the consequences. Sociologists such as Jason Moore (2016), who has championed replacing the notion of the Anthropocene with that of the Capitalocene, and philosophers such as Slavoj Žižek (2011), who insists that capitalism continues to provide the key to solving the ecological crisis of climate change, have fiercely criticized the narrative of species-wide agency that has typically accompanied the Anthropocene. Posthumanist theories, by contrast, no matter what their specific assumptions might be, tend to converge in questioning the conceptual foundations for human agency, whether it is postulated at the level of the individual, the social class, or the species.

The study of culture is today caught up in the tension between the new emphasis on the centrality of human agency in the Anthropocene debates and its sustained questioning in theories of posthumanism. Julia Adeney Thomas's work shows at least implicitly how the two might be connected to each other if the Anthropocene is understood as more than just climate change. The future study of cultures will need to be multiscalar, reconceptualizing cultural practices with different definitions of human collectivity in mind that range from the microscopic to the geological scale. And it will need to re-envision the human in a context of multispecies networks that take culture beyond the human. Several of the new areas of study that have emerged over the last two decades, including food studies, human-animal studies, and the environmental and medical humanities, have begun to develop the concepts, methods, and tools for such a study of cultures across species.

3 Multispecies Cultures

In my rough-and-ready sketch of posthumanist theorists in the last section, I already mentioned human-animal studies or critical animal studies, as it is sometimes called. This field has its historical roots in the animal liberation movement as it was initiated by Peter Singer in the mid-1970s and developed by philosophers such as Tom Regan and Mary Midgley later on. Certain types of moral consideration that are usually only extended to humans, these philosophers argued, should also be applied to nonhuman species that share particular characteristics with humans, whether it be the ability to suffer or to function as the subject of one's own biography, for example. The debate over which kinds of moral consideration, including certain "rights," should be extended to which species, continues to this day. But in the 1990s, animal welfare thinkers and activists were sometimes criticized by theorists in the poststructuralist tradition – Jacques Derrida and Cary Wolfe, for example – for still privileging human subjectivity by considering only those animals as deserving of rights who shared certain characteristics with humans. The blurry boundary between humans and animals, they argued, should rather encourage us to take a critical look at the implied integrity and exceptionality of human identity itself. "Critical" animal studies, then, tend to be skeptical of conventional beliefs about human identity, extending poststructuralist critiques of meaning and subjectivity into the consideration of biological species.

These debates are clearly crucial for the study of cultures, since they open up for analysis the way in which species boundaries have been historically and regionally variable, and how they have functioned to legitimate or criticize particular regimes of power. Giorgio Agamben has famously reminded us that biology's first modern taxonomist, Carl von Linné, who invented the binomial system of species designations, hesitated over how to classify humans in relation to other primates (Agamben 2004, 23–27). Cary Wolfe has argued that racism and other forms of social oppression are historically closely related to speciesism, which continues to underwrite social discrimination today (Wolfe 2003, 43). Graham Huggan and Helen Tiffin have explored how species distinctions function to legitimate colonial regimes by relegating colonial subjects to the less-than-human category of the animal (Huggan and Tiffin 2010, 18–19).

These explorations of the cultural and political work that the species concept does in different contexts and communities has in recent years been complemented by two other strands of research. Anthropologists and philosophers such as Vinciane Despret, Eben Kirksey, Stefan Helmreich, Roberto Marchesini, and Anna Tsing have developed approaches variously called multispecies ethnography, etho-ethnology, or zooanthropology, which analyze what we normally understand as *human* societies and cultures as, in reality, assemblages of many species, ranging from the microbes inhabiting our gastro-intestinal tracts and disease-carrying viruses to food plants and animals, pets, and those plants and animals that figure in ritual and religious practices. Tsing has observed that "human nature is an interspecies relationship" (quoted in Kirksey, Schuetze, and Helmreich 2014, 2) in that human life is inconceivable without its dependence on a wide variety of bacteria, microbes, plants, and animals. On this basis, multispecies ethnography seeks to redefine what 'the human' means individually and collectively:

> Ethnographers are now exploring how 'the human' has been formed and transformed amid encounters with multiple species of plants, animals, fungi, and microbes. Rather than simply celebrate multispecies mingling, ethnographers have begun to explore a central question: Who benefits, cui bono, when species meet? To answer this question, multispecies ethnographers are collaborating with artists and biological scientists to illuminate how diverse organisms are entangled in political, economic, and cultural systems.
> (Kirksey, Schuetze, and Helmreich 2014, 1–2)

In this vein, Deborah Bird Rose (2011) has analyzed the relations between dogs, dingoes, and humans in Aboriginal and white Australian communities; Anna Lowenhaupt Tsing has investigated the cultivation, harvesting, distribution, and consumption of *matsutake* mushrooms in communities on several continents; and Thom van Dooren (2016) has explored processes that lead to species extinction and conservation efforts in different regions. My own recent work has focused on particular communities' relationships to endangered or extinct species from the perspective of narrative analysis. Building on the work of multispecies ethnographers, I have suggested the concept of "multispecies justice" as a way of thinking together the concerns for environmental justice and biodiversity conservation (Heise 2016, 162–168).

In all of these case studies, the study of human cultures includes nonhuman species as a constitutive element without which the meaning of 'culture' itself could not be established. In future, such research projects may need to consider a different but related broadening of the culture concept that has come from ethology. The idea that animal communities themselves have cultures, in the sense of knowledge and skills that are transmitted from adults to juveniles not through genes but through learning, of locally specific practices by a population that is not shared by the species as a whole, and even of a sense of aesthetics, is no longer new. Ornithologists and cetologists have for several decades documented the existence of "dialects" in the vocalizations of different populations of birds and whales, and such distinctive characteristics were also discovered in different whale populations' foraging and migration traditions. Primates are by now well known for tool usage, complex social structures and relationships, and behaviors that would be difficult to call by any name other than politics, as Frans de Waal's (1998) classic study of power relations among chimpanzees has shown. Richard Prum (2017), finally, has forcefully argued that a good deal of animal behavior cannot be explained without postulating a sense of aesthetics. In studies such as these, the concept of culture migrates beyond the human realm even as it sheds new light on human practices. The future of cultural studies, especially but not only if culture is envisioned from a multispecies perspective, will need to situate itself in this broadened context of cultural structures and practices that humans share with other animal species.

4 A "More Than Human" Future

I'd like to briefly explore this path forward for cultural studies through one suggestive example: music. Literature and art have engaged with multispecies communities and conflicts across a wide spectrum of genres and media. Animated films often feature a variety of speaking animals, sometimes in the absence of any human characters, as in *Bambi* (1942), and sometimes in conflict or collaboration with them, as in Isao Takahata's *Pom Poko* (1994), Andrew Stanton's *Finding Nemo* (2003), or Vincent Patar and Stéphane Aubier's *Panique au village* (2009). Comic books, which have conventionally often featured animals that were simply humans in a more light-hearted guise, have over the last few decades metamorphosed into graphic novels with serious themes, complex plots, and three-dimensional characters. Some of them have addressed relations between different species in sophisticated ways, as Alan Moore's reinvention of the *Swamp Thing* comic (1984), Grant Morrison's reconceptualization of *Animal Man* (1988–1990), and the ongoing *Saga* series by Brian K. Vaughan and Fiona Staples (2012–) demonstrate. Short stories, novellas, and novels throughout the twentieth century have similarly engaged with questions of relations between species, from the critiques of domestication and captivity in works by Franz Kafka and Jack London to novels such as Bernard Werber's trilogy *Les fourmis, Le jour des fourmis*, and *La révolution des fourmis* (1991, 1992, 1996) and Barbara Gowdy's *The White Bone* (1999), which integrate scientific knowledge about the cognition, perception, and social behavior of nonhuman species such as ants and elephants into fictional scenarios that reach beyond conventional realism to imagine new multispecies worlds.

But the translation of multispecies visions into aesthetic form does not occur only through image and text. Sound artists, too, have explored multispecies networks in innovative ways, by recording, recreating, or musically transforming the vocalizations of individual species, as well as entire natural soundscapes. Well known in this genre is Bernie Krause's *Wild Sanctuary Audio Archive*, first initiated in 1968, which includes "marine and terrestrial soundscapes representing the voices of living organisms from larvae to large mammals and the numerous tropical, temperate and Arctic biomes from which they come. [...] 4,500 hours of wild soundscapes and in excess of 15,000 identified life forms" (http://www.wildsanctuary.com). Krause's goals are mostly archival and documentary, but soundscapes also feature in works that are equal parts documentation and composition. The Spanish sound artist Francisco López' *La selva* (1997/2001), the American composer Steven Feld's *Rainforest Soundwalks* (2006), and the Italian composer David Monacchi's monumental project *Fragments of Extinction* (2001–2015) all combine sound recordings of multiple species in the natural world with original sound creations.

Monacchi's "environmental sound-art project" focuses, according to his own explanations, on the acoustic biodiversity of rainforests and seeks to collect "three-dimensional sound portraits of entire circadian cycles. The complex network of inter- and intra-specific communication found in these recordings is [...] *the sonic heritage of millions of years of evolution.* We must save fragments of it in order to study, understand, experience, enjoy, and conserve it, *preserving for future generations* imprints of the disappearing sonic intelligence of nature" (http://www.fragmentsofextinction.org/mission/; original emphasis). For fifteen years, Monacchi traveled to the Amazon, to the Congo, and to Borneo to record the sounds of intact equatorial rainforests. In the process, he developed innovative microphones and recording techniques to capture sounds from all the different levels and directions of a given rainforest location, as well as to withstand extreme humidity. Ecologists such as E. O. Wilson have argued that the current global biodiversity crisis eliminates species before humans have had a chance to find and name them. Monacchi, analogously, emphasizes that extinction silences natural voices and along with them entire "'*eco-symphonies' we have not even heard or recorded*" (http://www.fragmentsofextinction.org/mission/; original emphasis). The soundscapes he focuses on have not been documented in their entirety, and they occasionally include individual voices that are unknown to current science.

Monacchi intends his sound art to communicate the beauty of the natural world to the audience, to influence public discourse, and thereby to support conservation efforts. To this end, he creates continuous twenty-four-hour recordings of particular rainforest locations, which he then submits to meticulous analysis in terms of its spatial information, progressing moments in time, and occupied frequencies. For his performances, he condenses the twenty-four hours into ninety minutes and accompanies the recorded sounds with a sound spectrogram that translates the voices of different species into dynamically moving neon-colored lines, a visual spectacle with a distinctive aesthetic all its own. As part of his analysis during the performances, Monacchi includes the insights of acoustic ecology: the study of how individual species use particular niches of sound and frequency; how they adapt to different sonic environments; and how they work around cross-species similarities in calls and frequencies that impede territorial defense or mating. Just as topographical and climatic conditions, vegetation, and the presence or absence of other species open or foreclose particular possibilities for a species, in other words, so does the sonic environment, where certain frequencies and types of calls – acoustic niches – are occupied and others not.

Acoustic ecology also informs how Monacchi himself intervenes into his rainforest recordings. He manipulates existing sounds and adds others in what he calls "eco-acoustic composition" (http://www.fragmentsofextinction.org/eco-acoustic-music/), following the principles of acoustic ecology in that the human-generated sounds he adds to the soundscapes cannot overlap in time, location, frequency, or type with those of other species. Through this procedure, Monacchi goes beyond the documentary goal of registering nonhuman voices and ecologically significant silences. He adds a human presence to the ecosymphony, but meticulously respects the rules that also guide the sound behavior of other members of the multispecies rainforest community. The imagination that informs eco-acoustic composition is therefore, in the end, in equal parts nostalgic, technoscientific, and utopian, in that it technologically generates a sonic environment in which human voices are perfectly integrated with the species around them – presumably in contrast to the perceived separation of modern humans from nature.

Whether one agrees with this narrative about the reintegration of modern humans and nature through art that underwrites Monacchi's compositions or not, his work highlights one of the many forms that multispecies approaches to the creation and interpretation of culture take today. From the perspective of multispecies theory, the cultural studies of the future would explore everyday cultural practices as well as works of art and literature as imbricated in networks that always reach beyond humans themselves to involve other species, often in locally or regionally distinctive ways. In some cases, this perspective may combine with the skepticism of humans' singularity and exceptionality that characterizes posthumanist theories; in others, it may emphasize humans' impact on the global environment in the way debates about the Anthropocene do. But in either case, the future of cultural study will be, in David Abram's (1996) phrase, "more-than-human."

References

Abram, David. *The Spell of the Sensuous: Perception and Language in a More-than-Human World.* New York: Pantheon, 1996.

Ackerman, Diane. *The Human Age: The World Shaped by Us.* New York: Norton, 2014.

Agamben, Giorgio. *The Open: Man and Animal.* Trans. Kevin Attell. Stanford: Stanford University Press, 2004.

Alaimo, Stacy. *Bodily Natures: Science, Environment, and the Material Self.* Bloomington: Indiana University Press, 2010.

Chakrabarty, Dipesh. "The Climate of History: Four Theses." *Critical Inquiry* 35 (2009): 197–222.

Crutzen, Paul J. "Geology of Mankind." *Nature* 415 (2002): 23.

Crutzen, Paul J., and Eugene F. Stoermer. "The 'Anthropocene.'" *Global Change Newsletter* 41 (2000): 17–18.

De Waal, Frans. *Chimpanzee Politics: Power and Sex among Apes.* Rev. ed. Baltimore/London: Johns Hopkins University Press, 1998.

Ellis, Erle C., and Navin Ramankutty. "Putting People in the Map: Anthropogenic Biomes of the World." *Frontiers in Ecology and the Environment* 6.8 (2008): 439–447. https://esajournals.onlinelibrary.wiley.com/doi/epdf/10.1890/070062 DOI 10.1890/070062 [accessed: 27 June 2019].

Fitzpatrick, Kathleen. "The Humanities, Done Digitally." *Chronicle of Higher Education* (8 May 2011). https://www.chronicle.com/article/The-Humanities-Done-Digitally/127382 [accessed: 27 June 2019].

Heise, Ursula K. *Imagining Extinction: The Cultural Meanings of Endangered Species.* Chicago: University of Chicago Press, 2016.

Huggan, Graham, and Helen Tiffin. *Postcolonial Ecocriticism: Literature, Animals, Environment.* London: Routledge, 2010.

Jamieson, Dale. "The Anthropocene: Love It or Leave It." *The Routledge Companion to the Environmental Humanities.* Eds. Ursula K. Heise, Jon Christensen, and Michelle Niemann. Abingdon: Routledge, 2017. 13–20.

Kareiva, Peter, Sean Watts, Robert McDonald, and Tim Boucher. "Domesticated Nature: Shaping Landscapes and Ecosystems for Human Welfare." *Science* 316 (2007): 1866–1869.

Kirksey, Eben, Craig Schuetze, and Stefan Helmreich. "Introduction: Tactics of Multispecies Ethnography." *The Multispecies Salon.* Ed. Eben Kirksey. Durham, NC/London: Duke University Press, 2014. Kindle edition.

Krause, Bernie. *Wild Sanctuary.* https://www.wildsanctuary.com/ [accessed: 27 June 2019].

Marris, Emma. *Rambunctious Garden: Saving Nature in a Post-Wild World.* New York: Bloomsbury, 2011.

Marris, Emma, Peter Kareiva, Joseph Mascaro, and Erle C. Ellis. "Hope in the Age of Man." *New York Times* (7 December 2011).

McKibben, Bill. *The End of Nature.* New York: Anchor, 1989.

Monacchi, David. *Fragments of Extinction*. 2001–2015. http://www.fragmentsofextinction.org/ [accessed: 27 June 2019].

Moore, Jason W., ed. *Anthropocene or Capitalocene? Nature, History, and the Crisis of Capitalism*. Oakland, CA: PM Press, 2016.

Nordhaus, Ted, and Michael Shellenberger. "The Death of Environmentalism: Global Warming Politics in a Post-Environmental World." 2004. https://grist.org/article/doe-reprint/ [accessed: 27 June 2019].

Nordhaus, Ted, and Michael Shellenberger. *Break Through: From the Death of Environmentalism to the Politics of Possibility*. Boston: Houghton Mifflin, 2007.

Penley, Constance. "Brownian Motion: Women, Tactics, and Technology." *Technoculture*. Eds. Constance Penley and Andrew Ross. Minneapolis: University of Minnesota Press, 1991. 135–162.

Prum, Richard O. *The Evolution of Beauty: How Darwin's Forgotten Theory of Mate Choice Shapes the Animal World – and Us*. New York: Doubleday, 2017.

Radway, Janice. *Reading the Romance: Women, Patriarchy, and Popular Literature*. Chapel Hill, NC: University of North Carolina Press, 1984.

Rose, Deborah Bird. *Wild Dog Dreaming: Love and Extinction*. Charlottesville: University of Virginia Press, 2011.

Schwägerl, Christian. *Menschenzeit: Zerstören oder gestalten? Die entscheidende Epoche unseres Planeten*. Munich: Riemann, 2010.

Thomas, Julia Adeney. "History and Biology in the Anthropocene: Problems of Scale, Problems of Value." *American Historical Review* 119 (2014): 1587–1607.

Tsing, Anna Lowenhaupt. *The Mushroom at the End of the World: On the Possibility of Life in Capitalist Ruins*. Princeton: Princeton University Press, 2015.

Dooren, Thom van. *Flight Ways: Life and Loss at the Edge of Extinction*. New York: Columbia University Press, 2016.

Wolfe, Cary. *Animal Rites: American Culture, the Discourse of Species, and Posthumanist Theory*. Chicago: University of Chicago Press, 2003.

Žižek, Slavoj. *Living in the End Times*. Rev. and updated ed. London/New York: Verso, 2011.

Peter L. Galison and Jens Kugele

Future Trading Zones for the Study of Culture: An Interview with Peter L. Galison

5 September 2019

Jens Kugele: Thank you very much, Peter, for agreeing to this interview. I truly appreciate this opportunity to continue our conversations on possible futures of the study of culture and to include your perspective as a scholar and filmmaker in this dialogical form. In addition to your academic work as a physicist and historian of science, you have also been involved in the production of several documentary films. In your and Robb Moss's documentary *Containment* (2015), for instance, you raise questions about possible futures when you shed light on governments' practices in their efforts to (safely) contain overwhelming amounts of radioactive sludge for the next ten thousand years. Your film addresses the question of how we can communicate with future generations and, indeed, future cultures about these containment attempts. In your view, how can we in the academic study of culture make sure that we foster communication with future generations and with future cultures? What kind of questions, topics, and concerns are of central importance in this context? What kind of (new) genres, formats, and media might be helpful or even necessary in your view?

Peter Galison: An interesting double question! On the one side, in the United States and in Europe, for some years a mix of physicists, futurists, astronomers, anthropologists, and material scientists have been grappling with the problem of how to warn the far future about the dangers of our buried nuclear waste. It seemed to the American Congress that it would constitute a plain moral hazard to bury and forget such dangerous materials. But then how does one communicate across 10,000 years or more during which the radioactive waste remains dangerous? Should one proceed by burying an image, icon, or image sequences? By a contemporary version of the Rosetta Stone? By entombing samples of the waste itself? By encoding scientific formulae and descriptions of the state of our knowledge of radiological medicine, nuclear physics, and geology? These are hard questions indeed, and they push to the limit – to the breaking point – our capacity to imagine the societies ten millennia from now. But the very act of trying to grapple with this necessary but nearly-impossible task is itself a great good thing. My hope in *Containment* was that to convey the idea that the very act of thinking far ahead – backed by nations – could help dislodge us from the presentism that threatens us everywhere. Most dramatically, we must resist our avaricious

https://doi.org/10.1515/9783110669398-018

present moment: The future of the planet depends on addressing the unfolding planetary catastrophe of global warming.

Turning to the other side of your good question: How might we, scholars of culture today, address matters of consequence, starting with finding ways to bring our concerns to an engaged public even today? When the futurists had to decide what to do, they felt inexorably drawn to the idea that no single modality could be relied upon – our best chance of communicating with the future would necessarily encompass many forms: images, texts, samples, ceramic plates, monuments, and scientific information. In some sense I think we face something similar in the human sciences today: It behooves us to think not only of a multitude of *sources* for our research, but a multitude of productive forms in which to present it. In my work, I've tacked back and forth between film, material culture, and text. Text is adept at cutting across times and places; it can, more easily than other media, take a topic and follow it through a multitude of countries, times, cultures. Books, physical and digital, travel relatively easily and can be read in any order. Film may more easily register the density of specific circumstance; it generalizes through its particularity and unfolds across time. Focus on an individual or family or kind of work, and the density of affect, the physicality of circumstance, and the volatility of relationships can emerge in ways that are not so easily ignored. Exhibits do yet other things, insofar as they can establish new and unexpected kinds of juxtapositions among objects, images, sounds. Done well, they can make immediate a distant time and place: I think of the power of the National Museum of African American History and Culture in Washington, DC; artefacts compel a confrontation with the physicality of slavery in a way that text alone cannot. My own sense is that the human sciences could do much more with these other (non-textual) forms of address: Culture is material and visual as well as textual. I strongly believe that we will need to be adept at using these and other modalities to present our work.

Over the last years, this kind of concern has driven me, for example – from writing on state secrecy, e.g., "Removing Knowledge" (2004) and "Secrecy in Three Acts" (2010), to making the film, with Robb Moss, *Secrecy* (2008) – all of these addressed the question of how historical (and present) attempts to block understanding help us frame how knowledge works. Grappling with university knowledge meant also confronting the historical inequity of the university – I made a film about a disputation that occurred in 1773 over the moral legitimacy of slavery itself – at a time when there was slavery at many American universities. The short film I made, *No More, America* (2017, with Henry Louis Gates), was accompanied by a textual exploration of the role of the eighteenth-century debate, its antecedent forms, and how academic disputation crossed with slavery on the eve of revolution: "Disputation, Poetry, Slavery" (2019). A work in the history and philosophy of science, my book *Einstein's Clocks, Poincaré's Maps* (2003), led me to a collaboration with South African artist William Kentridge in the multi-screen installation *Refusal of Time* (2012) and that then carried over to its accompanying chamber opera *Refuse the Hour* (2012).

Jens Kugele: In your research as well as in your editorial work, you have built bridges between the natural sciences and other disciplines, particularly the humanities: Several of your publications in the 'history of science' field explore its relation to neighboring fields such as cultural history and art history. Moreover, you have also served, among others, on the editorial board of *Critical Inquiry*, one of the leading journals in the humanities. What kinds of methods and strategies have you found helpful to bridge the gap between these disciplinary formations and the different forms of knowledge construction?

Peter Galison: Many of the societal problems that face us just now can *only* be addressed with a concerted effort by natural sciences working *with* the social and human sciences. Above all, we need to see science and technology as part of culture, not exterior to it. To name three such arenas: global warming; digital privacy; and artificial intelligence. Each of these is all at once a technical problem-cluster *and* a concatenation of ethical, social justice, and political issues. Who lives next to the major sources of carbon production: natural gas, methane production, factory-scale animal plants? In country after country they tend to be the poorer, the less privileged of society: In these questions at the intersection of climate-altering substances and demographics is a zone of culture implicated by environmental justice. On computer science: Even to ask a seemingly most basic question, "What counts as online privacy?" takes us beyond pure technicalities. So too does the query, "Do Algorithms carry weight beyond the data bases on which they draw?" – I've written on this in "Algorists Dream of Objectivity" (2019). It is a technical circumstance that a DNA swab of a criminal suspect also delivers information about that suspect's blood relatives. But it is a social and ethical matter for us to deal with what that means for our society: How do we balance criminal inquiry with genetic privacy? This sort of question may be grounded in the gene-sequencing techniques but no amount of biological reasoning in isolation will confront these broader issues. So it is with the steady stream of digital exhaust captured by governments and multinational corporations: siloed disciplines (computer science, ethics, political science). The pure code technicalities of Python or HTML leave open some of the most pressing issues of our time, issues that bear on our freedom to think and act, our ability to vote, our ability to be treated equally under the law.

Collaborating across boundaries requires a certain kind of attentive listening. When I first started to work in film, I tried to import the structure of a text – a periodization that had done good work in understanding the moral-political history of the hydrogen bomb. It was a total disaster – in a text you can scan in a moment a five-part structure. In a film such a thematization comes across as unwatchable pedantry. Working with William Kentridge (I was the dramaturg) taught me something else – we learned to think together through episodic stories, the laying of pipes carrying compressed air to sync clocks, for example. More than that, the logic of the work often proceeded through an associative rather than inferential or deductive arc. Entering the trading zones between fields demands a certain suspension of our confident, go-to forms of reasoning in order to hear our collaborators.

Jens Kugele: How do you envision the future academic work on the intersections between the humanities, the social sciences, and the (natural) sciences?

Peter Galison: I envision future work where shared topics, concepts, and methods will offer deep and deeply consequential sites for inquiry. Some of the great questions about politics and culture in this century, it seems to me, will emerge not from classical party politics but from the driving force of seemingly technical issues, like the multi-national exchange of data and its impact on privacy. The history of our present culture – how we got here – may offer us a platform of resistance to the passivity of technological determinism. Things need not be the way they happen to be.

Jens Kugele: With regard to the possible futures of the interdisciplinary study of culture, what forms of individual and collaborative research do you see as central? Where do you see (new) responsibilities of this academic field?

Peter Galison: Over much of the twentieth century, anthropology, from the work of Franz Boas to that of Clifford Geertz, has held the study of culture to be urgent. Up until the 1970s and 1980s, it was held (obviously, I'm oversimplifying) that cultures had a certain structural integrity to them; they were not to be ranked hierarchically. Each particular culture was supposed to carry its own validity through the interrelated and co-dependent use of meanings, symbols, and values; each had its own account of origins, reproduction, and relation to the outside. There was good reason for anthropologists to have treated cultures as self-sustaining, quasi-autonomous entities: These ideas had a powerful anticolonial force; it offered, in its relativism, a bulwark against racism, subordination, and genocide. Cultural holism and relativism seemed necessary: If every culture was the equal of every other one, the imperial subordination of groups outside of white, European Christian ones in Western Europe (or North America) could not be justified.

Since the 1980s, in the highly interpenetrated world in which we live, we have come, more and more, to find it impossible to see cultures as isolated, crystalline structures; more and more, we see them as overlapping, tied together through the movements of people, ideas, objects, and struggles. We no longer see a homogeneous culture (singular), not in the nation, not in 'pre-contact' third world countries, not in former imperial capital cities.

In our cacophonous and interrelated world, we need to form new modes of understanding these shifting and ever-crossing boundaries of cultures. In my sector of inquiry – the study of the physical sciences in its broader cultural frame – it is clear that people and ideas are always on the move, there are no strict borders between previously separate disciplines and subdisciplines: String theory shares techniques with what used to be called condensed matter physics. Biology and physics share major spheres of interest as they address the nature of DNA and other biological materials. No longer are the sectors of civilian, military, and commercial science so distinct. A GPS chip is in your running watch, in a smart bomb, in drones, and in more apps on your smartphone than you can remember. Buying a GPS chip is cheap and they are ubiquitous. But the impact of these little objects – some just a few millimeters long, wide, and high – is vast: as they can report back on our positions; as data combines to report who we are with, where we idle, our objects of attention. Indeed, understanding our emerging technical world will take the collaboration of many disciplines, from economics and anthropology to physics, engineering, and surveillance studies. Even this little example of the GPS chip suggests that we will need a myriad of approaches to characterize, understand, and intervene in the great issues we face now.

The study of culture going forward necessarily must address the technological and scientific domains, it cannot retreat to a belle-lettrist self-definition. But the study of culture can and should be more than the study of the "impact" of the scientific-technical, the study of culture is needed to understand how we, as society ought to handle these intersections.

Jens Kugele: Where do you see key challenges for interdisciplinary work in light of the academic publishing market, tenure reviews, and the high value attributed in most Humanities disciplines to single-authored publications, a concept rooted in eighteenth-century discourses of individual geniality.

Peter Galison: I am very worried about the institutional frame that faces a new generation of scholars. Right now, I see early-career researchers whose interdisciplinary, collaborative work is highly valued, even celebrated, as they are chosen for postdoctoral fellowships across North America, Europe, and elsewhere. But then, when these same scholars go on the job market, they suddenly face obstacles. They confront resistance to cross-disciplinary work alongside discord about the validity of multi-authored books and articles. I see ambivalence about promoting to tenure someone who works in teams or steps over the disciplinary border. This switcheroo of values – telling a generation of scholars interdisciplinarity is good, until they are told it is bad – is, in my view, misguided academically. It ignores some of the best work produced today. But I would say more and with a certain degree of anger and frustration: By alternately encouraging and then disvaluing collaboration and interdisciplinarity, our institutions are behaving unethically, betraying an emerging generation of scholars.

The natural sciences have not solved the manifold problems of collaboration and interdisciplinarity, but they are ahead of the human sciences: Teams of hundreds, now (at CERN) thousands of physicists work together toward goals of the first importance, including the discovery of the Higgs. Over the last four years, I have been a member of one such team, with some 207 scientists and engineers, distributed over 18 countries and 59 institutions. Some of the collaborators come from computer science, some from theoretical astrophysics, yet others are experts in electrical engineering, or radio-telescope observation. I come from a mix of history and philosophy of science – focused to a certain degree on image-making, and theoretical particle physics. Together the many of us constitute the Event Horizon Telescope Collaboration, formed to assemble a world-spanning network of radio-telescopes capable of imaging a black hole 53 million light years away. We are constantly grappling with questions of credit and the advancement of early career scientists – and by no means has this or any other big collaboration solved the problem. But there are substantive things one can do to promote the visibility recognition of rising PhDs, postdocs, and assistant professors: They can be promoted to give academic and public talks; they can take on recognized roles in working groups; they can report at collaboration meetings; they can be leads on white papers. We ought to be thinking now about ways to do such things in the growing number of interdisciplinary collaborations in the human sciences.

Jens Kugele: In your *Image and Logic* (1997), among several others of your publications, you introduce the notion of "trading zones." Is the concept of "trading zones," as you understand it, applicable to the study of culture as an interdisciplinary and international field, i.e., can it help us to shed light on dynamics, potentials, and challenges?

Peter Galison: I introduced the notion of trading zones because I was frustrated with the false choice we were offered: Either there was a universal reduction basis to all of the sciences, as the notion of a physical thing-language would have it (logical positivists). Or cultures depicted as so disjunct that passing between them would be like a Gestalt shift, a radical linguistic translation, or a religious conversion. That did not (so it seemed to me) correspond to the real, partial, ever-developing coordination that is constantly in play between and among cultures. Or put another way, in many fields, certainly in science and technology studies, we are used to focusing on local practices (the focus on the local conditions of knowledge production seems to me the single most important innovation in STS). By the 1990s, it seemed to me far too late in the day to call into play a *global* notion of languages and cultures and join it (as was common in Kuhnian-inflected studies) with *local* conceptions of scientific work. Instead, I wanted to see language itself as an evolving *locally* inflected formation, of which exchange languages (jargons, pidgins, and creoles) were only the most notable manifestation. So too at the boundaries between fields, there is always coordination that can, over time, develop into fields themselves (think of biochemistry, algebraic geometry, physical chemistry, just to name a few).

Jens Kugele: If you think about the possible futures of the study of culture, where do you see institutional "trading zones" for this interdisciplinary field of research? What kind of "trading zones" should be explored further? What kind of competencies should we foster, e.g., in our Ph.D. training, if we are convinced that the ability to make meaning in more than one discipline is based on more than "interactional expertise" (Collins and Evans 2002)?

Peter Galison: Institutions that can best assist the formation of trading zones are not entirely abstract. Instead, they have a focus. Artificial Intelligence offers an example. AI systems are being used to determine who, among those accused of having committed a crime, should be granted pre-trial release. Here is an arena where a technical, computational concern needs to cross with cultural concerns: constitutional and ethical questions – questions of social justice. Similarly, we have a growing number of institutes that study the cross of genetics with politics and ethics – modifications of crops, animals, and ultimately humans raise pressing issues that demand a way of reasoning that is more than just technical drives constrained by cultural constraints. Instead, we need to develop a way of teaching, researching, and applying genetic knowledge that is reasoning about ethics all the way down, so to speak.

Perhaps our Ph.D. training should include at least some work in an interdisciplinary team. Instead of focusing all our energy on the production of a thesis, we could have at least one project, or a chapter within a thesis built on collaborative research, where, say a literary scholar, a historian, and an economist could address a body of literature not only in its associations of structure, meaning, and allusion, but in the materiality and finances of production and distribution. Book history, media theory, and literary analysis could work together rather than squaring off as antagonistic approaches. With filmmaker/anthropologist Lucien Castaing-Taylor, we set up at Harvard a program called "Critical Media Practice," where advanced graduate students from across the university could learn, develop, and supplement their thesis work in film, interactive online sites, audio, installation, and other digital work (http://cmp.gsas.harvard.edu/). This inter-field, inter-modal mix aims to be an institutional trading zone.

Jens Kugele: To what extent is the concept of "trading zones" helpful to think about future exchange and collaboration between the study of culture, the natural sciences, and the life sciences? Where do you see trading partners who, as you phrased it, "can hammer out a local coordination, despite vast global differences" (Galison 1997, 783)?

Peter Galison: The central concern of trading zones is that it is possible for different domains to work out a local, specific, common form of action and reasoning, even if the larger disciplinary demands remain quite disjunct. Nanotechnology does require that virology, surface chemistry, and atomic physicists learn to work together, create techniques, and develop ways of speaking that are common enough for them to generate new work. But productive nanoscience work does *not* require making physicists into biologists or biologists into chemists, much less all these groups into an undifferentiated morass. Instead, the human, social, and natural sciences come into consequential interaction through a sufficiently developed, *specific* common language and set of actions.

For example, computer science needs to be taught and practiced with concepts of privacy built in; not added as an afterthought. And here the humanities and interpretive social sciences have much to contribute: What notions of privacy do we want to protect? What does the history of the concept reveal? What is understood by taking onboard the way it is understood in psychology, political science, or critical feminist theory, to name but a few examples?

Jens Kugele: Scholars such as Mieke Bal have suggested the idea of a concept-based approach to the study of culture with a notion of "traveling concepts" at its core. How do you see the potential of concept-based research in the study of culture in the future? Do concepts bear the potential to facilitate such trading zones? Do concepts bear the potential to serve as building blocks of a "contact language," a "system of discourse" as you phrase it in the context of your notion of "trading zones"?

Peter Galison: I take concepts to be quasi-stable entities, holding practices and meanings in a form that is recognizable over some region of time and place, but not in any sense absolute. We fight over concepts because they organize so much: think of political concepts like liberty, property, rights, nation, citizenship. Understanding their contingency, their remit, their history, is essential to moving forward. Marriage may have had certain meanings (who can marry whom) but in hard-fought battles, that notion is changing – opening up – across many countries. So too is it in science (energy, mass, time, entropy): Einstein's main contribution was to level a critical re-evaluation of space and time. And much of my work has been organized around an understanding of how certain scientific concepts shifted under the pressure of scientific and philosophical engagement: objectivity, simultaneity, secrecy, containment – to name a few.

I certainly agree with Mieke Bal that concepts move – they *travel* as she has called it in her persuasive studies – as she stresses, the points of intersection and coordination can be generative. My understanding of trading zones – as a local and coordinative venture – sets concepts, material objects, and manipulations in historically-shifting syntactic frames. What we do with concepts (how they relate to each other) is also essential. Rules of combination and exclusion figure vitally in the notion of a trading zone: There is no building structures from bricks alone. We need the mortar binding the bricks; we need guiding principles so to speak: For example, you are better off intercalating bricks from one level to another if you want the house to stand. This is why I come back to the more-than-metaphor of trading languages in which jargons (highly restricted coordinative structures), pidgins (more extensive than jargons but still specific), and creoles (interlanguages sufficiently rich in structure and metalinguistic development that one can grow up in them). There is a semantics, to be sure (the concepts and meanings), but also a syntax.

Of course, not all jargons become pidgins on the way to creoles, some jargons and pidgins persist as such, or vanish altogether. This much we know from the anthropological linguists. Indeed, I reject the idea that there are pure disciplines as opposed to hybrid fields: Today's "pure" languages are just the more consolidated and elaborated forms of yesterday's hybrids. So it is in disciplinary fields. Take what many mathematicians would consider the purest of pure mathematics: algebraic geometry. The hybrid antecedents of that field (algebra and geometry) are worn on its name-sleeve. Purity in disciplines or languages is a *product*, not an essence. For this reason, we ought to be highly suspicious when interdisciplinarity is derogated. But *which* interdisciplinary forms will persist? That is an open question.

Jens Kugele: In your newspaper article "Self-censorship in the Digital Age" (*Frankfurter Allgemeine Zeitung*, 2014), you point to ways that surveillance and harvesting of communications has and continues to reshape culture and ultimately the self. What are some of the central challenges and potentials that you see for the academic study of culture in this digital age? Do you see a changing role for the scholar in the field of culture studies?

Peter Galison: It is too late in the day to bemoan the new forms of cultural production, circulation, and consumption as if they can be driven back into the keyboards, cameras, and microphones from whence they came. Instead, those of us studying digital cultures have openings. We can study the evolving forms: games, tweets, postings, sites, apps, seriality, blogs. We can use them under critical pressure: What kind of subject and object is constructed by these genres? And we can take an active role in reshaping them, asking how might they be turned to other ends, as in an earlier epoch, film, neon lights, and theatre were bent away from classical structures. My hope for the future of cultural studies is that it will have all of these elements: a critical history *and* a productive taking-up of the forms to other, adventurous, and generative ends.

References

Bal, Mieke. *Travelling Concepts in the Humanities*. Toronto: University of Toronto Press, 2002.

Collins, Harry, and Robert Evans. "The Third Wave of Science Studies: Studies of Expertise and Experience." *Social Studies of Sciences* 32.2 (2002): 235–296.

Galison, Peter (dramaturg) for William Kentridge. *Refusal of Time* and *Refuse the Hour*. 2012.

Galison, Peter, and Henry L. Gates. *No More, America*. 2017. 13 minutes.

Galison, Peter, and Lorraine Daston. *Objectivity*. Boston: Zone, 2007.

Galison, Peter, and Robb Moss. *Secrecy*. Premiered Sundance 2008. 81 minutes.

Galison, Peter, and Robb Moss. *Containment*. Premiered Full Frame 2015. 81 minutes.

Galison, Peter, Kazunori Akiyama, and The EHT Collaboration. "First M87 Event Horizon Telescope Results, I–VI." *The Astrophysical Journal Letters* 875.1 (2019).
Galison, Peter. "21 July 1773: Disputation, Poetry, Slavery." *Critical Inquiry* 45 (2019): 376–404.
Galison, Peter. "Algorists Dream of Objectivity." *Possible Minds: Twenty-Five Ways of Looking at AI*. Ed. John Brockman. New York: Penguin Press, 2019.
Galison, Peter. "Removing Knowledge." *Critical Inquiry* 31 (2004): 229–243.
Galison, Peter. "Secrecy in Three Acts." *Social Research* 77 (2010): 941–974.
Galison, Peter. *Einstein's Clocks, Poincaré's Maps*. New York: W.W. Norton, 2003.
Galison, Peter. "Self-censorship in the Digital Age: We Won't be able to Recognize Ourselves." *Frankfurter Allgemeine Zeitung* (07 April 2014). <https://www.faz.net/aktuell/feuilleton/debatten/self-censorship-in-the-digital-age-we-won-t-be-able-to-recognize-ourselves-12885374.html?printPagedArticle=true#pageIndex_2> [accessed: 10 August 2019].
Galison, Peter. *Image and Logic: A Material Culture of Microphysics*. Chicago: University of Chicago Press, 1997.
Further information on films at <https://galison.scholar.harvard.edu/>.

Notes on Contributors

Nicole Anderson is Professor of Cultural Studies, and Head of the Department of Media, Music, Communications, and Cultural Studies at Macquarie University. She is the co-founder and chief editor of the journal *Derrida Today* published by Edinburgh University Press, and the Founder and Executive Director of the Derrida Today Conferences. Her research fields and interests include: cultural theory and practice, deconstruction, Derrida Studies, posthumanism, animal and environmental studies, bioculture, biopolitics, poststructuralism, and continental philosophy. As well as numerous book chapters and articles in various academic journals such as *Parallax*, *Journal of Southern Philosophy*, *Social Semiotics*, *Oxford Literary Review*, etc., she is the author of *Derrida: Ethics Under Erasure* (Bloomsbury, 2014); *Cultural Theory in Everyday Practice* (Oxford University Press, 2009, with Katrina Schlunke), *Political Animal(s)* (forthcoming, 2020), and is currently writing a book for Routledge entitled *Culture* (which explores the intersection between, and co-implications of, science and culture).

Doris Bachmann-Medick is Adjunct Senior Research Fellow at the International Graduate Centre for the Study of Culture (GCSC) of the Justus Liebig University Giessen. She held numerous appointments as a visiting professor, recently at the universities of Graz, Göttingen, UC Irvine, Cincinnati, and Georgetown University/Washington, DC. Her main fields of research are cultural theory, *Kulturwissenschaften*, literary anthropology, and translation studies. Her recent book publications include *Cultural Turns. Neuorientierungen in den Kulturwissenschaften* (6th ed. Rowohlt, 2018 [2006]), transl. into Polish, Russian; revised English edition *Cultural Turns: New Orientations in the Study of Culture* (De Gruyter, 2016) and her edited volumes *The Translational Turn*, a special issue of the journal *Translation Studies* (2.1, 2009), *The Trans/National Study of Culture: A Translational Perspective* (De Gruyter, 2014), *Migration: Changing Concepts, Critical Approaches* (De Gruyter, 2018, co-edited with Jens Kugele). She serves on the editorial board of *Translation Studies* (since 2008).

Hubertus Büschel is Professor for Modern History at the University of Kassel. He was Juniorprofessor for Cultural History at the International Graduate Centre for the Study of Culture (GCSC) of the Justus Liebig University Giessen between 2009 and 2015 and Professor for Contemporary History (Chair) at the University of Groningen (The Netherlands) between 2015 and 2019. His main fields of research include cultural history of Germany, Europe, and Africa in the nineteenth and twentieth centuries, historical anthropology, global history, history of development, and history of psychiatry. His recent publications include *Hilfe zur Selbsthilfe. Deutsche Entwicklungsarbeit in Afrika* (Campus, 2014), *Hitlers adliger Diplomat. Der Herzog von Coburg und das Dritte Reich* (Campus, 2016). He is currently working on a book project about 'glocal' histories of psychiatry in Africa and the experiences of African psychiatric patients.

Tom Clucas completed his B.A., M.St., and Dr. phil. in English Literature at the University of Oxford, where he also taught at several of the Colleges. Subsequently, he worked as a Postdoctoral Researcher at the International Graduate Centre for the Study of Culture and Deputy Professor of English and American Literature and Culture at Justus Liebig University Giessen. He co-edited the book *Futures Worth Preserving: Cultural Constructions of Nostalgia and Sustainability* (transcript, 2019) and has published articles on Romantic Poetry in journals

https://doi.org/10.1515/9783110669398-019

including *Notes and Queries*, *The Wordsworth Circle*, *The Cowper and Newton Journal*, *Textual Cultures*, *The Review of English Studies*, *European Romantic Review* and *The Coleridge Bulletin*. He now lives and works in London.

Peter L. Galison is a physicist, historian of science, and filmmaker at Harvard University, where he is the Joseph Pellegrino University Professor. In 1997, he was named a MacArthur Fellow; with his Event Horizon Telescope colleagues, Galison shared in the 2020 Breakthrough Prize in Fundamental Physics for the first image of a black hole. He is the author of several books, including *How Experiments End* (University of Chicago Press, 1987), *Image and Logic: A Material History of Microphysics* (University of Chicago Press, 1997), *Einstein's Clocks, Poincaré's Maps* (W.W. Norton, 2003), and *Objectivity* (with Lorraine Daston, Zone Books, 2007). Galison partnered (as dramaturg) with South African artist William Kentridge on a multi-screen installation, *The Refusal of Time* (2012), and an associated chamber opera. His documentary film *Ultimate Weapon: The H-bomb Dilemma* (with Pamela Hogan, 2000) probed the moral-political debates over the H-bomb. He and Robb Moss co-directed *Secrecy* (2008) on national security secrecy, which premiered at Sundance. The two also co-directed *Containment* (2015), about the need to guard radioactive materials and warn the 10,000-year future. Galison's current film, still in production, is on black holes: *The Edge of All We Know*.

Isabel Maria de Oliveira Capeloa Gil is a Full Professor of German and Culture Studies at UCP's School of Human Sciences (Lisbon), and holds a PhD in German Studies. Her scholarly work pivots around the exploration of the disciplinary boundaries between literature, the arts, and other disciplines. Her research has been published in Portuguese, English, German, French, Italian, and Spanish. She is the author of over 140 publications, bridging cultural theory, interarts studies, visual culture, culture, and conflict. Recent publications include *Hazardous Future: Disaster, Representation and the Assessment of Risk* (De Gruyter, 2015), *The Cultural Life of Money* (De Gruyter, 2016, co-edited with Helena Gonçalves da Silva), and *The Ballets Russes: Modern Times after Diaghilev* (Documenta, 2018). With an extensive experience in research development, she founded the Research Centre for Communication and Culture (CECC) at UCP focused on understanding the role of contemporary cultural and media production in societal development. In 2019, she was considered amongst Portugal's top women in science by the Ciência Viva Foundation. She is a Fulbright, DAAD, and Gulbenkian Scholarship recipient. She is currently the Rector of the Catholic University of Portugal. In 2019, she was awarded an Honorary Doctorate in Humane Sciences by Boston College and gave the university's Commencement Speech. In addition, she sits in the Advisory Board of the Gulbenkian Foundation and is a member of the European Council of Foreign Relations.

Richard Grusin is Distinguished Professor and Director of the Center for 21st Century Studies at University of Wisconsin-Milwaukee. He has published numerous articles and book chapters and authored five books: *Transcendentalist Hermeneutics: Institutional Authority and the Higher Criticism of the Bible* (Duke, 1991), *Remediation: Understanding New Media* (MIT, 1999, with Jay David Bolter), *Culture, Technology, and the Creation of America's National Parks* (Cambridge, 2004), *Premediation: Affect and Mediality After 9/11* (Palgrave, 2010), and *Radical Mediation: Cinema, Estetica, e Tecnologie Digitali*, ed. and transl. Angela Maiello (Cosenza, Italy: Pellegrini Editor, 2017). He has edited three volumes of essays: *The Nonhuman Turn* (Minnesota, 2015), *Anthropocene Feminism* (Minnesota, 2017), and *After Extinction* (Minnesota, 2018).

His work has been translated into several languages including Italian, Korean, French, German, Spanish, Portuguese, and Polish.

Ursula K. Heise is Chair of the English Department at UCLA. She holds the Marcia H. Howard Chair in Literary Studies, and is co-founder of the Lab for Environmental Narrative Strategies (LENS) at UCLA's Institute of the Environment and Sustainability. Her research and teaching focus on contemporary literature and the environmental humanities, environmental literature, arts, and cultures in the Americas, Western Europe, and Japan, literature and science, science fiction, and narrative theory. Her books include *Chronoschisms: Time, Narrative, and Postmodernism* (Cambridge University Press, 1997), *Sense of Place and Sense of Planet: The Environmental Imagination of the Global* (Oxford University Press, 2008), *Nach der Natur. Das Artensterben und die moderne Kultur* (After Nature: Species Extinction and Modern Culture) (Suhrkamp, 2010), and *Imagining Extinction: The Cultural Meanings of Endangered Species* (University of Chicago Press, 2016), which was written with the help of a Guggenheim Fellowship and won the 2017 book prize of the British Society for Literature and Science. She is editor of the series *Natures, Cultures, and the Environment* with Palgrave, and co-editor of the series *Literature and Contemporary Thought* with Routledge. She produced and wrote *Urban Ark Los Angeles*, a documentary created as a collaboration of LENS with the public television station KCET-Link.

Jens Kugele is Head of Research Coordination at the International Graduate Centre for the Study of Culture (GCSC) of the Justus Liebig University Giessen. He is one of the research centre's Principal Investigators and has been a member of the centre's Executive Board since 2014. Previously, he held appointments as Visiting Researcher and Assistant Professor at the LMU Munich, the Berkley Center for Religion, Peace and World Affairs at Georgetown University, and the Institute for German Cultural Studies at Cornell University. His research interests include intersections of the literary and cultural history with the history of religion, German-Jewish literature and culture, constructions of belonging, memory, and space. Recent publications include the volume *Migration: Changing Concepts, Critical Approaches* (De Gruyter, 2018, co-edited with Doris Bachmann-Medick), the volume *'Heimat': At the Intersection of Memory and Space* (De Gruyter, 2012, co-edited with Friederike Eigler), and a special issue of the *Journal of Religion in Europe* with a focus on *Relocating Religion(s)* in museal spaces (Brill, 2011, co-edited with Katharina Wilkens). One of his current research projects looks at 'sacred space' as a concept for the interdisciplinary study of culture. He is co-founder and co-editor of the peer-reviewed interdisciplinary open access journal *On_Culture*.

Dirk van Laak is Full Professor of German and European History from the nineteenth to the twenty-first century at Leipzig University. He studied German Literature and History at Essen University and finished his Ph.D. (on Carl Schmitt) at Hagen University in 1993. From 1991 to 1993, he was employed at a State Archive in Düsseldorf. From 1993 to 2007, he was scientific assistant at the Historical Institute of Jena University, supplemented by visiting professorships at universities in Chicago, Tübingen, and Freiburg. From 2007 to 2016, he was Full Professor of Contemporary History at Giessen University. He is co-editor of Docupedia-Zeitgeschichte, Themenportal Europäische Geschichte, and 'Geschichte der technischen Kultur.' His main research interests are German and European history and the history of globalization, colonialism, planning and technology, also the history of historiography and its relationship to fiction. His most recent book addressed the everyday and cultural history of infrastructures (*Alles im Fluss. Die Lebensadern unserer Gesellschaft – Vergangenheit und Zukunft der Infrastruktur* (S. Fischer, 2018).

Andreas Langenohl is Professor of Sociology with a focus on Comparative Studies and Head of Graduate Studies at the International Graduate Centre for the Study of Culture (GCSC) at Justus Liebig University Giessen. His research and teaching cover the areas of economic sociology and the social study of finance, transnationalism, the epistemology of the social sciences and economics, and social and cultural theory. He is co-editor of the journal *Behemoth: A Journal on Civilisation* and elected board member of the section "Economic Sociology" within the German Sociological Association. Recent and forthcoming publications include *Town Twinning, Transnational Connections, and Trans-local Citizenship Practices in Europe* (Palgrave Macmillan, 2015), *Conceptualizing Power in Dynamics of Securitization: Beyond State and International System* (Nomos, 2019, co-edited with Regina Kreide), *Security Infrastructures*, a special issue of the journal *Politikon: South African Journal of Political Studies* (47.1, 1–3, 2020, co-edited with Gideon Van Riet).

Laura Meneghello holds a postdoc research and teaching position in Modern European History of Knowledge and Communication at the University of Siegen. She held a scholarship at the International Graduate Centre for the Study of Culture (GCSC) and received her PhD from Justus Liebig University Giessen. Her dissertation was published in 2017 under the title *Jacob Moleschott – A Transnational Biography: Science, Politics, and Popularization in Nineteenth-Century Europe* (transcript, 2017). She is also co-author of the collective monograph *Library Life. Werkstätten kulturwissenschaftlichen Forschens* (Meson Press, 2015). Her current research and teaching activities focus on the history of science and technology, as well as on the history of knowledge and scientific translation, urban history, and the history of infrastructures and media. Her habilitation project deals with the history of pneumatic mail networks.

Ansgar Nünning is Professor of English and American Literature and Cultural Studies at Justus Liebig University Giessen, and holds an honorary doctorate from the University of Stockholm. He is the founding and managing director of the International Graduate Centre for the Study of Culture and the European PhD Network "Literary and Cultural Studies," both pioneering models for the reform of postgraduate research training and international collaboration. One of the leading narratologists, he has published widely on English and American literature, cultural memory studies, genre theory, narratology, and literary and cultural theory. His most recent English book publications include *The American Novel in the 21st Century* (ed. with Michael Basseler, in collaboration with Nico Völker, WVT, 2019), *The British Novel in the Twenty-First Century* (ed. with Vera Nünning, WVT, 2018), *Literature and Crises* (ed. with Elizabeth Kovach and Imke Polland, WVT, 2017), *Travelling Concepts for the Study of Culture* (ed. with Birgit Neumann, De Gruyter, 2012), *Cultural Ways of Worldmaking* (ed. with V. Nünning and Birgit Neumann, De Gruyter, 2010), and *A Companion to Cultural Memory Studies* (ed. with Astrid Erll, De Gruyter, 2008). He is the editor of *Metzler Lexikon Literatur- und Kulturtheorie* (5th ed., Metzler, 2013) and of the book series *Uni Wissen Anglistik/Amerikanistik*, *WVT-Handbücher zum literaturwissenschaftlichen Studium* and *ELCH: English Literary and Cultural History* (both with Vera Nünning) as well as *MCM: Media and Cultural Memory/ Medien und kulturelle Erinnerung* (with Astrid Erll) and co-editor of the peer-reviewed journal *Germanisch-Romanische Monatsschrift* (*GRM*).

Andreas Reckwitz is Professor for Social Theory and Cultural Sociology at Humboldt-Universität zu Berlin. Before, he was professor at Constance University from 2005 to 2010 and at European University Viadrina in Frankfurt/Oder from 2010 to 2020. He received the Leibniz prize of the DFG in 2019. He held numerous visiting professorships and fellowships, among others in St. Gallen, Vienna, Berkeley, London, Heidelberg, Freiburg, and Bielefeld. Book publications include *Das Ende der Illusionen. Politik, Ökonomie und Kultur in der Spätmoderne* (Suhrkamp, 2019; English translation Polity, 2021), *Die Gesellschaft der Singularitäten* (Suhrkamp, 2017, English translation Polity, 2020), *Die Erfindung der Kreativität* (Suhrkamp, 2012, English translation Polity, 2017), *Das hybride Subjekt* (Velbrueck, 2016, new edition Suhrkamp, 2020).

Andressa Schröder is a PhD researcher at the Giessener Graduiertenzentrum Kulturwissenschaften (GGK) at Justus Liebig University Giessen. The working title of her thesis is "Affective Sustainabilities in Alternative Art Festivals: An Aesthetic Approach to Sustainability Ethics." She was the main editor of the recent volume *Futures Worth Preserving: Cultural Constructions of Nostalgia and Sustainability* and she has also written several articles focusing on the role of arts-based research and environmental aesthetics for sustainability debates, among them: "Aesthetics as a 'Middle Way' in Sustainability Ethics" and "Aesthetic Strategies to Explore beyond the Models of Sustainable Development: An Analysis of Lisa Simpson's Musical Sewing." She has also written and coordinated different artistic projects funded by the Brazilian Ministry of Culture and worked as a curator for various art exhibitions at the Brazilian Embassy in Berlin in several occasions. Her research interests include environmental aesthetics, arts-based research, sustainability ethics, alternative art, and psychedelic culture.

Uwe Wirth is Professor of German Literature and Cultural Theory at the Department of German Literature of the Justus Liebig University Giessen. From 2005 to 2007 he coordinated the Center for Advanced Studies in Literature and Culture (ZfL) in Berlin. Since then he was visiting professor at the University of California, Berkeley, at Cornell University, and at the Imperial University of Tokyo. He had various fellowships at the Exzellenzcluster *Kulturelle Grundlagen von Integration* at Konstanz University and at the *Deutsche Haus* at New York University. He is on the editorial board of the handbook series *Kulturwissenschaftliche Philologie* (De Gruyter) and published widely on concepts of philology, authorship, and narratology (as for instance with the monograph: *Die Geburt des Autors aus dem Geist der Herausgeberfiktion. Editoriale Rahmung im Roman um 1800: Wieland, Goethe, Brentano, Jean Paul und E.T.A. Hoffmann* (Fink, 2008). Other research interests concern theories of humor – here he edited the handbook *Komik – ein interdisziplinäres Handbuch* (Metzler, 2017) and a volume on *Komik der Integration* (Aisthesis, 2019, co-edited with Deniz Göktürk). Following a research project on cultural theories "after hybridity," he is currently working on a monograph on grafting as cultural technique and as a cultural model which is a part of a research project that will be published by Fischer 2021.

Index

Adelson, Leslie A. 10, 66–78
Adorno, Theodor W. 67–69, 72, 115, 226–227
affordance 45
Alexievich, Svetlana 156, 158, 164
Anderson, Benedict 158, 205
Anthropocene 5, 7, 10, 32, 41, 114–115, 215, 276–280, 286
anthropology 3, 8, 9, 11, 22, 52, 91, 123, 127, 128, 131, 171, 274, 275, 282, 291, 292
apocalypse 90
Appadurai, Arjun 3, 22, 23, 66–67, 70–73, 75, 77, 78, 99, 159
area studies 52, 203, 213, 246
art 4, 32, 33, 38, 55, 91, 96, 103–106, 110–114, 144–148, 152, 157, 160, 163, 165, 167–180, 182, 188, 189, 223, 228, 229, 246, 251, 262, 267, 274, 276, 283–285, 290
– arts-based research 9, 169–180
artificial intelligence 9, 21, 60, 290, 294

Bachmann-Medick, Doris 3–15, 47, 51, 224, 243, 245, 246, 270
Bakhtin, Mikhail 22, 185, 186, 199
Ballard, James G. 165
Bal, Mieke 44–46, 51, 52, 296
Barthes, Roland 163
Baudrillard, Jean 101
Bauer, Fritz 69, 70
Bauman, Zygmunt 41, 100, 101
Beck, Ulrich 5, 7, 9, 11, 14, 115, 116
Belsey, Catherine 54
Benjamin, Walter 68, 112, 182, 184, 185, 199, 226
Bentham, Jeremy 95, 98, 99, 105
Bhabha, Homi 183, 184, 192
big data 13, 21
biology 49, 195, 215, 250, 276, 281, 292
– microbiology 279
body 95, 96, 98, 102, 105, 161, 193, 195, 196, 244, 251–255, 257–266, 279
border thinking 11, 123–134
Bourdieu, Pierre 124, 127–130, 133, 225
Bowie, David 83–85, 87, 90, 91
Braidotti, Rosi 4, 8, 10
Breines, Wini 74, 75
Brekhus, Wayne 267
Brexit 85–88, 111
Bruner, Jerome 34–37
Brus, Günter 265, 266
Buc, Philippe 127, 132
bullshit 49, 50, 55

capitalism 34, 41, 43, 58, 70, 72, 73, 77, 89, 117, 122, 141, 149, 151, 242, 245, 253, 256, 280
– capitalist culture 32–35
Chakrabarty, Dipesh 10, 46, 60, 131, 213, 246 , 278
Chamayou, Grégoire 96, 101, 102
change 4, 7, 9, 17, 33, 34, 39, 42, 43, 55, 71, 167, 184, 213, 215, 243, 274–286
Chartier, Roger 125, 127
Citton, Yves 43, 48
climate 6, 10, 42, 83, 88, 89, 115–117, 213, 290
– climate change 7, 9, 13, 21, 32, 39, 41, 43, 49, 55, 66, 89, 113, 115–118, 277, 278, 280
Cohn-Bendit, Daniel 85
Cold War 203, 206, 207
collaboration 4, 10, 11, 13, 14, 18–20, 23, 29, 51, 52, 54–56, 105, 110, 145, 173, 270, 283, 290, 292, 293, 295
– collaborative research 17–23, 46, 291, 295
colonial 5, 11, 123–134, 192, 196, 203, 208, 209, 213, 239, 243, 245, 246, 281
– colonial history 203, 208
– colonialism 124, 126, 128, 129–131, 182, 192, 208, 209, 240, 258, 269, 274
– decoloniality 131–133
– German colonialism 208–209
– postcolonialism/postcolonial 46, 129, 130, 134, 159, 182, 183, 194, 205, 246, 274
Comaroff, Jean 132, 133
Comaroff, John 124, 126, 132, 133
commercialization 21, 40, 259, 260
concepts 3–12, 14, 15, 18, 30–34, 37–39, 43–49, 53, 56, 59, 66–71, 75, 76, 86, 97,

https://doi.org/10.1515/9783110669398-020

115, 124, 126–131, 133, 134, 143, 144, 155, 157–159, 161, 172, 175, 182–196, 200, 203–205, 210, 211, 214, 224, 225, 230, 236–246, 252–255, 258, 266–269, 274–277, 280–283, 291–296
– concepts of culture 9, 39, 224, 252, 283
– concepts of history 205
– conceptual tools 13, 18
– travelling concepts 44–46, 51, 53, 237, 243–245
concerns 4, 5, 7, 10, 13, 16, 21, 23, 29–47, 54–60, 67, 75, 78, 103, 111–118, 149, 165, 169, 250, 251, 258, 260–262, 266–268, 282, 288, 289, 294, 295
conjunctures 32–36, 39–44
consumption 40, 41, 101, 142, 148, 150, 152, 157, 158, 164, 227, 275, 282, 297
control 14, 88, 94–106, 117, 118, 121, 150, 161, 211, 253
Covid-19 pandemic 13, 17
Crary, Jonathan 41, 94, 96
crisis 4, 13, 14, 17, 33, 34, 36, 37, 41, 42, 47–51, 66, 87–89, 126, 169, 179, 228, 230, 241, 242, 280, 284
– financial crisis 42, 224, 228, 230
Crutzen, Paul 276, 277
culture 3, 9, 17, 18, 23, 29, 33, 34, 35, 41, 42, 44, 47, 50, 53–55, 57, 68, 86, 99, 106, 114–116, 122, 146, 152, 155, 156, 167, 182–188, 191, 193, 195, 196, 198, 205, 209, 212, 213, 223, 225, 227, 251, 276, 282, 283, 288, 292
– corporate culture 225
– culturalization 142, 143, 146, 147, 150–152
– cultural facts 70, 71, 75, 77, 78
– cultural frame 183, 292
– cultural history 9, 11, 52, 123–134, 215, 236–247, 278, 290
– cultural market 150
– cultural memory 44
– cultural representation 101, 155, 156
– cultural studies 3, 18, 19, 30, 31, 38, 39, 51–60, 88–91, 96, 124, 132–134, 191, 225–227, 233, 237, 251–257, 261–270, 274–276, 283, 285, 286, 297
– cultural translation 10, 182–200
– cultural turns 6, 51, 268
– Kulturwissenschaften 51, 52, 54, 57
– study of culture 3, 17, 29, 66, 90, 91, 94, 106, 110, 131, 146, 155, 156, 167, 168, 169, 223, 224, 232, 237, 251, 274, 288, 291–297
Cusa, Nicholas of 96, 97, 105

Darwin, Charles 194, 195, 211, 241
Davidson, Cathy 30, 32, 54, 56
Davis, Natalie 123, 125, 131, 132
Day, Richard 67, 74, 75
delinking 7, 11, 12, 123–134
Derrida, Jacques 83, 86, 87, 90, 191, 192, 199, 280, 281
digital 10, 41, 43, 113, 142, 148, 151, 161–163, 275, 290, 295, 297
– digitalization 5, 14, 40, 55, 150
– digital age 32, 33, 41, 54, 297
– digital information technologies 40, 41, 42, 48
disciplines 3, 8, 10, 22, 29, 38, 39, 44–46, 49, 51, 52, 55, 56, 90, 105, 106, 112, 114, 117, 124, 155, 156, 160, 169, 171, 175, 215, 225, 229, 236–238, 246, 262, 275, 276, 290, 292, 294, 296, 297
– cross-disciplinarity 55, 56, 274, 293
– disciplinarity 22
– interdisciplinary 8, 11, 18, 20, 22, 23, 31, 38, 44, 46, 52–55, 58, 110, 112–115, 228, 233, 237, 243, 246, 255, 261, 291, 293–295, 297
– transdisciplinarity 7, 8, 10, 12, 23, 40, 52, 55, 56, 76, 106, 114, 177
discourse analysis 46, 124, 236, 238, 246
diversity 8, 22, 88, 134, 215
drone 101–104, 292

ecology 9, 40, 43, 48, 52, 162, 167, 178, 276, 285
– acoustic ecology 285
economy 33, 39, 42, 48, 55, 58, 85, 86, 142, 147–153, 155, 157, 158, 213, 223, 225, 227–229, 233, 236–238, 245, 246, 275
– economic 30, 34, 42, 43, 46, 48, 73, 87, 111, 114, 115, 117, 133, 148, 149, 150, 169, 188, 205, 207, 211, 228–232, 236, 237, 239–246, 260, 280
– global economy 9, 141, 236–247

Edelman, Lee 111, 112
elective affinity (Wahlverwandtschaft) 250–270
environmental 5, 73, 116, 117, 170, 173, 174, 177, 209, 247, 277, 278, 280, 282, 284, 290
– environmental humanities 9, 55, 58, 275, 276
epistemology 23, 132, 133, 155, 164–166, 253, 254, 267
– epistemic spaces 159, 160
– epistemological crisis 48–51
– epistemological framework 8
ethics 11, 40, 71, 72, 86, 87, 111, 114, 169, 237, 250, 252–255, 261, 262, 266, 267, 269, 290
– bioethics 252, 254–257, 261–266, 268
– ethical values 36, 236

Facebook 37, 40, 95, 141, 148, 151, 226
facts 19, 46, 48, 165, 253, 257
– alternative facts 48, 49, 165
– social facts 67, 78, 253
Faludi, Susan 36, 47
Felski, Rita 54, 57
fiction 19, 32–38, 48–51, 103, 156, 162–167, 190, 250, 251, 265, 274, 278, 284
– fictional narrative 36, 37
– ideological fiction 49, 50
– science fiction 250, 251, 265
film 20, 38, 83, 84, 89, 91, 96, 103, 114, 148, 150, 164, 226, 246, 247, 250, 275, 283, 288, 289, 291, 295, 297
Fleck, Ludwik 18, 19, 124, 238, 242
Foucault, Michel 73, 74, 87, 98–99, 105, 115, 124, 129–131, 157, 188, 189, 237
fragmentation 33, 224
– fragmented novels 33, 34
frame 4, 8–10, 35, 47, 103, 157, 159, 169, 174, 289, 292, 296
– frameworks 4, 7, 8, 9, 14, 29, 31, 38, 39, 43–45, 47, 54, 60, 90, 111, 123, 134, 145, 152, 155, 156, 166, 167, 190, 196, 204, 251, 253, 256, 258, 266, 276
– framing 5, 129, 132, 156–158, 257, 261
Frankenstein 250, 253
Fridays for Future 30
futures 3, 17, 18, 22, 23, 31, 53, 56, 58, 90, 91, 122, 123, 139, 223, 232, 233, 274, 288, 291, 294
– futurism 111, 115, 116
– futurist 72, 288, 289
– futurity 3, 17, 66–68, 74, 75, 78, 111
– Futurium 17, 18
– futurology 31
– future sense 10, 66–78

Galison, Peter L. 11, 14, 19, 20, 22, 246, 288–297
Geertz, Clifford 21, 22, 251, 291
Gennep, Arnold van 125
geography 86, 158–160, 167, 210, 275
– geography of literature 158
global 5, 7–10, 14, 30, 39, 46, 49, 50, 52, 59, 60, 66, 84, 94, 96, 111, 115–117, 122, 134, 141, 147, 148, 159, 160, 203, 206, 209–215, 226, 236, 238, 254, 257, 259, 260, 276, 284, 285, 289, 290, 294, 295
– globalization 8, 40, 42, 45, 51, 53, 59, 117, 159, 203, 207–210, 215, 216, 258, 274
– global justice 6
– glocalization 203, 213, 258, 259
Goethe, Johann Wolfgang von 158, 253
Goodman, Nelson 36–38
Graeber, David 54, 55, 67, 74, 75
grafting 10, 12, 182–200
Grossberg, Lawrence 3, 19, 20, 21, 30–32, 39, 45, 46, 54, 58, 96
Gumbrecht, Hans Ulrich 4, 7–8

Habermas, Jürgen 87, 165, 227
habitus 124, 127–129, 133, 134
Hall, Stuart 3, 59, 159, 223, 225, 229
Harari, Yuval Noah 32, 35, 49, 51, 60, 95
Harrison, Helen M. 169–180
Harrison, Newton 169–180
Hegel, Georg Wilhelm Friedrich 210
Heise, Ursula K. 9, 14, 21, 29, 30, 274–286
Herder, Johann Gottfried 111, 196–199
higher education 17, 20, 21, 23, 86–89, 112, 115, 226
history 4, 11, 37, 49, 56, 85, 91, 113, 116, 123–125, 127, 129–132, 156, 159, 203–215, 224, 236–247, 275, 278, 289–291, 293, 297
– big history 203, 215

– cultural history 9, 11, 52, 123–134, 182, 215, 236–247, 278
– global history 203, 210–213, 215
Hobsbawm, Eric 159, 204, 205
Hoggart, Richard 226
humanities 4–6, 17, 18, 21, 38, 48, 51, 55, 57, 58, 83–92, 95, 99, 110–115, 123, 133, 169, 171, 173, 179, 206, 226, 231, 236, 237, 246, 264, 275, 276, 278, 280, 290–292, 295
– public humanities 10
human rights 6, 9, 13, 133, 207, 210
Hunt, Lynn 123
hybridity 182–199, 274
– hybridization 159, 166, 182, 186–189, 193, 195, 209

institution 12, 18–23, 29, 30, 32, 46, 49, 54–56, 73, 74, 87–89, 95, 98, 101, 106, 117, 157, 166, 167, 206, 236, 257, 293, 294
– academic institutions 21, 85, 88, 89, 117
– institutional framework 29, 31, 53, 54, 60

Jameson, Fredric 89, 223–225

Kittler, Friedrich 161
Klein, Naomi 70
Kluge, Alexander 67–70, 78
knowledge production 8, 11, 29, 53, 57, 103, 237, 294
Koselleck, Reinhart 66, 77, 224
Kubrick, Stanley 84

Lagoon Cycle 169, 173–179
Landauer, Gustav 75
Langenohl, Andreas 3, 10, 66–78, 232
language 11, 19, 44, 51, 53, 66, 68, 71, 91, 103, 104, 106, 158, 159–161, 166, 167, 178, 182–187, 191, 196–200, 205, 206, 223, 224, 226, 227, 230–232, 240, 242, 245, 246, 251, 254, 256, 263, 264, 267–269, 294, 295–297
– exchange language 11, 294
Latour, Bruno 5, 38, 116, 117, 165, 186, 193, 237, 246, 279
life sciences 55, 57, 250, 255, 256, 268, 269, 295
"liquid spaces" 203–216
literature 33, 38, 50, 53, 54, 68, 91, 96, 114, 155–168, 171, 194, 196, 228, 229, 251, 262, 265, 274, 275, 276, 283, 285, 295
– literary studies 57, 66, 159–161, 228, 230, 246, 274
– literature and economics 228, 229
– technology of literature 160
Luhmann, Niklas 66, 76, 77, 279
Lyon, David 96, 99, 100, 101

MacIntyre, Alistair 50
management 12, 14, 42, 126, 149, 211, 224, 225
– manageability 12
March for Science 51
market 6, 21, 41, 42, 60, 70, 73, 77, 85, 142, 148, 150, 152, 153, 160, 165, 223–231, 232, 233, 241, 246, 257, 258–260, 292, 293
– academic market 6, 224, 227
– financial market 6, 70, 77, 224
– marketization 226
Marx, Karl 75, 85, 208, 223, 274
Mbembe, Achille 104, 123, 124
McAdams, Dan P. 33–35
media 4, 5, 35–38, 40, 46, 48, 58, 83, 84, 95, 113, 114, 116, 141, 142, 148, 149, 152, 160, 162, 165, 246, 251, 259, 262, 269, 283, 288, 295
– media ecology 40, 163, 167
– media frameworks 156
– social media 41, 48, 49, 91, 95, 141, 148, 151, 275
medicine 9, 10, 55, 60, 193, 215, 237, 250–270, 275, 276, 288
Medick, Hans 123
Medusa 96–98
memory 44, 53, 68, 274, 276
metamorphosis 5, 9, 10, 11, 14
metaphors 35, 36, 38, 42, 45, 52, 68, 96, 98, 105, 121, 173, 176, 177, 178, 184, 186, 190–192, 194, 196, 199, 223, 224, 230–232, 244, 251, 296
methodology 170, 172, 203, 209, 216, 255, 269, 270

– methodological 5, 12, 14, 21, 49, 52, 57, 68, 69, 123, 124, 127, 132, 133, 161, 172, 175, 179, 180, 204, 214, 251, 254, 257, 266, 269, 274
Mignolo, Walter 131–134
migration 7, 13, 40–42, 47, 113, 205, 210, 211, 214, 283
Mirzoeff, Nicholas 94
Mitchell, W. J. T. 37, 38
modernity 46, 58, 67, 72, 77, 94–96, 98–100, 103, 115, 116, 123, 133, 141–145, 147–149, 151–153, 193
– modernism 72, 76, 142
Moretti, Franco 161
multispecies (futures) 14, 274–286

narrative 3, 32, 33–38, 45, 50, 58, 68, 69, 70, 102–105, 124, 129, 130, 132, 33, 145, 146, 152, 166, 177, 205, 250–252, 262, 270, 275, 277, 280, 282, 285
neoliberalism 73, 74, 88, 116
Nietzsche, Friedrich 84
nonhuman 110, 114, 115, 117, 193, 275, 277, 278, 279, 281, 282, 283, 285
Nünning, Ansgar 14, 18, 29–61, 110, 230
Nussbaum, Martha 21, 169, 226

Obama, Barack 117, 118

pandemic v, 7, 13–14, 17, 21
panopticon 95, 98–101, 103, 105
Paul, Jean 190, 191
Phelan, James 33
pluralism 22, 51, 52, 224
Poitras, Laura 103–105
politics 17, 33, 36–39, 47, 48, 53, 57, 58, 67, 74–76, 85, 87, 99, 101, 103, 111, 155, 165, 166, 192, 207, 211, 223, 236, 237, 239, 251, 276, 283, 291, 294
– necropolitics 104
– prefigurative politics 10, 67, 73–78
(the) present 5, 7–10, 12–14, 17, 22, 23, 32–34, 41, 54, 66–78, 90, 102, 105, 111–113, 139, 145, 158, 165, 167, 170, 209, 213, 232, 239, 245, 250, 276, 289, 291
provincializing 11, 46, 60

Quine, Willard Van Orman 198, 199

Rancière, Jacques 158
reading 4, 57, 68–70, 72, 76, 102, 134, 155, 161–163, 166, 171
– distant reading 161
research 3–13, 17–23, 31, 36–40, 42–58, 61, 66, 90–92, 95, 113, 124, 126, 131, 167, 169–175, 177–180, 203, 204, 207, 209, 211–213, 215, 224, 226, 227, 236, 243, 246, 247, 253, 255, 256, 275, 291, 293–296
– cutting-edge research 4, 44
– research areas 39, 47, 56, 275
resilience 14, 46, 47, 117
responsibility 6, 10, 14, 23, 29–61, 86, 169, 254, 255, 265–269
risk 4, 5, 10, 20–22, 72, 78, 115, 116, 167, 170, 256, 260, 263, 270
– risk society 5, 115, 116
ritual 46, 124, 125–127, 129, 132, 133, 146–148, 161, 147, 282
Rushdie, Salman 159, 164, 194, 195

Said, Edward W. 129, 130, 205
Schleiermacher, Friedrich 185–187, 195, 196, 199
science studies 236–247
security 66, 94–106, 113, 116, 117, 128, 149, 206, 232
Self, Will 161, 162
Shelley, Mary 250
Simmel, Georg 226, 231, 232
singularity 6, 95, 97, 142–145, 147, 149–153, 285
– singularization 142, 143, 145–147, 149, 151
– "society of singularities" 6, 141–153
social science 4, 13, 18, 22, 55, 57, 111, 113, 114, 170, 174, 275, 280, 291, 295
sound 104, 231, 247, 284, 285, 289
Spivak, Gayatri C. 130, 160
standards; see also language standards
stories 33–37, 50, 58, 102, 132, 148, 149, 167, 174, 208, 209, 213, 270, 283, 291
– life stories 35, 102
– storytelling 33, 34, 37, 104, 106, 165
surrogacy 259–263, 268

surveillance 4, 5, 13, 21, 41, 95, 96, 98, 99–106, 114, 116, 117, 292, 297
sustainability 7, 170, 269
– sustainable future 43, 56
Szeman, Imre 52, 53, 59

Tawada, Yoko 199, 200
Thompson, E.P. 225
thought collective (Denkkollektiv) 18, 19, 20
Tolkien, J.R.R. 94
Toulmin, Stephan 255
trading zone 11, 14, 291, 293–296
transformation 4, 5, 9, 10, 14, 20, 40, 41, 43, 96, 100, 103, 104, 132, 142, 148, 149, 153, 156, 158, 162, 166, 167, 184, 188, 225, 230, 237, 239, 274
translation 10–12, 14, 46, 47, 51, 53, 159, 167, 178, 182–200, 238, 244–246, 264, 268, 284, 294
– cultural translation 10, 182–200
– translational 10, 11, 14, 53, 182, 185, 196, 246, 264
– translation studies 10, 11
transplantation 185, 186, 190, 193, 195, 196, 198, 200, 250, 253, 256
Trump, Donald 110
Tsing, Anna Lowenhaupt 282
Turner, Victor W. 124–126, 128, 130
Twitter 40, 48, 95, 148, 151
Tylor, Edward B. 127

university 6, 12, 19, 60, 87, 88, 90, 204, 289, 295
– corporate university 6, 60
utopia 23, 67, 71, 72, 75, 76, 78, 96, 179, 285

visuality 94, 103, 106

Weber, Max 76, 87, 115, 143, 191, 226, 253, 269
Welzer, Harald 30, 60
Wolfe, Cary 280, 281
worldmaking 33, 35–39, 48, 49, 55, 67
writing 68, 69, 86, 90, 104, 129, 130, 155, 190, 212, 231, 289

Young, Edward 195
YouTube 48, 95, 148, 151

Žižek, Slavoj 89, 280

Bei Fragen zur Produktsicherheit wenden Sie sich bitte an:
If you have any questions regarding product safety,
please contact:

Walter de Gruyter GmbH
Genthiner Straße 13
10785 Berlin
productsafety@degruyterbrill.com